Electrical Engineering at Manchester University

Dedicated to the memory of Marguerite

Electrical Engineering
at
Manchester University

125 years of achievement

by

T. E. Broadbent

Published by
The Manchester School of Engineering
University of Manchester

Published by The Manchester School of Engineering
The University of Manchester
Oxford Road
Manchester M13 9PL

© T.E. Broadbent 1998

No part of this book may be reproduced, stored in a retrieval system, or transmitted in any form, or by any means electronic, mechanical, photocopying, recording or otherwise, without the prior permission of the copyright holder.

ISBN 0 9531203 0 9

Printed and bound by Campus Print Ltd
Linac Building
Wilton Street
Manchester M13 9LN

Contents

List of illustrations	iv
Acknowledgments	vi
Picture credits	viii
Preface	ix

Chapter:

1.	1800-1851:	Early science in Manchester and the founding of Owens College	1
2.	1851-1868:	Science at Owens College in its early years	12
3.	1868-1896:	Reynolds, Schuster, and the start of electrical engineering at Owens	26
4.	1896-1907:	The new physical laboratory and the end of the Schuster era	44
5.	1907-1912:	Electro-technics in the Rutherford years	67
6.	1912-1914:	The founding of the Electro-technics Department	80
7.	1914-1918:	The First World War	96
8.	1918-1930:	Consolidation in the post-war years	115
9.	1930-1938:	Productive years, and the retirement of Beattie	130
10.	1938-1945:	Willis Jackson and the war years	146
11.	1945-1950:	Williams, Kilburn, and the world's first stored program computer	173
12.	1950-1964:	Golden years	192
13.	1964-1977:	Changing times and the end of the Williams era	220
14.	1977-1994:	4-year courses, widening research and a new School of Engineering	245
15.	1994 onwards:	The present and future	277

Appendices:

1. Members of staff in electrical engineering, 1896-1997	292
2. Staff lists at intervals, 1897-1997	296
3. List of first-degree graduates in electrical engineering, 1915-1997	302
4. Numbers of electrical engineering graduates year by year, 1915-1997	321
5. Papers on electrical subjects published by Owens College staff, 1851-1900	324
6. Robert Beattie's published papers	329
7. An electrical examination paper from 1904	330
8. M.Sc. and Ph.D. projects in electrical engineering, 1920-1946	332
9. Syllabuses of the electrical engineering lecture courses, early 1930s	334
10. Undergraduate laboratory work, 1950	339
11. Student and graduate engineering societies at the university	341

Index	342

List of Illustrations

	Page
John Owens, whose gift made possible the founding of the College	10
The Quay Street building which was the first home of Owens College	14
View of Alfred Waterhouse's new Owens College building, c. 1876	28
Professor Osborne Reynolds in 1896, aged about 54	34
Plan of Owens College, c. 1890	39
Professor Arthur Schuster in 1906, aged about 55	40
Advertisement for the Queen's Jubilee Exhibition, 1887	45
Dr John Hopkinson	53
The New Physical Laboratory at Owens College, completed in 1900	56
Two views of the dynamo house, early 1900s	61
Physics group, 1905	64-65
Physics groups, 1909 and 1912	73
Front cover of the brochure for the opening of the Electro-technics extensions in 1912	81
Plans of the new extensions	84-85
The Vice-chancellor's letter to Robert Beattie, 1912	89
Plan of the university in 1914	98
Beatrice Shilling, a notable Electrical Engineering graduate of 1932	133
Engineering graduation photograph, 1937	141
Electrical Engineering graduates, 1937	143
Professor J.C. West's receipt for his graduation fee, 1943	155
Professor Willis Jackson, as a young graduate and in mid-career	171
Tom Kilburn and F.C. Williams at the control panel of the Manchester University Mark I computer	179
The Manchester University Mark I computer and some of the design team	182
Engineering Staff, 1950, including Electrical Engineering	193

List of illustrations

The new Electrical Engineering building, completed in 1954	198
200kV research laboratory, the first working room in the new building	199
High-voltage research group, 1954	200
Michael Lanigan and Tom Kilburn in 1959 with part of 'Muse'	208
Professor Williams explains a variable-speed a.c. motor	213
Four studies of Professor Williams	224-225
Three scenes from the Open Day of 1975	232-233
Professor Williams prepares for a test lap at the Harris Stadium, early 1970s	236
Staff of the Electrical Engineering Department, 1975 and the team which won the IEE Student Challenge, 1977	239
Analogue and digital computers, late 1970s, and an R.M. Nimbus computer in the new Microprocessor Laboratory, early 1980s	250
Two scenes from the 1978 Open Day	252
Two scenes from the 1981 Open Day,	254
The new Information Technology Laboratories, and the Princess Royal being shown the new Laboratories, 1988	260
Clean room facility in the Electrical Engineering Department, late 1980s, and the staff of the Electrical Engineering Department, 1990	271
The Manchester School of Engineering	284

Front cover: Tom Kilburn (left) and Professor F.C. Williams in the late 1940s, at the control panel of the Manchester University Mark I computer designed and built by them. It was the world's first stored-program computer.

Back cover: The main building of the university, viewed from the Manchester School of Engineering

Acknowledgments

Help has been received over many years from numerous people in response to my requests for information of a historical nature. Deserving special mention are Mrs Janet E. Paterson (daughter of the late Professor R. Beattie), Messrs P.M. and T.O. Gerrard and Mrs Ann Burton (sons and daughter of the late Harold Gerrard) and the late Mr Joseph G. Higham and Mrs Jean Oliver (son and daughter of the late Joseph Higham), all of whom have allowed me to read their archives of family history and to use appropriate material in this book. Mrs Paterson also took part in a tape-recorded interview. Their contributions have been invaluable in establishing facts relating to the early years of electrical engineering in the university.

Thanks are due to the following former colleagues who allowed themselves to be subjected to tape-recorded interviews of a reminiscent nature: Professor R. Cooper, Professor Tom Kilburn, Professor J.C. West and the late Mr G.A. (Albert) Cooper. Professor Eric Laithwaite sent a tape of recollections and anecdotes relating to his days in the department.

The late Mr Herbert Shackleton, who graduated in Electrical Engineering in 1924, contributed interesting written recollections of his student days. Dr Alan Fairweather (graduated 1933), who was F.C. Williams's first research student in the department, has provided, in correspondence starting in 1991 and continuing to the present time, copious information about the department in the 1930s. Professor Cooper has been similarly helpful concerning his student days and the war years (1939-45) when he was engaged in research work in the department. Thanks are also due to Professor Cooper for reading and commenting on the draft of Chapter 10, and Professor Tom Kilburn for doing likewise with Chapters 11 and 12. The late Professor J. Lamb kindly wrote a seven-page account of his days in the department as a student and young research worker, in response to my request. He was very ill at the time, a fact of which I was unaware when I contacted him, but he still summoned up the will to respond. His handwritten notes were sent to me by his wife after his death. The recollections of the late Mr G.A. Cooper, whose association with the university began in 1929, have been particularly valuable as he was closely concerned with much that occurred in the department over a lengthy period.

Former colleagues from the Electrical Engineering Department not already named who have helped in various ways (passing on documents, photographs, anecdotes, etc.) include Messrs M. Atherton and J.K. Birtwistle, Drs W.W. Clegg and M.J. Cunningham, Professor D.B.G.

Acknowledgments

Edwards and P.G. Farrell, Drs D.R. Hardy, C.G.M. Harrison, the late Professor G.R. Hoffman, Dr D. King, Professor S.H. Lavington, Dr C.N.W. Litting, Mr J. McCormick, Mrs L. MacGill, Professor B.K. Middleton, Mr D. Moore, Drs G.F. Nix, J.P. Oakley, R.M. Pickard, I.E.D. Pickup, Mr L.S. Piggott, Drs R.S. Quayle and R. Shuttleworth, Mr D.J. Tait, Dr D. Tipping and Mr C. Walker.

Other members of the university who have helped in various ways include Mr. D. Camm (former Secretary to the Faculty of Science), Mrs Ruth Hicks (Formerly Alumni Secretary), Drs V. Knowles (former Registrar) and D.J. Sandiford (Physics Department), Mr M. Koiston (Computer Science Department), Mr B.A. Dawson and Mrs Sharon B. Wilkinson (Faculty of Science Office) and Professors J.D. Jackson and P. Montague and Mr D.G. Walton (Engineering Department).

Mr G.A. Naylor (husband of the late Beatrice Shilling) kindly provided interesting details, including newspaper cuttings, about his late wife's career, and also loaned the original of the photograph shown on page 133.

In finding out all I needed to know about Edward Stocks Massey, the benefactor whose gift led to the setting up of the Chair of Electro-technics, I am greatly indebted to Mr K.G. Spencer, local historian, Burnley; Mr Brian Whittle, Chief Executive and Town Clerk to Burnley Borough Council; and Mr J.D. Hodgkinson, District Librarian, Lancashire County Council. All were unstinting in their efforts to provide the necessary facts.

Staff of the Manchester City Council and Central Library helped me to establish when various street names around the university changed.

The Superintendent Registrar of West Somerset at Williton provided the death certificate of Osborne Reynolds, and the death certificate of Frank Roberts was obtained from the Superintendent Registrar at Whitehaven.

Brochures outlining the early history of their companies were kindly donated by two local firms, Ferranti Ltd. and Mather & Platt.

I may have omitted people or organisations whose names should have been included. If so I apologise for any such unwitting omission.

The work of my three proof-readers for nobly plodding through the typescript, noting misprints and suggesting improvements, is acknowledged, with many thanks. They are my ex-colleague Dr Dennis Tipping and my daughters, Miss Judith M. Broadbent and Mrs Susan V. Clews.

Finally, I must thank my dear late wife, Marguerite, for all her love and support over the years and for unfailingly cheering me up when things in the department went wrong. She accepted without complaint my strange interest in the history of the Electrical Engineering Department and I hope she would have been happy with the result, as it appears in this book.

Picture credits

Page 34. Donated by Professor J.D. Jackson.
Page 45. Published by courtesy of Mather & Platt.
Page 53. From *John Hopkinson: Original Papers.* (Vol. 1). Cambridge University Press, 1901.
Pages 64, 65, 73. Courtesy of the Department of Physics, Manchester.
Pages 89, 141, 143. Loaned by Mrs J.E. Paterson and published with her permission.
Page 133. Loaned by Mr. G.A. Naylor and published with his permission.
Page 155. Published by courtesy of Professor J.C. West.
Page 171. *Left:* Photograph provided by Professor J.D. Jackson
Right: From the Proceedings of the B.T.H. Summer School, July 1948.
Page 199. From *Electrical Review*, July 1955, p. 26.
Pages 213, 224, 225. From *International Science and Technology*, February 1964.
Page 236. Picture taken and donated by Mr J.K. Birtwistle.

The pictures on pages 198, 200, 232, 233, 239, 252 and 254 are from the author's collection.

The remaining illustrations are Electrical Engineering Department pictures or University of Manchester ones which have appeared in various university publications over the years. The prints of the photographs appearing on pages 179, 182, 193, 208 and the front cover were obtained from the Computer Science Department which currently holds the original negatives.

Preface

When did teaching and research in electrical engineering at Manchester begin? That was one of the first problems to be faced when this book was being planned. An obvious starting point would be the year 1912 when the Department of Electrical Engineering (known then as "Electro-technics") was set up. But before then, Robert Beattie had been teaching electrical engineering in the Department of Physics for sixteen years. And even earlier, substantial amounts of electrical technology were taught by Professor Schuster and others. Indeed, it could be argued that the origins of electrical engineering go back as far as 1851 when Owens College was founded, as there was evidence of electrical teaching in the physics courses from the very earliest days of the college.

So I decided after much thought to follow the example of Dylan Thomas in *Under Milk Wood* - "To begin at the beginning". The book therefore opens with a very brief summary of the state of science in Manchester in the first half of the 19th century, culminating in the foundation of Owens College. Next, the earliest elements of electrical teaching and research at the college are identified, after which the more formal electrical instruction and the research in related topics that took place in the college during the 1870s and 1880s are followed through in more detail. By the end of the 1880s electrical technology was an established part of the syllabus in Physics at Owens and remained so until Robert Beattie took over responsibility for the subject in 1896. Consequently, this book's subtitle, "125 years of achievement", is not, I feel, an exaggeration from the chronological point of view, because electrical technology has occupied a place at Owens College, and subsequently at the university, for at least that length of time.

When considering the 'terms of reference' of this book it was decided at the outset to follow through the story of electrical engineering in whichever department it happened to be taught; departments are not isolated castles and it would have been pointless to restrict the account to events in the Electrical Engineering Department when so much had been done in the Physics Department in earlier years.

My qualification for writing this book, if any is needed, is that I have been fascinated by the subject for years. The view is often taken that the past is over and done with and should be forgotten; we should look to the future, which is all that matters. My feeling is that as we are moving along on a uniform time scale, the past, present and future are all equally important. So the past should not be forgotten. I believe that electrical engineering at Manchester has a proud history and it is right that its story should be set down in reasonable detail whilst there is still someone here to do it. It has long been my ambition to achieve this end and with this object in mind I have been gathering information for many years. At long last the book is finished. Few people will be interested in its contents unless they have a connection not only with Manchester University, and the Electrical Engineering Department in particular, but also an interest in what happened in days gone by. I hope there are a few people 'out there' who will find the story worth reading.

The information in these pages has come from many sources. A vast amount is available in libraries and such sources have been utilised to the full. Also, a great deal of help has been given by a large number of people, some of whom have no connection with the electrical engineering department and many who do. As a student I was taught by, amongst others, Harold Gerrard and Joseph Higham who graduated in Physics in 1909 and 1911 repectively. They could have told me a lot about the university of that era, had I thought to ask them when I was a student. But young people rarely have an interest in such history so the opportunity went by. Fortunately the relatives of those two staff members and those of Professor Beattie have very kindly provided much valuable detail which has been of great assistance in the preparation of this book. Other people have also been unstintingly helpful in response to my requests for information. Conversations have been tape-recorded and many people with whom I have corresponded have given their recollections. Consequently it has been possible to amass more than enough facts for the purposes of this book, even concerning the distant past. The details coming in from different sources act like the many pieces of a large jigsaw. When put together, a fascinating story emerges of what the university, and electrical engineering in particular, was like in former days. Sadly, some of the people who helped me have since died, but the information they gave has all been used in painting what I hope is an interesting portrait.

In selecting the years covered by the various chapters I decided that the ends of the chapters should coincide with a significant event, such as the appointment or departure of a particular professor, the setting up of a department, the start or end of a war, etc. This method, adopted by every novelist or soap-opera writer, seemed to me much more sensible than splitting the story arbitrarily into fixed periods of time. As the events of the kind quoted above rarely occur at the ends of calendar years but at some point within the year, the last year corresponding to a particular chapter is the same as the first year of the next. That is the reason for the apparent chronological overlap of successive chapters.

Preface

The final remark I will make here is a personal one. My association with the Electrical Engineering Department at Manchester University goes back to 1949 when I arrived as a student, having been told by my schoolmasters that Manchester University was "a good place", a view I have always held with an unshakeable conviction. I have been a member of the Department in one capacity or other ever since, except for a two-year break in the 1950s, and even in those two years I often called in to discuss technical matters. So I have had the enjoyable privilege of observing events for nearly 50 years. It has been a pleasure to be a part of the department for so long and to work with the friendly and talented people who have comprised its members.

T.E. Broadbent
1997

Note: It is appropriate to mention here that this book is concerned only with the teaching of and research in electrical engineering at Owens College (for the years 1851-1903) and in the Faculty of Science of the University (for the period after 1903). Anyone familiar with scientific education in Manchester will know of the existence of a separate establishment which is now known as UMIST. It evolved from the Manchester Municipal School of Technology (Manchester Municipal College of Technology from 1918) which developed from the Manchester Mechanics' Institution founded in 1824. In 1905 it became the Faculty of Technology of the University and was empowered to award the degrees of B.Sc.Tech and M.Sc.Tech. Electrical Engineering was taught in the Faculty of Technology from an early period; the first professor in the subject was A. Schwartz, who was appointed in 1905 and resigned in 1912, to be followed by Miles Walker who held the post for 20 years. There have been many other professors since then and, as in the Faculty of Science's Electrical Engineering Department which forms the subject of much of this book, there has been steady growth. The two departments have operated in parallel over the years, collaborating from time to time, and have occasionally offered joint post-graduate courses. On 1st August 1994 UMIST became virtually autonomous and independent of the University.

Chapter 1
1801-51: Early science in Manchester and the founding of Owens College

Science in early Victorian Manchester

The setting up of Owens College, followed within a few years by the introduction of teaching and research in electrical engineering, were direct results of the rapid growth of Manchester as an industrial town in the 19th century. By the time Owens College became the Victoria University of Manchester in 1903, engineering, including branches of the subject that might be termed electrical engineering, was an established part of the college's structure. Let us trace briefly the course of events which led to the foundation of Owens College and the establishment of courses in physics and engineering.

In the late 18th century Manchester was little more than a country town with a population estimated at around 25,000. But in 1801 the first official census determined the inhabitants to number well over 70,000, and by 1840 the population of Manchester and Salford together had grown to over 300,000. [1] The increase was caused, primarily, by the success of the cotton industry. There was little regard in those days for the quality of life of those who lived in the town, most of whom were poor, with the result that as Manchester grew bigger it also grew smokier and dirtier, with many people living in squalid surroundings. At the same time there were others who were wealthy, to a greater or lesser degree, and many educated, cultured people were numbered amongst these. The wealth and prestige of Manchester lay in the hands of its clergy, professional men and merchant-manufacturers. Thus amidst the smoke and grime, there existed a powerful bourgeoisie together with, but in marked contrast to, a powerless but unrevolutionary proletariat.

As would be expected, the leaders of the cultural and scientific movements in Manchester in the early 19th century were the professional class - the doctors, apothecaries, ministers of religion, and some businessmen. It was they who were responsible for the creation of a number of societies and institutions which reflected their cultural and scientific interests, the latter limited by the primitive condition of scientific knowledge at the time. First and foremost was the Manchester Literary and Philosophical Society, founded in 1781. The membership was originally limited to 50, all of whom were professional men or merchants with wide interests. The society was quite unlike modern professional groups such as the Institution of Electrical Engineers or the Institute of Physics. It was, so far as science was concerned, a society of amateurs, a league of gentlemen with cultural aspirations who met regularly to discuss and present papers on whatever took their interest. The society had a proper structure with an elected president, vice-presidents and officers. But it remained basically a society of amateurs until the arrival of one notable figure. John Dalton came to Manchester around 1790 in order to teach mathematics, physics and chemistry at New College, an institution which offered lectures on these and other subjects to anyone who wished to attend, provided they had the financial resources to do so. In 1794 he was proposed for membership of the Literary and Philosophical Society by three members; Thomas Henry, apothecary, Thomas Percival, physician, and Robert Owen, a wealthy Manchester textile manufacturer. Dalton was appointed secretary of the society in 1800 and remained in that post until 1809 when he was elected vice-president. He was re-elected vice-president at regular intervals until 1817 when he became president.

Dalton was in a different category from the other members of the Literary and Philosophical Society for he was a professional scientist of great ability. He did not remain in his teaching post at New College for long. In 1800 he resigned and began a career of private teaching and research. Moreover the Literary and Philosophical society, recognising his outstanding talents, offered him a room for teaching and experimentation, and he equipped his rent-free laboratory from the proceeds of his private earnings. Thus he was able to operate in Manchester in much the same way as Humphry Davy and Faraday did at the Royal Institution in London, and he set out on a scientific career that brought him international fame and acclaim.

Dalton was thus Manchester's famous resident scientist and the results of his research and that of others was reported in the proceedings of the Manchester Literary and Philosophical Society. He was Manchester's first truly great scientist. He was elected a Fellow of the Royal Society in 1822, received its royal medal in 1826 and was elected a foreign associate of the French Academy of Sciences in 1830. In 1832 the degree of Doctor of Civil Laws was conferred on him by Oxford University and he was awarded a civil pension in 1833. Soon afterwards a public subscription fund was opened for the creation of a John Dalton statue in Manchester. [2]

There were other important societies in Manchester in the early part of the 19th century, notably the Natural History Society (founded in 1821) and the Royal Manchester Institution (founded 1823), devoted to lecture series and conservation, mainly in the field of fine arts. Under its aegis an art gallery was set up. Meanwhile, the growing use of steam power in factories was leading to a burgeoning interest in matters relating to mechanics. Consequently the Manchester Mechanics' Institution was set up in 1824; its avowed aim was to extend the benefits of learning to the working man, though after about 1840 non-scientific subjects took up an increasing proportion of its function. The New Mechanics' Institution, a breakaway group of the above, was founded in 1829, followed by the Salford Mechanics' Institution in 1839. The Statistical Society of Manchester was founded in 1834 but it was not, as the name might suggest, a mathematical society - but in fact an institution for collecting data relating to urban and industrial affairs. In 1840 the Manchester Society for the Promotion of Natural History was set up with W.C. Williamson (later to become well known as an early luminary of Owens College) as its first curator.

In the 18th century the textile industries of Manchester had played a major role in the tremendous impact of the industrial revolution on working lives through the inventive minds of local men like Arkwright, Crompton and Hargreaves. By the 1830s we see not only a commercially-thriving town (Manchester did not become a city until 1853) but also a flourishing intellectual and scientific community. The increasing importance of Manchester as a scientific and cultural centre was due in part to the existence of its societies, in particular the Literary and Philosophical Society; it was due probably even more to the influence of one man, John Dalton, who strode like a Colossus above his fellow citizens of the scientific community in the earlier years of the 19th century. But by the 1830s John Dalton's scientific work was virtually completed. In 1837 and in the following few years he suffered a series of strokes, after which he was not well enough to work again.

The 1840s

The British Association was founded in 1831, the year of Faraday's discovery of electromagnetic induction. Its aim was "To give a stronger impulse and a more systematic direction to scientific enquiry". It met annually in a different British town or city in order to survey progress and to present current papers. Science in Manchester received a boost in 1842 when it was the town's turn to play host to the Association. The president for the year was customarily elected from the local scientific worthies. John Dalton was the obvious choice but he was by then too ill to assume the duties other than as a figurehead so Lord Francis Egerton, M.P., president of the Manchester Geological Society and the Manchester Agricultural Society, was elected president. Dalton agreed to serve as vice-president. The scientific community of Manchester and its region was excited at hosting this prestigeous event. The Royal Institution was used as a centre for meetings, other venues being the Athenæum (opened in 1836), the premises of the Literary and Philosophical Society, and the Mechanics' Institution. More than 200 new members from Manchester and district were recruited for the British Association, and exhibitions, soirées and dinners were held. The engineer J.F. Bateman designed a membership ticket embossed with a medallion showing a likeness of Dalton. [3] The occasion provided the opportunity to bestow further honours on Dalton. He was presented with a medallion sponsored by George Faulkner, editor of Bradshaw's Journal. Other eminent people active in the scientific life of Manchester were also honoured, among them J.F. Bateman and Eaton Hodgkinson F.R.S. (mechanical science), Edward Binney (geology), Rev. William Herbert (Dean of Manchester), Edward Holme (president of the Association's medical section), Lyon Playfair (chemistry) and G.W. Wood (president of the statistics section). The meetings of the British Association in Manchester were a great success, but it was not the established 'old guard' of Dalton and his contemporaries whose papers formed the basis of the meetings, it was the newcomers including such men as Joule and Fairbairn. The old names were the ones nationally known, but it was the youngsters who were changing the face of Manchester science.

    Another event of significance at this time was the arrival in Manchester of William Sturgeon, a noted experimenter and inventor in electrical science. His appearance on the scene was due to the founding of the Royal Victoria Gallery in 1840. This was created as a result of a meeting of enthusiasts in 1839 and was intended to be modelled on the Adelaide Gallery in London which was opened in 1830. William Fairbairn added his support to the proposal for the creation of the gallery which he said would have "the very best effects upon the population at large, particularly among the younger classes". Eaton Hodgkinson said he particularly wanted to have exhibits of models, machinery and instruments. "As we in Manchester possessed such great ingenuity in our workshops, we should have some institution in which all these machines might be collected and shown and the elements in them explained." [4] The proposal was not to everyone's liking but the gallery was duly set up and was described by early visitors as "elegant, commodious and well-lighted". Amongst the exhibits were dial weighing machines, electrical and philosophical apparatus, mathematical instruments, fossils from the excavations of the Manchester and Leeds railway, a model of the "electromagnetic telegraph of Wheatstone and Cooke", electromagnets, Sharp, Roberts & Company's ball and socket valve, and cast iron surface plates deposited by Joseph Whitworth & Co. [5]

William Sturgeon was 57 years of age when he came to Manchester and was already famous as the inventor of a practical electromagnet (1825), the amalgamation of zinc plates in batteries (jointly with Kemp) (1828), an electromagnetic engine (1832), the two-part commutator (1835) and the use of a suspended coil in a galvanometer (1836). Moreover, at the Adelaide Gallery in London he had edited their Journal, the *Annals of Electricity*. To secure the services of Sturgeon must have been regarded as quite a coup for Manchester. Sturgeon found many sympathetic colleagues in the town, among them the youthful James P. Joule, who was already well known to Sturgeon because he had made six contributions to *Annals of Electricity* in the form of letters concerning his own experiments on Sturgeon's electromagnetic motor. Soon Sturgeon arranged for Joule to lecture at the gallery. A small close-knit group of friends and enthusiasts formed a social and intellectual link with Sturgeon, amongst them, in addition to Joule, were John Leigh, a young medical man employed as a chemist by the Manchester Gas Works, and Edward Binney, an attorney who was a major figure in the Manchester Geological Society and eventually became an important officer and president of the Literary and Philosophical Society.

In spite of the enthusiasm of Sturgeon and his friends the Royal Victoria Gallery did not fare well. As so often happens, the vigour shown by people in encouraging the creation of a new enterprise was not matched by their necessary support once it had been established. In 1942, only two years after its foundation, the gallery foundered through lack of interest. Fifteen years later Joule recalled, bitterly, that "the indifference to pursuits of an elevated character which too frequently marks wealthy trading communities destroyed this, as it has many other institutions." [6] So Sturgeon found himself out of a job but in 1843 he announced the formation of a new body, the Manchester Institute of Natural and Experimental Science. Under its auspices Sturgeon announced courses of lectures and conversaziones. But it fared no better than its predecessor and was defunct within months. Again out of work, the talented Sturgeon had to resort to becoming an itinerant lecturer, travelling from village to village in the Manchester area, taking his demonstration equipment with him on a cart. In 1844 he was elected, belatedly, to membership of the Literary and Philosophical Society but that of course did nothing for his finances and by the late 1840s he was destitute. Owing to the vigorous efforts of his friends the sum of £200 was awarded to him in 1847 and again in 1849, and he was awarded an annual government pension of £50. But in 1850 he died, the pension continuing to his widow. [7] In delivering a eulogy James Joule, still only 33 himself, made a plea for the adequate recognition of a scientist's talents when such a man fell on hard times: "The sum which would be necessary to succour the needy man of science, and so to enable him to continue his researches, would appear trifling if regard were made to the important objects to be realized. But he appeals not to his country as a pauper. He asks it to discharge the debt it owes for labours which have contributed to the common weal, a debt which cannot be left unpaid." [8]

In Manchester the death of William Sturgeon passed almost unnoticed. This was in marked contrast to the death six years earlier (in 1844), of John Dalton, whose body lay in state in the town hall in a large room overhung with black drapery. The melancholy scene was lit by a large candelabra and no less than 40,000 people filed past. The funeral was organised by a committee of civic dignitaries, and the coffin was taken in a cortège almost a mile long to Ardwick Cemetery. [9] It was as though a king had died.

The work of James Joule in Manchester

By the 1840s James Prescott Joule had become a major figure on the Manchester scientific scene. Born on Christmas Eve 1818 in Salford, his father was the owner of the Salford Brewery. James was a sickly child who suffered from a slight spinal deformity; perhaps for these reasons he spent a lot of his time reading. From the age of 11 he studied with a series of private tutors and in 1834, at the age of 15, began studies with John Dalton, probably at the instigation of his father who thought that the study of the physics and chemistry of gases with so famous a chemist would be of value to the brewing business. At first James, along with his elder brother, studied mathematics with Dalton but just as lessons in chemistry were about to start in 1837 Dalton suffered the first of his strokes and their education with him terminated abruptly. Joule later wrote: "Dalton possessed a rare power of engaging the affection of his pupils for scientific truth; and it was from his instruction that I first formed a desire to increase my knowledge by original researches." [10]

Joule followed in the master's footsteps by displaying a burning desire to carry out research. His father provided him with a room and the youthful Joule "filled it with voltaic batteries and other apparatus". Early in 1838 he built an electromagnetic engine in order to try to improve the efficiency of motors, which led to his series of letters to Sturgeon's *Annals of Electricity*. In 1839 he began a course of study in chemistry with John Davies of the Mechanics' Institution and Pine Street Medical School, not only studying with him privately but also attending Davies's lectures to medical students. On Sturgeon's arrival in Manchester Joule had become his ardent disciple.

1800-1851: Early science in Manchester and the founding of Owens College

Joule lectured at the Royal Victoria Gallery at Sturgeon's invitation and was starting to submit papers to the Royal Society in London. His paper "On the production of heat by voltaic electricity" was rejected in part but appeared in an altered form in the *Philosophical Magazine*. In it Joule showed that the 'calorific effect' produced by the passage of current in a wire is proportional to the square of the magnitude of the current multiplied by the resistance of the wire, "whatever may be the length, thickness, shape or kind of metal". Joule then began to study combustion and developed an analogy between electrical heating and chemical combustion which started him on a series of experiments which ultimately led to his determination of the mechanical equivalent of heat. A paper he submitted on the subject to the Royal Society was rejected. Joule later remarked, "I was not surprised. I could imagine those gentlemen in London sitting round a table and saying to each other, 'what good could come out of a town where they dine in the middle of the day?'" [11] The mutual mistrust that often exists between north and south was clearly as evident then as it is now.

As the 1840s progressed Joule's researches continued with increasing vigour. His private researches took on a more publicly-recognised aspect and after election to the Literary and Philosophical society he was elected its secretary in 1846. In 1843, still working in his laboratory in his father's house, he concentrated on the problem of finding the mechanical equivalent of heat and read a paper on the subject to the Manchester Literary and Philosophical Society. At about the same time he investigated the mechanical and heating effects of electrolysis and in August 1843 at the British Association's meeting in Cork he gave a paper, "On the calorific effects of magneto-electricity and the mechanical value of heat". The paper showed that heat is generated and not merely transferred from a source in the electromagnetic machine, and that a constant ratio exists: "The quantity of heat capable of increasing the temperature of a pound of water by one degree of Fahrenheit's scale is equal to, and may be converted into, a mechanical force capable of raising 838 lb to the perpendicular height of 1 foot."

Meanwhile Joule's father moved to Whalley Range in 1843 and built his son a larger and better-equipped laboratory. At the same time as he was continuing his experiments on the mechanical equivalent of heat, Joule collaborated with the chemist Lyon Playfair (born in 1818, the same year as Joule) in research on atomic volumes and specific gravity. Never one to be cautious, Joule wrote to his colleague in 1845: "Don't be afraid of being called rash, it is a natural thing for successful young theorists to be called so. And remember what Sedgwick said: 'The way to make no errors is to write no papers.'" [12]

Between 1843 and 1847 Joule evolved his famous paddle-wheel experiment to make an accurate determination of the mechanical equivalent of heat, and gave a paper in which he exhibited an apparatus consisting of a paddle wheel operated by weights and pulleys. The work expended in revolving the paddle wheels produced a rise in temperature of the water, which was duly measured. For this purpose some improved thermometers were specially made for him by J.B. Dancer, his friend and fellow-member of the Literary and Philosophical Society and partner in the firm Abraham and Dancer. By 1847 Joule's experiments on the mechanical equivalent of heat were virtually completed and on 28th April, 1847, he gave his famous lecture "On matter, living force and heat" at the St. Ann's Church reading room. In this, Joule described clearly his views on the conservation of what would later be called "energy". When Joule's findings were presented in London at the Royal Society some members found the results difficult to accept. One member said he did not believe Joule because he had nothing but hundredths of a degree to prove his case.

But in spite of criticism from some die-hards Joule's brilliant work was gradually accepted by the scientific establishment and honours were soon bestowed on him. In 1850 he was elected a Fellow of the Royal Society (he was still only 31) and in 1857 became a member of its council. In 1857 he also received an honorary doctorate from Trinity College, Dublin and Oxford conferred a degree on him in 1860. From then on his national and international reputation rose year by year. In the few years following his determination of the mechanical equivalent of heat he refined and developed his experiments to obtain greater accuracy, and in 1857 Joule and his friend William Thomson (later Lord Kelvin) embarked on researches leading to a series of papers including some on the thermal effects of fluids in motion, delineating the Joule-Thomson effect. It is a remarkable fact that from the age of 15 Joule worked in the family brewery and spent the most part of each day in commerce; although scientific research was his passion he was only able to devote his 'spare time' before breakfast and after business hours to it. It was only in 1854, with many of his major discoveries behind him, that James's father sold his brewery, whereupon, at the age of 36, James was able to retire from the world of commerce to devote himself full time to his scientific researches.

Although we are concerned here with the scientific scene in Manchester as the mid-point of the century approached, it is perhaps worth noting, in passing, that the scientific events described were matched by those in the artistic, social and cultural field. In 1806 the Portico Library was opened. The Literary and Philosophic Society was by its nature concerned as much with the Arts as with Science, as were some of the other active societies in the town. Before the middle of the century Elizabeth Gaskell was writing her locally-based novels, Engels was trudging the Manchester streets and calling attention in his writings to the condition of the poor, whilst in the world of music there was plenty of activity. Concerts were given regularly and many notable visiting musicians performed in the town. Paganini played 10 times in Manchester in 1832-33. Liszt gave two recitals as a 12 year old in 1825, two more the following year, and played twice, at the Athenæum*, in 1840. In 1847 Mendelssohn conducted *Elijah* and in 1848 Chopin gave a recital. In the same year Charles Hallé settled in Manchester; he formed his famous orchestra nine years later. Manchester in that period was no more a cultural backwater than it was a scientific one.

The Manchester scene in the first half of the 19th century was thus characterised by several institutions dedicated to the dissemination of knowledge, there was a lively enthusiasm on the part of many citizens wishing to further their scientific and cultural knowledge, and present in their midst were a number of major figures in the world of science, in particular John Dalton and James Joule. But the efforts of the various organisations and individuals were largely unco-ordinated and one major amenity was missing. The town possessed neither a university nor a college for the arts and sciences. The subject of this omission had been broached at the inaugural meeting of the Athenæum in January 1836. The president, James Heywood, remarked . . . "Such a college is not only wanted, but actually called for in Manchester . . . a general interest is already excited in favour of a college in Manchester." [13] These sentiments were supported by many others. The idea of a university in Manchester was by no means new and Henry Longueville Jones (1806-1870) had proposed and described such a project in a pamphlet entitled *A Plan of a University of Manchester* published in 1836 at Heywood's expense. [14] A number of meetings of influential men took place to discuss the details of the proposals but despite initial optimism it seemed that nothing would come of the idea, partly because of the failure of different parties to agree on relevant matters but more importantly, because of the difficulty of raising the necessary funds, for it was obvious that the founding of a university, even a small one, would require considerable financial investment. But a chance event came to the rescue of the scheme when, on 29th July 1846, John Owens died, aged 55.

---

* The building on Princess Street where Liszt played (the Athenæum) still stands and is now part of the City Art Gallery, but the room where he performed has been partitioned into smaller rooms.

Owens, a bachelor, was a wealthy textile manufacturer who was interested in education. By the terms of his will drawn up on 31st May 1845 he bequeathed a considerable portion of his estate for the establishment of an institution "for the instruction of young men in such branches of learning and science as now and may be hereafter taught in the English universities."[15] The trustees were left with a sum of slightly under £100,000 for the purpose. These trustees were drawn from diverse political and religious lines and were open to advice from anyone on how best to set up the college. They drew as their models the Scottish universities and the colleges of the Metropolitan University with which the proposed new college was to affiliate. At their second meeting the trustees decided to appoint professors in classics, mathematics and natural philosophy, mental and moral philosophy, and English language and literature. The posts were widely advertised and Faulkner, as chairman of the trustees, bought Richard Cobden's house in Quay Street to provide premises for the college, leasing it back at £200 per annum.[16] Thus Owens College was born.

John Owens, whose gift made possible the founding of the College (from a medallion by T. Woolner, R.A.)

1800-1851: Early science in Manchester and the founding of Owens College

References: Chapter 1

General:

Joseph Thompson. *The Owens College: its foundation and growth and its connection with the Victoria University, Manchester.* J.E. Cornish, Manchester, 1886.

P.J. Hartog. *The Owens College, Manchester: A brief history of the College and description of its various departments.* J.E. Cornish, Manchester, 1900.

Edward Fiddes. *Chapters in the History of Owens College and of Manchester University, 1851-1914.* Manchester University Press, 1937.

H.B. Charlton. *Portrait of a University.* (Written to commemorate the centenary of the founding of Owens College). Manchester University Press, 1951.

R.H. Kargon. *Science in Victorian Manchester.* Manchester University Press, 1977.

D.S.L. Cardwell. *James Joule: A biography.* Manchester University Press, 1989.

Specific to the text:

1. W. H. Chaloner. *The Birth of Modern Manchester*, in *Manchester and its Region.* Manchester University Press, 1962, p. 134;
   B.R. Mitchell and P. Deane. *Abstracts of British Historical Statistics.* Cambridge University Press, 1971, pp. 24 - 26.
2. F. Greenaway. *John Dalton and the Atom.* Ithaca, New York, 1966, p. 92.
3. *Manchester Guardian*, 18th June 1842, p. 3.
4. *Manchester Guardian*, 27th March, 1839, p. 3.
5. Bradshaw's Manchester Journal, **1** (1841), pp. 9 -11, 134-35.
6. *Manchester Literary and Philosophical Memoirs*, 2nd series, **14** (1857), p.77.
7. ibid, p. 83.
8. ibid, p. 82.
9. Elizabeth Patterson. *John Dalton and the Atomic Theory.* Garden City, 1970, pp. 277-83.
10. *Manchester Literary and Philosophical Society Memoirs* (reporting an autobiographical note of Joule), **75**, 1930-31, p. 110.
11. A. Schuster. *Biographical Fragments.* London, 1932, p. 201.
12. Joule/Playfair correspondence, Manchester City Reference Library.
13. *Manchester Guardian*. 16th January 1836, p. 3.
14. See E. Fiddes (general references, above), p. 21.
15. John Owens' will is quoted in B. W. Clapp; *John Owens, Manchester Merchant.* Manchester University Press, 1965, p. 173.
16. *Minutes of the Trustees' Meetings for Educational Purposes*, Owens College. **3**, pp. 9-10. 30th January 1849. Manchester City Reference Library.

Chapter 2

1851-1868: Science at Owens in its early years

Owens College opened officially on 11th March 1851, six weeks before the start of the Great Exhibition, for a shortened session as it was then more than half way through what would be the normal academic year. Five professors had been appointed, contrary to the initial plan to appoint four. One of them was also the college's principal. They were:

Alexander J. Scott:	Principal of the College; also professor of Hebrew, Grammar, English Lang. and Lit., Logic, Mental and Moral Philosophy
J.G. Greenwood:	Greek, Latin, History
Archibald Sandeman:	Mathematics, Natural Philosophy
W.C. Williamson:	Botany, Zoology, Geology
Edward Frankland:	Chemistry

Clearly these men were expected to be all-rounders compared with their modern counterparts but the subjects were less well compartmentalised than they were later to become and the body of knowledge available in any discipline was much less than is now the case. The old title "natural philosophy", in common use in the 19th century, encompassed the science of the physical properties of bodies; i.e. physics, or physics and dynamics. Thus, science was well represented from the start in the persons of Frankland, Sandeman and Williamson (three of the five professors were scientists) though engineering was absent. However, some basic instruction in mechanics, magnetism and electricity etc. would be given in the natural philosophy courses. Although five professors had been appointed instead of the four initially proposed, the appointments of Frankland and Williamson were officially part-time ones, though it is doubtful whether their duties left them much time for anything else. In those days a substantial portion of a professor's remuneration came directly from the fees paid by students. For instance, Scott received £200 as Principal, £350 as professor plus two-thirds of the class fees, Greenwood's professorial salary was £350, whilst Frankland and Williamson received £150 plus fees. [1]

The professors gave inaugural addresses which, in the case of the scientists, formulated policies which most good scientists would approve of to this day. Frankland, for instance, in his address entitled "On the educational and commercial utility of chemistry", observed:

"[It] can scarcely be over-rated; as an analytical science it is not greatly excelled by mathematics, whilst the close and minute observations, and the careful analogies which it requires, form an invaluable discipline for the faculties of youth that can scarcely fail to give a high degree of tone and energy to the mind, and to exercise a most important and salutary influence in future life even should the student not be destined for any profession in which the knowledge acquired can be rendered practically useful." [2]

Frankland also outlined his own plans, the nature of which augured well for the teaching of science in the new college, stating that each student would have his own working table and set of apparatus and would himself conduct his own analytical experiments. [3]

There was an air of great optimism; Frankland later confided to a friend that his enthusiasm for Owens was so great in his early years there "that nothing would have induced me to leave the college." [4] A trustee who had visited other universities reported that the facilities were "second to none in the kingdom". [5] At the close of the first full session in July 1852 the chairman of the Board of Trustees for Owens, J.F. Foster declared that "We have every reason to rejoice and congratulate ourselves upon the present state of the college." Numbers of students were of course tiny by present-day standards. Frankland's classes in systematic chemistry (lectures) and practical chemistry (laboratory) were attended by 18 and 17 students respectively, or, in all, 25 young men. This represented a considerable fraction of the 62 students currently enrolled at the college. [6]

By the end of the second full academic year Frankland's initial optimism had waned. The college's student population had risen to 99 but Frankland's enrolments, particularly those in his laboratory classes, had declined. He reported "mixed feelings of pleasure and disappointment." Some students, he said, had distinguished themselves, but others, had been "remiss in their exertions", an observation that will strike a chord to this day with anyone who has taught in a university. Nevertheless Frankland planned to increase the range of classes available, but his plans were thwarted through lack of support and most of the planned courses were not given. This uneasy state of affairs continued during the next couple of sessions leaving Frankland somewhat disillusioned.

Meanwhile the three science professors, Williamson, Sandeman and Frankland, were elected to the Literary and Philosophical Society, as was the Principal, Scott, but Greenwood did not join. Frankland and Williamson did publish papers in the society's memoirs but for the most part published in national journals, the papers including one by Frankland read before the Royal Society. Frankland, an enthusiastic and far-seeing man, saw Owens not as a northern imitator of Oxford and Cambridge but as the provider of first-rate teaching and research which would benefit the local community in particular and Britain in general.

Electrical Engineering at Manchester University

The building in Quay Street, the first home of Owens College

Frankland's chemistry courses, along with the other disciplines in the college, continued to struggle to attract students. In March 1856 he confided in a letter to the eminent chemist Robert Bunsen in Heidelberg:

"It must be in the highest degree satisfactory to you to be surrounded by such a number of students so many of whom are engaged in original research. Unfortunately this is far from being the case here and there is rarely any further desire for knowledge than the testing of "Soda-ash" and "Bleaching Powder". [7]

The next few years, following the initial euphoria when the college was founded, were difficult ones for Owens. The college had begun with only 25 day students; by the 1852-53 academic year the numbers had increased to 72 with an additional 28 evening pupils, and in 1854 there were 71 day students and 73 evening ones. But in 1854-5 the number of day students dropped to 48. The disillusioned Principal blamed the city, remarking on its "distaste for thorough and systematic teaching." The decline continued and in 1856-57 the number of day students fell to a mere 33. The alarming nature of the decline occasioned much debate, not only within the College but in public. Scott remarked bitterly in 1856:

"Names of greater weight than mine might easily be cited for the uncertainty of the experiment made in offering to Manchester not popular lectures, not an education of the school or of the workshop but a College education; an education not professional but general. If there is disappointment, it is no more than just to inquire whether Manchester had cause to be disappointed in Owens College or Owens College in Manchester . . . the demand has yet to be created for the kind of instruction which it is our duty to furnish."

Scott was not the only one to express himself in virulent terms. Frankland observed:

"It is impossible to reside in Manchester for even a much shorter time than five years without becoming painfully conscious of the very slight appreciation which the Manchester people in general possess for a higher education."

Clearly Owens College was in deep crisis. Following a faculty meeting at Frankland's house where a few policy proposals were made that were clearly aimed at Scott, the Principal resigned "owing to ill health" on 28th May, 1857. J.G. Greenwood, the college's professor of Classics, was appointed to succeed him. Frankland felt he had had enough and in August 1857, tired of being a pioneer, he too resigned. A measure of Frankland's pique at the way events were turning out may be gauged by the fact that his letter of resignation was sent from Zurich. [8]

In Greenwood's first year as Principal the College's fortunes failed to improve, the number of students enrolled remaining about the same as in the previous year.

The plight of Owens College was taken up by the press. A leader in the *Manchester Guardian* was critical of the college's policies, stating:

"Explain it as we may, the fact is certain that this college, which eight years ago it was hoped would form the nucleus of a Manchester university, is a mortifying failure." This prompted a lively exchange in the newspaper's letters pages. "A former student" remarked: "If the College is the mortifying failure you make it out to be, the fault is in the parents of Manchester for not justly valuing education for their sons." The recently-departed Scott, still smarting from his wounds, retorted in similar vein: "That love of information which leads to a 'marvellous development of the cheap press' is not identical with a taste for systematic study." The *Manchester Guardian* continued to proffer its opinions: "Mr Scott thinks and writes of Owens College as a rival of Oxford and Cambridge; we speak of it as an institution for supplying the middle classes of Manchester with suitable education . . . we want Mr Scott to take society as he finds it." [9]

The dispute rumbled on. Opinions were entrenched and the sort of accusation and counter-accusation on the philosophy of education that were being exchanged then would not be unfamiliar today. Following Frankland's sudden resignation the college's priority was to appoint a new professor of chemistry as quickly as possible and Henry Enfield Roscoe was selected for the post in 1857. He arrived when the fortunes of the college were at their lowest ebb. In his recollections published in 1906, Roscoe told a story which illustrated the pessimism pervading at that time:

"The institution was at that time nearly in a state of collapse, and this fact had impressed itself even on the professors. I was standing one evening, preparing myself for my lecture by smoking a cigar at the back gate of the building, when a tramp accosted me and asked me if this was the Manchester Night Asylum. I replied that it was not, but that if he would call again in six months he might find lodgings there! That this opinion as to the future of the college was also generally prevalent is shown by the fact that the tenancy of a house in Dover Street was actually refused to me when the landlord learnt that I was a professor in that institution." [10]

Fortunately, the years 1856-58 turned out to be a nadir in the fortunes of the college. Following the appointment of Roscoe, and partly because of his lively presence, matters began to improve; indeed, Owens College was never again to experience the crises of those early years.

Owens College rises in stature

Henry Roscoe, the son of a Lancashire barrister, was a bright young chemist who had studied in London and with the famous chemist, Bunsen, in Heidelberg where he had obtained his Ph.D. When appointed to the Professorship in Chemistry at Owens he was still only 24, which demonstrates the forward-looking wisdom of those who appointed him in opting for ability and potential rather than experience. In his inaugural address Roscoe defended experimental physical science as a mental discipline. He put forward the view that the method of teaching suited to Owens was a middle way between the professorial system of the German universities and the tutorial system of Oxford and Cambridge. He aimed to convince Manchester's commercial and industrial society of the wisdom of a systematic experimental training in sciences using common sense. In stating these views Roscoe cleverly showed that his intentions for academe at Owens were entirely consistent with the requirements of Manchester commerce. Owens, or at any rate the part over which he presided, was to be no ivory tower. [11]

This book is concerned with electrical engineering and Roscoe was a chemist, but his contribution to our story is relevant and significant, for in furthering the pattern of experimental science at Owens that had been started by Frankland he set the scene for all the other disciplines in science that were to follow. Roscoe was a clever man, not only in the technical sense as a chemist, but also in his ability to set out his views on teaching and research clearly, and to convince the city of Manchester that Owens would progress along the right lines. His arrival marked a turning point in the fortunes of the college.

When Roscoe came to Manchester the morale of staff and students was low and only 19 new students entered in 1857. The number of day students was only 34 and the enrolment of evening students had dropped to 59. But within a year of Roscoe's arrival there was a steady improvement. In 1859, 37 new day students entered the college and there were 77 evening students despite a 50 per cent increase in the evening class fees from the previous year. In 1860 Robert Clifton was appointed as Professor of Natural Philosophy, thereby allowing Sandeman to devote his time to mathematics alone. Clifton instituted a new course in physics, and enrolments increased even more dramatically. For the 1861-62 session there were 88 day students and 235 evening class ones. In 1864-65 the corresponding figures were 127 and 312. In 1863 the college reported that it had "every reason to hope that the progress was not of a transient character but that it would be permanent." [12] Morale of staff and students had markedly improved. The college had, it seemed, turned the corner. A former student wrote later: "We had a kind of feeling that we were the fore-fathers of a great race to be." [13]

Recognition of the value of degrees

Although the skill, vigour and enthusiasm of Henry Roscoe played a significant part in the revival, there were other factors. The years 1856 to 1858 when Owens reached rock bottom coincided with an economic depression. The college had been founded at a time of economic optimism as the Great Exhibition had demonstrated, but the country's economic fortunes had plummeted in the next few years and it would not be an exaggeration to state that Owens' prospects had been brought down correspondingly. Owens was not alone in having difficulty in attracting students. All other universities and colleges were experiencing the same sort of problems, but for Owens the difficulties were particularly acute because the institution was new, which made it more susceptible to external problems. After the early 1860s the country's economy improved, and with it Owens' fortunes. But there was another important reason for the sudden increase in the quest for education by the country's young. Beginning in the early 1850s, British society was demanding qualifications for careers. Thus from the late 1850s onwards, emphasis would be placed increasingly upon courses, examinations and degrees.

In the 1850s Owens had awarded its own certificates to successful candidates but at the end of the decade an important decision was made which proved to be of great help in establishing the stability of the college. London matriculation examinations would be held at Owens starting in July 1859; the matriculation was much sought after and was a nationally-recognised qualification. In October of the same year London Bachelor of Arts examinations were held at Owens. The Principal, Greenwood, had argued strongly for this to happen, considering it critical for the college. [14] The measures certainly helped. The college was aided further by the University of London's decision of 1858 to alter its charter to enable it to introduce new science degrees. These changes at London University came about after much discussion and the deliberations of committees which included in their membership such men as Michael Faraday and Sir James Clark. [15] London's new B.Sc. degree, introduced in 1858, required a knowledge of mathematics, chemistry, biology, physics, logic and ethics. The D.Sc. introduced at the same time required the first degree, further training in a principal and a subsidiary subject, and research for further examination. The new measures received general support throughout the country. The *Lancet*, in an editorial of 1859, observed: "that the engineer, the miner, the architect of the next generation will, as well as the chemist, recommend themselves to their clients by the guarantee of training and acquirements which the initials B.Sc. and D.Sc. will afford." [16]

At Owens there was a modest, but not spectacular, initial response to the introduction of the new degrees. In 1861 five London B.Sc. degrees were awarded from Manchester compared with 55 B.A.s and 20 Bachelor of Medicine. In 1862, 13 B.Sc.s were awarded and for the remaining years of the 1860s the number of B.Sc. degrees stabilised at about 10 to 15 per cent of the number of B.A.s. The usefulness of the new degrees seems to have been accepted outside the college. At the opening of the autumn term in 1860 a trustee, William Nield, acknowledged the importance of the public examinations and the changing professional requirements of the medical and the engineering professions.

Further progress

The appointment of the new Professor of Natural Philosophy, R.B. Clifton (then aged only 24), in 1860, on the recommendation of a committee headed by Roscoe, had generally been regarded as a step forward for the scientific interests of Owens. Hitherto the teaching of both mathematics and natural philosophy (i.e. physics) had been in the hands of Archibald Sandeman who was said to be insensitive to the needs of such people as aspiring engineers, and furthermore, though he was an amiable man, had a reputation as a poor teacher whose lectures were a mystery to most of the students who attended. He was interested in the more abstract branches of mathematics and published a work on his favourite subject - pelicotetics - in 1868. (Pelicotetics is defined as the science of quantity.) He looked upon his obscure work as being of the utmost importance. The story is told that when two of his former pupils visited him in old age he told them of the deep pain he had felt in having once misled them on some abstruse principle involved in the differential calculus, whereupon one of them replied, "Well, Sir, it may comfort your mind to know that we didn't understand a word of what you said, so it hasn't done us any harm." [17]

Sandeman resigned in 1865 and moved north to Perth to oversee his family's calico manufacturing business. The new courses in experimental physics which had been set up on Clifton's arrival comprised mainly statics and dynamics, heat, light and sound but also included some electrical content - the first elements of electrical teaching at Owens. The syllabus for 1862-63 (the first year that an Owens College Calendar was published though the college had existed for 11 years) mentioned Magnetism and Electricity. One of the examination papers set in June 1863 included a question in which students were asked to enunciate Ohm's Law and to "find the intensity of the current" in a circuit in which a number of cells with specified internal resistances were interconnected. Clifton did very little research in his short period at Owens (he left for Oxford in 1865) but established a reputation as a fine teacher. Because of this it is no surprise that his classes grew in size. He was an energetic man; he instituted an evening course in natural philosophy and during his second year he began courses in experimental mechanics and mathematical and natural philosophy. By June 1864 he had drawn 44 students to his experimental mechanics course, 28 to experimental physics and 10 to his mathematical physics courses out of a total Owens pool of only 110 full time students. The physics syllabus of 1865 included "Magnetism and Dynamical Electricity".

Whilst Clifton was building up the physics courses, Roscoe's combination of academic and political skills, together with his unbounded energy and enthusiasm, was pushing him to the forefront of the Owens' professors in the now expanding college. Frankland had done well; Roscoe seemed to be doing even better and furthermore it was he who had persuaded Clifton to come to Owens. Roscoe knew that his effectiveness and that of the college generally was dependent on his relationship with the local community. He made himself available to industrialists and local government as a consultant, a course of action which brought about much technical collaboration, not only in proffering technical advice but in performing chemical analyses. He also gave many outside lectures. In short, he was very effective in every possible way and was a great benefit to Owens. Not only did he co-operate with and help those in the higher echelons of business and commerce, he was active at all levels of society, helping to establish a branch of the Working Man's College in Manchester.[18] Roscoe was elected to the Literary and Philosophical Society in 1858 and became an officer within two years. He also assumed a major role when the British Association held its meetings in Manchester in 1861. William Fairbairn proposed Roscoe for Fellowship of the Royal Society in 1861 and he was elected in 1863 at the age of 30.

Roscoe's success in helping to build up Owens depended on his convincing Manchester's merchants and industrialists of the necessity for a good scientific education in the training of their sons for managerial positions in industry and commerce. At the time Roscoe was building up his chemical school about one half of the college's income was drawn from Owens' endowment, a sixth from the interest from gifts and other bequests, and one third from student fees. The average age of students at that time was about 18 but the youngest entered at 14. In 1868 the average fee paid was £15 p.a. but laboratory classes cost £21 p.a. for six days a week. About one half of the students in the day school were drawn from the sons of merchants and manufacturers, a quarter from the children of the professional class, one eighth from the sons of shopkeepers, and the origin of the rest was unknown. Those in evening classes were in the main young men employed in warehouses and factories who wished to better themselves, and there were also a few teachers, artisans and factory foremen.[19]

The courses taken by B.Sc. students seem to have been thorough, as can be judged by the following schedule:
First year: Latin, Greek, mathematics, natural philosophy, history, inorganic chemistry and either French or German.
Second year: Mathematics, natural philosophy, the junior chemistry class, chemical laboratory (2 days/week), anatomy, physiology and French or German.
Third year: Courses including logic, mental and moral philosophy, mathematics, natural philosophy, organic chemistry, chemical laboratory (2 days/week), geology and botany.[20]

Students usually took the 1st B.Sc. examination at the end of the second year and the Final B.Sc. at the end of the third year. The quality of the students and the teaching must have been high for the Owens students did well in the London examinations from the very beginning. In 1867 between a quarter and a third of the successful London candidates were Owens men.

Like Clifton, Roscoe was a good teacher. He explained his teaching philosophy as follows:

" The personal and individual attention of the professor is the true secret of success; it is absolutely essential that he should know and take an interest in the work of every man in his laboratory, whether beginning or finishing his course. The professor who merely condescends to walk through his laboratory once a day, but who does not give his time to showing each man in his turn how to manipulate, how to overcome some difficulty, or where he has made a mistake, but leaves all this to the demonstrator, is unfit for his office, and will assuredly not build up a school." [21]

Roscoe explained his approach to teaching as follows:

"I feel I am doing the best for the young men who, wishing to become either scientific or industrial chemists, are placed under my charge, in giving them as sound and extensive foundation in the theory and practice of chemical science as their time and abilities will allow, rather than forcing them prematurely into the preparation of a new series of homologous compounds, or the investigation of some special reaction, or of some possible new colouring matter, though such work might doubtless lead to publication. My aim has been to prepare a young man by a careful and complete general training, to fill with intelligence and success a post either as teacher or industrial chemist, rather than to turn out mere specialists who, placed under other conditions than those to which they have been accustomed, are unable to get out of the narrow groove in which they have been trained." [22]

The success of Owens students in the London B.Sc. examinations continued throughout the 1860s and several of the best went on to gain the London D.Sc. Some went to Germany to take a Ph.D. Many became distinguished scientists who would form the nucleus of the newer universities. [23] This record was unmatched anywhere in Britain, and Roscoe was justifiably proud of what was being achieved.

Moves to 'extend' the College

The 1860s, then, were years of consolidation and expansion for Owens College, and in particular for Roscoe in his chemical school and to a lesser extent (until his departure for Oxford) for Clifton in physics. Owens had started with five professors; in 1865 the number had increased to nine and there were six assistants in the delightfully-named 'teachership'.[24] The student numbers had increased correspondingly, as already outlined. The college was bursting at the seams and if any further expansion was to take place it would clearly be necessary to leave the premises in Quay Street which were by now totally inadequate to house the number of students enrolled. So began the 'Extension of Owens College' movement, the aim of which was to involve not an 'extension' in the usual sense where existing premises were added to, but the finding of a new site and the raising of a new building. This would require a lot of money and the solving of many legal battles but as the end of the decade drew near the movement was rapidly gaining momentum.

The introduction of an engineering course

What of engineering in this expansion? It is perhaps surprising that Owens took so long to introduce engineering as a formal discipline, bearing in mind the Manchester district's wonderful record of invention and commercial success in so many branches of technology, particularly in textiles. Moreover, in the country at large, engineering discoveries had been made and other developments had taken place at a rapid rate during the first 60 years of the 19th century. Disregarding mechanical and civil engineering which are not the area of concern of this book, but taking only electrical technology as an example, here are a few of the developments that had occurred between 1825 and 1868:

    1825 Sturgeon produced practical electromagnets
    1831 Faraday discovered electromagnetic induction
    1832 Sturgeon constructed an electromagnetic engine
    1832 Schilling's electromagnetic telegraph
    1833 Faraday set down the laws of electrolysis
    1833 Gauss and Weber operated the electric telegraph commercially
    1835 Sturgeon adopted the two-part commutator
    1836 Sturgeon used a suspended coil in a galvanometer
    1836 Morse devised a simple relay
    1837 Cooke and Wheatstone used the telegraph on the L.& N.W. railway
    1840 Davidson demonstrated an electric car
    1842 The Slough murder case showed the value of electric telegraphy
    1843 Wheatstone announced his bridge
    1844 Hand operated arc lamps used at the Paris Opera House
    1845 Morse opened a public telephone service
    1845 Matteui invented the mica capacitor
    1846 Bain invented the automatic telegraph
    1848 Swan started experiments on filament lamps
    1854 Hughes invented his printing telegraph
    1856 Formation of the Atlantic Telegraph Company
    1858 Holmes' first electric lighthouse at South Foreland
    1858 Second Atlantic cable completed and the first message sent
    1858 Electrical firm of Siemens Bros. established
    1859 Henley built his cable factory at North Wooland on the Thames
    1860 Planté produced a lead storage battery (a secondary cell)
    1863 Wilde patented the separately-excited dynamo
    1864 Clerk Maxwell postulated electromagnetic waves

1866 Great Eastern lays a new Atlantic cable and established ocean telegraphy
1866-67 Werner von Siemens, Wheatstone and Varley all announced self-excitation of dynamos
1868 Stated that there are 16,000 miles of telegraph line in Britain
1868 Leclanché invented his famous cell
1868 Wilde announced the synchronisation of A.C. machines

There is surely enough evidence in this list to show that by 1868 electrical technology was very much a thriving, commercially successful business; the study and practice of electricity was no longer merely the pursuance of interesting but commercially-useless experiments by privately-financed dilettantes. Electro-plating was now being undertaken commercially, the electric telegraph was used extensively, electric lighting was beginning to be put into practice, successful storage batteries were being manufactured, and a variety of relatively efficient electric motors and generators, d.c. and a.c., were on the market. Thus, electrical technology was clearly an up-and-coming successful branch of engineering. Not only that, but one of the 19th century's leading scientists, Joule, lived and worked 'on the doorstep', as did the practical electrical engineer Henry Wilde. It is perhaps surprising that the Owens College authorities took so long to seek funds to introduce engineering, in particular electrical engineering, as a formal academic discipline.

However, on 11th December 1866 a group of leading Manchester engineers with Joseph Whitworth as secretary met at the town hall and drafted a resolution to the effect that "it is expedient to establish a professorship of civil and mechanical engineering, together with a special library, a museum of models, a drawing class etc. in connection with and under the management of Owens College." [25] They aimed to raise £10,000 for the purpose. By the following November gifts had been received totalling £8,750, including £3,000 from Beyer, Peacock & Co. and personal gifts of £1,000 each from Whitworth, C.P. Stewart and John Robinson. The sum finally given to the college was £9,505. [26] It was this same committee which solicited candidates for the chair by advertisement. Candidates of the necessary quality were found wanting until Charles F. Beyer offered to supplement the post's salary for a time to bring it to £500 p.a. and new candidates were then sought. In March 1868 Osborne Reynolds was appointed Professor of Engineering.[27] His tenure was to bring great distinction to the college in the years that lay ahead.

References: Chapter 2

General:

The general references listed in Chapter 1 (page 11) and the following in addition:

W.H. Chaloner. *The Movement for the Extension of Owens College, 1863-73*. Manchester University Press, 1973.

Percy Dunsheath. *A History of Electrical Engineering*. Faber and Faber, London, 1962.

Specific to the text:

1. *Minutes of the Trustees Meetings for Educational Purposes*, Owens College, 3, 20-21, Manchester Central Reference Library.
2. *Manchester Guardian*, 22nd March, 1851. See also *Introductory Lectures on the Opening of Owens College*, Manchester, 1852, p. 96.
3. *Manchester Guardian*, 22nd March, 1851, p. 8.
4. *Sketches from the Life of Edward Frankland*, London, 1902.
5. *Manchester Guardian*, 7th July 1852.
6. Figures on student numbers, etc. are from Owens College Annual Reports, various years.
7. Letter from Frankland to Bunsen, 3rd March 1856. Bunsen papers, Heidelberg University.
8. As ref. 1, but pages 53 to 55.
9. Correspondence in the *Manchester Guardian*, 1858: 9th July, p. 2; 12th July, p. 4; 14th July, p. 3; 15th July, p. 2; 22nd July, p.2.
10. H.E. Roscoe. *The Life and Experiences of Sir Henry Enfield Roscoe*. (Autobiography). Macmillan, London, 1906.
11. *Manchester Guardian*, 15th October 1857, p. 3.
12. Owens College, Miscellaneous Materials, Manchester Central Reference Library; *Manchester Guardian*, 29th June, 1861, p. 5 and 4th July, 1863, p. 4.
13. *Record of the Owens College Jubilee*. Manchester, 1902.
14. J. Thompson (see general references, above), p. 211.
15. D.S.L. Cardwell. *Organisation of Science in England*. London, 1972, pp. 92-93.
16. *Lancet*, No. 4 (January 1859), p. 564; ibid., No. 29 (October 1859), p. 442.
17. p. 44 of E. Fiddes's book on the history of Owens College and Manchester University (see general list on p. 11). Story told by Rev. F.W. Macdonald.
18. J. Thompson (see general references, above), pp. 229 - 236.
19. *Report of the Select Committee on Scientific Instruction* (London, 1868), p. 277.
20. *Report of the Royal Commission on Scientific Instruction and the Advancement of Science* (1872), p. 12.
21. H.E. Roscoe. *Record of Work Done in the Chemical Department of the Owens College, 1857-1887*. Manchester, 1887, p. 2.

22. *British Association for the Advancement of Science Report*, 1884, p. 668.
23. R. H. Kargon. *Science in Victorian Manchester.* Manchester University Press, 1977; page 180.
24. This and similar information is from Owens College calendars.
25. J. Thompson (see general references, above), p. 295.
26. Ibid, p. 295 and *Manchester Guardian*, 29th November 1867.
27. Jack Allen. *The Life and Work of Osborne Reynolds.* Introductory paper in the Osborne Reynolds Centenary Symposium, *Osborne Reynolds and Engineering Science Today*, ed. J.D. Jackson and D.M. McDowell. Manchester, 1970.

Note: *The Manchester Guardian*, quoted several times in this chapter and in chapter 1, was founded as a weekly newspaper (published on Saturdays) on 5th May 1821 at the substantial cost of 7d, of which 4d was tax (stamp duty). The stamp duty was condemned by contemporary critics as a 'tax on knowledge'. From 1836 it was published twice weekly (Wednesdays and Saturdays) at 4d per copy, the tax having been reduced to 1d. It became a daily paper (excluding Sundays) on 2nd July 1855 at a cost of 2d which was reduced to 1d in 1857 when the stamp duty on newspapers was finally abolished. It has remained a daily newspaper ever since. The name was changed to *The Guardian* on 24th August 1959.

Chapter 3
1868-1896: Reynolds, Schuster, and the start of electrical engineering at Owens

The years covered in this chapter have been chosen because they span the period from the appointment of Osborne Reynolds as the first Professor of Engineering in 1868 to the appointment of Robert Beattie to teach electrical engineering in 1896. In order to see how electrical engineering came to be introduced as an academic discipline we need to look at events in the Engineering Department and also in the Physics Department, as both are relevant to our story. These will be considered in turn in this chapter. But first it is advantageous to look at the Owens scene in general, as the period under consideration was one of great change within the college and was the environment within which the engineering and physics departments were operating.

Extension of Owens College and the formation of the Federal University

As stated towards the end of the previous chapter Owens College was bursting at the seams by the mid 1860s and ideas were afoot to move into new, larger premises. The forward-seeing Henry Roscoe enlisted the help of Thomas Ashton, a lay member of the college's council known for his vigour in executing projects he thought were worthwhile. At a meeting at the town hall on 1st February 1867 Manchester was awakened to the exciting possibilities ahead which would lead eventually, it was said, to the creation of a true University of Manchester. An appeal to the public of Manchester and its surrounding towns was made for money with which to buy land and to erect there 'commodious and stately buildings' so that Owens College would provide 'the foundations of . . . the University of South Lancashire and of the neighbouring parts of Cheshire and Yorkshire.' Ashton was elected chairman of an executive committee whose function would be to raise funds and steer the project through the legal and other minefields which would assuredly lie ahead. One of the difficulties was the proposal to admit women to the new 'extended' college. The terms of John Owens' will referred to the "education of young men" and legally it mattered not whether he objected to the higher education of women in principle or whether he had never considered the question. The constitution of the college would need to be altered not only in this but in other respects.

A way round the difficulties was found through an ingenious plan. A new college, in which there were to be vested the newly raised funds, was to be incorporated by a private Act of Parliament. In its constitution distinct powers were to be given to the Court, Council, the Senate of Professors, and the Associates (who were the alumni; the equivalent of the present-day convocation). It would be permissible to admit women. The name of the college would be the "Owens Extension College". Once this new college was established, the old Owens College was to be incorporated with it and, at the same time, its name, "The Owens Extension College" was to be changed to "The Owens College". [1] Thus, by the use of this clever scheme, Owens College would still be there but with a new constitution which in certain respects, i.e. the admission of women, was inconsistent with John Owens' will. As would be expected the scheme drew criticism from those who said it violated Owens' intentions. Nevertheless, the supporters of the scheme had their way. The plan was shepherded through parliament by Acts of 1870 and 1871 though not without opposition. [2]

Meanwhile a committee was looking for a new site, the Quay Street area being considered unsuitable for expansion for a number of reasons. The chairman of the site committee was Murray Gladstone, chairman of the Manchester Royal Exchange and a cousin of W.E. Gladstone. A number of sites were inspected but the committee found that as soon as the intended purpose of the land became known the asking price increased. Finally, in March 1868, Gladstone purchased secretly a large and cheap out-of-town site of about 4 acres "which was being offered for private sale on favourable terms" in his own name and at his own risk. It cost £29,100 and comprised 19,164 square yards forming an 'oblong plot with a frontage of 127 yards to Oxford Street* and extending about 152 yards down Burlington Street on the south and 177 yards down Coupland Street on the north.' [3] Gladstone's enterprising speculation was approved by the executive committee and the land was duly conveyed from Gladstone's private ownership to that of Ashton as chairman. Alternative sites in Ardwick (Stockport Road), Cheetham Hill and Nelson Street had already been inspected and rejected. Thus was the present site of Manchester University acquired.

The building sub-committee did their homework thoroughly, having obtained and studied plans of newly-erected academic buildings in Glasgow, Oxford, Cambridge, London, Cork and Belfast and in the summer of 1868 Roscoe and the Principal (Greenwood) made a tour of inspection of ten European universities and colleges, taking particular note of the layout of those with a good reputation for science. It had been decided that the sum of at least £100,000, and preferably £150,000 would be required to set

---

*The main road was then Oxford Street from St. Peter's Square to the start of Wilmslow Road at Moss Lane East. Oxford Road did not exist until 1902-03 when the main road was designated Oxford Street from St. Peter's Square to Medlock Bridge and Oxford Road to the south of that.

up the new college. Money was coming in well, and early in 1868 the great Victorian architect Alfred Waterhouse was appointed to design the new Extension College. After much debate it was decided the building should be set some distance back from but parallel to Oxford Street. The construction of the first part of the college, now usually known as the 'main building' or, more recently, 'The John Owens Building', then went ahead.

The first stone of the new college building was laid ceremoniously by the seventh Duke of Devonshire (who had accepted the invitation to become President of the College) on Friday, 23rd September 1870 in the presence of a large and distinguished company. Thomas Ashton stated that the new buildings would accommodate about 600 day students and a much larger number of evening students. The cost of the buildings, fittings and site was £90,000 and £30,000 was still needed. The architect Alfred Waterhouse presented the Duke with an engraved trowel while Thomas Clay, the contractor, then deposited in a cavity below the foundation stone a glass bottle containing current coins, copies of the three Manchester daily newspapers, the London *Times* and printed documents issued by the Extension Committee. The cavity was covered with a lead plate bearing on its lower side an inscription:

"The first stone of this building erected for the Owens College, Manchester, was laid by his Grace the Duke of Devonshire, KG, FRS, etc., the first president, September 23rd, 1870. Architect: Alfred Waterhouse, Esq." [4]

View of Alfred Waterhouse's fine new building seen from Oxford Street, c. 1876. The whole of the college was housed in this building

The building, completed in 1873 and opened for the new term on 6th October of that year, must have been a fine sight viewed from Oxford Street. Other buildings, also designed by Waterhouse, were added in stages over a 30-year period to complete the main quadrangle, the last of them being the Whitworth Hall block completed in 1902. [5]

Although the college had succeeded, through a change in the constitution, in its aim to admit women, they were enrolled for the next few years only into what was termed the 'Women's College' which acted rather as a separate department of the University. However, this was only an interim measure.

With the college now housed in its fine new buildings expansion continued not only in student numbers but in staff and the introduction of new academic disciplines. Owens was still bound by Acts of Parliament to the restriction of awarding the London degree and its sights were set on becoming a university in its own right - 'a university of the north'. But there were objections from many quarters. Durham already had a university so Owens could hardly aspire to the title of university of the north. Furthermore, the attainment of university status was hard to come by and if one aspirant succeeded it would be very difficult for others to achieve it. This fact was very clear to other colleges in the industrial north, namely the Yorkshire College in Leeds and the Liverpool College, both of whom saw themselves as strong candidates for university status in due course. They knew that if Manchester became a university in its own right their chances of following would be poor, at any rate for many years ahead. Leeds objected to the proposals that were afoot for Owens to become the University of Manchester and made a counter suggestion. They proposed that a federal university be instituted with incorporated colleges. Of these colleges Owens would be the first. Others could apply for admission later on a footing of equality. The Yorkshire College was aware that it was not yet ready to apply itself, but wished the door to be left open for the future. Owens was unhappy about this; they proposed that Owens should become the University of Manchester with representation of other colleges on its administrative and academic boards. But the other big towns and cities of the north did not relish the idea of being relegated to a subordinate position, believing this to be even worse than Manchester setting up independently with no connection to the other colleges. Their civic pride was at stake.

In 1878 a counter-proposal from the Yorkshire College requested:

"1. That a charter should not be granted to Owens College but to a new corporation with powers to incorporate the Owens College and such other institutions as might fulfil the necessary conditions;

2. The new University should not bear the name of a town or of a person of merely local distinction."

Owens reluctantly agreed. Under the new scheme, Owens would be a constituent college of the new federal university and if the Yorkshire College or any other college were to be admitted it would be the equal of, not subordinate to, Owens. Furthermore the new university would not be the University of Manchester or the John Owens University. It was finally agreed that the new university would be The Victoria University, empowered to award its own degrees. The inclusion of the name of the much-respected monarch probably helped to overcome obstacles to the creation of the University. A new charter was drawn up (dated 20th April 1880) which allowed the conferment of degrees in Arts, Science, Law and Music, though degree courses in music were to be some years away. For legal reasons no provision was made at this stage for degrees in medicine but the authority to introduce such degrees came in 1883. The first medical students graduated in 1892. Under the terms of the new Charter of 1880 the degree of B.A. or B.Sc. was offered to the Associates of Owens College (those who had completed their course to the satisfaction of their examiners) at the date of the Charter. Initially Owens was the only constituent college of the Victoria University but the University College, Liverpool was admitted on 5th November 1884 and the Yorkshire College, Leeds on 3rd November 1887. No other colleges were admitted subsequently. The federal university lasted from its inception in 1880 until October 1903 when Owens College became the Victoria University of Manchester in its own right.

Osborne Reynolds and the Engineering Department

As explained at the end of Chapter 2, Osborne Reynolds was appointed in 1868 on the recommendation of very famous engineers and had the full support of local industry. Equally, his appointment at the age of 26 was approved by the professors at Owens who saw him as a bright young academic. It was a bold and imaginative choice. Born in Belfast of an Anglican clerical family, engineering was in the blood; his father, also called Osborne Reynolds, was a competent engineer, despite making the church his vocation, and had four technical patents relating to agricultural machinery registered in his name. Young Osborne Reynolds already had practical and academic achievements at the time of his appointment to the Manchester chair. At the age of 19 he had been apprenticed to a mechanical engineer in order to become a working mechanic before entering Cambridge in October 1863. At Cambridge he excelled at mathematics, graduating B.A. as seventh wrangler in 1867. He was elected a fellow of Queens' College and simultaneously set out on a civil engineering career in London. He then heard of the new chair in Manchester and applied, stating in his letter:

"From my earliest recollection I have had an irresistible liking for mechanics, and the physical laws on which mechanics as a science are based. In my boyhood I had the advantage of the constant guidance of my father, also a lover of mechanics, and a man of no mean achievements in mathematics and their applications to physics . . . having now sufficiently mastered the details of the workshops, and my attention at the same time being drawn to various mechanical phenomena, for the explanation of which I discovered that knowledge of mathematics was essential, I entered the Queens' College, Cambridge, for the purpose of going through the university course, previously to going into the offices of a civil engineer." [6]

In his inaugural lecture, entitled "The progress of engineering considered with respect to social conditions in this country", Reynolds, like Roscoe before him, avowed to link the city and the college through his professional work. Like Roscoe he was keen to press home the need for well-trained scientists.

"Though another mile of railway should never be made, there will still be room for all the engineering skill the country can find. As the ladder of science gets higher it requires better training and greater dexterity to reach the top. The more perfect our mechanism becomes the more knowledge and labour it requires to improve it." [7]

The early years of Reynolds' career at Manchester were marred by personal tragedy. He had married in the June of 1868, the year of his Manchester appointment, but his wife died the following year after the birth of a son. This son died in 1879. In 1881 Reynolds married again; three sons and a daughter were born of this marriage. One of the sons graduated in engineering at Manchester in 1908.

Possibly due in part to the trauma of his wife's death Reynolds' Manchester career got off to a poor start. Enrolments to his course fell and in July 1870 the Owens Principal reported to the Trustees on "the non-success of the engineering classes". He concluded: "The Professor has a genuine enthusiasm for his subject and is, I am sure, sincerely devoted to the interest of his students. But it would be an affectation to conceal that our anticipations as to these classes have not been realised and that the great falling off in numbers in the second session is a cause of disappointment, and, I had almost said, of mortification." [8] Colleagues found him difficult, cynical, paradoxical and sometimes perverse but this was a surface veneer

---

N.B. The life and work of Osborne Reynolds are dealt with much more fully than is possible here in the excellent papers by Jack Allen and J.D. Jackson (see references on page 43).

and no one could mistake or underestimate Reynolds' ability and enthusiasm for engineering, for the college, and for the local community.

Perhaps because of these qualities things began to pick up. He participated actively in the Scientific and Mechanical Society (established in 1870). Through this and his links with local industry he soon became a well known and respected figure not only within the college but in the city as a whole. His research work began in earnest though in the first five years of his tenure, 1868 to 1873, he was hampered by having to work in the cramped, overcrowded premises in Quay Street which were totally unsuited to the needs of engineering. On his appointment Reynolds wasted no time in setting up his three-year engineering course. In the first two years the standard subjects of machines, strength of materials, structures, surveying etc. were covered. The lectures in Reynolds' senior class (third year) included the dynamics of machinery, balancing of machinery, gauges, brakes and governors, prime movers, the steam engine and hydraulic machinery. The recommended text books were Rankine's *The Steam Engine*, Baker's *Mechanics*, and Willis's *Elements of Mechanism.* It will be noticed that no electrical machines or any other electrical devices figured on this list. The course was devoted entirely to civil and mechanical engineering. More will be said about this later.

At first Reynolds had to lecture in all the courses himself as he comprised the entire staff in engineering, but in 1869 an assistant, J.B. Millar was appointed. A graduate of Queen's University, Belfast, Millar stayed with Reynolds for many years. One of Reynolds' early students was J.J. Thomson (knighted in 1905), who was to become one of the most famous pioneers of nuclear physics and who had a long and distinguished scientific career at Cambridge. Born in Cheetham Hill, he had joined Reynolds' class as a 14-year old in 1870. In his *Recollections and Reflections* published towards the end of his life Thomson described Reynolds' lectures:

"As I was taking the engineering course, the Professor I had most to do with in my first three years at Owens was Osborne Reynolds, the professor of engineering. He was one of the most original and independent of men and never did anything or expressed himself like anybody else. The result was that it was very difficult to take notes at his lectures so that we had to trust mainly to Rankine's text books. Occasionally in the higher classes he would forget all about having to lecture and after waiting ten minutes or so, we sent the janitor to tell him that the class was waiting. He would come rushing into the room pulling on his gown as he came through the door, take a volume of Rankine from the table, open it apparently at random, see some formula or other and say it was wrong. He then went up to the blackboard to prove this. He wrote on the board with his back to us, talking to himself, and every now and then rubbed it all out and said that was wrong. He would then start afresh on a new line, and so on. Generally towards the end of the lecture he would finish one which he did not rub out and said that this proved that Rankine was right after all." [9]

As the years progressed the course in engineering seems to have remained much the same, covering the basic aspects of civil and mechanical engineering. In 1880 the syllabus was given as:

1st year: Surveying and levelling, hydraulic surveying, surveying instruments, quantities and estimating.
2nd year: Applied mechanics, forces, strength of materials, structures, mechanical connexion, dynamics of machinery.
3rd year: Theory of machines, prime movers, theory of structures.

The Engineering Department was fortunate to benefit in this period from two major bequests. In 1875-76 Charles Clifton of New Jersey, U.S.A. left £21,500 to the department, and about the same time Charles Frederick Beyer of the firm of Beyer & Peacock bequeathed £100,243 for the endowment of professorships in Science. Consequently Reynolds' engineering chair became the Beyer Chair of Engineering, the other Beyer chairs going to mathematics and zoology.

By modern standards the department remained small but the proper study of engineering requires large machinery and even though the new Waterhouse building afforded much better accommodation than had Cobden's house in Quay Street, Reynolds' Department was cramped for space throughout the 1870s and much of the 1880s. After 1873 his department and the physics department occupied the basement of the original Waterhouse building with lecture and other rooms above, physics towards the southern end of the building and engineering at the northern end.

In 1884 Reynolds proposed the construction of a new engineering laboratory. The university authorities were sympathetic but were unable to proceed through lack of money. An appeal for funds was set up but the amount raised was inadequate for Reynolds' requirements; however, the need for new accommodation was so acute that in 1885 the college authorities relented and resolved to erect a new building for the department. The financial problem was alleviated when Sir Joseph Whitworth, who had long been a supporter of Owens College and of the teaching of engineering in particular, died in 1887, and his legatees generously offered to meet the outstanding costs. Reynolds helped with the planning of the building which was carried out under the supervision of the architect, Alfred Waterhouse. The plans included testing machines, an experimental compound steam engine, experimental boiler and furnaces, an Otto gas engine, dynamos, electric motors, pumps, precision instruments and workshops for carpenters, smiths and fitters.[10] The new laboratory, known as the Whitworth

Laboratory, was constructed in 1886-87 and opened in 1887. In 1890, 22 years after Reynolds' appointment, the staff consisted of Reynolds, a senior demonstrator (J.B. Millar) and a junior demonstrator (H. Bamford).

Turning to the matter of research in the engineering department during these years, it must be stated that Osborne Reynolds' research was outstanding, to the extent that his name became famous internationally. The area of research in which he achieved perhaps most distinction was hydraulics and heat transfer. Terms such as the Reynolds number, Reynolds equations, Reynolds stresses and Reynolds analogy became part of established scientific terminology. As a scientist and engineer he was one of the foremost figures of his generation; Jack Allen, who studied Reynolds'

Professor Osborne Reynolds, who established engineering as an academic discipline at Manchester and who was one of the most gifted men of his era. A photograph dating from 1896 when Reynolds was about 54.

life and work, describes Reynolds as a major master and ventures so far as to suggest that he was the most distinguished man ever to occupy a chair of engineering in any British University. [11]

A detailed account of Reynolds' research is clearly beyond the scope of this chapter. However, his contributions to science and engineering have been summarised by J.D. Jackson [12] as follows:

Reynolds the scientist
1. Papers on 'out-of-door physics' (1870 - 1881):
   Solar and cometary matters; natural phenomena.
2. Papers on the physics of gases, liquids and granular materials (1874-1903):
   Kinetics of gaseous fluids; physics of liquids; properties of granular media.
3. Papers on fluid motion and turbulence (1872-1894):
   Wave motion, vortex motion, laminar and turbulent flow in pipes; the Reynolds number; the dynamical theory of fluid flow; hydrodynamic lubrication; visualization of the internal motion of fluids.

Reynolds the engineer
1. Papers on applied fluid dynamics and hydraulics (1872-1894):
   Ship propulsion and dynamics; pumps and turbines; modelling of rivers and estuaries; cavitation.
2. Papers on heat transfer and thermal power (1873-1897):
   Condensation of steam; analogy between heat transfer and momentum transfer; thermodynamics and heat and work; thermodynamics of gas flow; steam engine trials; determination of the mechanical equivalent of heat.
3. Papers on the dynamics of machines and the mechanics of materials (1872-1902):
   Rolling friction, dynamics of machines, repeated stresses and fatigue.

This list illustrates the wide range of Reynolds' interests. From the electrical engineering viewpoint the most interesting of Reynolds' papers are the early ones concerned with a field of investigation described by J.J. Thomson as 'out-of-door physics', and some of those on the subject of the kinetics of gaseous fluids. The papers are listed in Appendix 5.

The first five papers on the list, on 'out-of-door physics', are concerned with various aspects of electrical discharge in gases, which is of great interest in certain branches of electrical engineering. Reynolds probably undertook this type of work at that time (the papers all date from the first few years of his professorship) because his department was still housed in the Quay Street building where experimental facilities would have been very sparse. The two papers on the communication of heat were on a somewhat related topic to the first group and led to an explanation by Reynolds of the modus operandi of the Crookes Radiometer. The paper on John Faulkner's electromagnet is pure 'electrical engineering'.

Reynolds moved into Alfred Waterhouse's fine new building in October 1873, along with all the other departments of the college, and it was in the new building that the greater part of his work was done. It is perhaps no surprise that after the move from the cramped quarters in Quay Street, where laboratory facilities were very limited, his interest in such areas as atmospheric electricity waned and he turned his attention to the various fields of mechanical engineering in which he became so famous.

A hundred years have passed since Osborne Reynolds was at the height of his powers. Looking back on his work there is no doubt that as a mechanical engineer he was a giant of his time. His achievements ensured that his name entered the Oxford English Dictionary which devotes a paragraph to him [13], and into the scientific terminology of botany and zoology, such is the generality of application of some of his discoveries. From the standpoint of electrical engineering, his worth must be valued in a different light because he hardly did more than dabble in this area. He and the physics professors were a close-knit group, each of whom knew what the others were doing, and he left electrical matters to them. His value to the electrical engineering discipline is that he did probably more than anyone to establish the status of the professional engineer in universities. Until he came on the scene engineers tended to be regarded as in a class below 'pure' scientists but Reynolds would have none of it. This is illustrated in an exchange of correspondence with the Professor of Mathematics at Owens, Horace Lamb (later Sir Horace), who, after reading the draft of one of Reynolds' papers, referred to 'the empirical formula adopted by engineers'. In a tetchy letter written to Lamb from his home in Lady Barn Road, Fallowfield, on 5th April 1895, Reynolds remarked: " . . . Nor is it polite or true to speak of 'the *empirical* formula adopted by engineers', since it is engineers who have done the scientific investigations which alone have given us accurate data". The letter concludes: "I am obliged for the trouble you [have] taken to bring my work in . . . I fear my criticism will bother you but you will take what notice of it you like." [14] Reynolds' pioneering work in establishing engineering as a true university discipline cannot be overestimated.

Arthur Schuster and the Physics Department

It will be recalled from Chapter 2 that the academic discipline of physics at Owens began when the college was created, with Archibald Sandeman as Professor of Natural Philosophy from 1851 to 1860 and Robert Clifton from 1860 until his resignation in 1866. Clifton was replaced by William Jack, a Scot, who remained in the post only until 1870 when he left to become editor of the *Glasgow Herald.* During the 1860s no electrical research was undertaken in physics but it is clear from the college calendars that there was instruction in such matters as Ohm's Law, the voltameter and electrolysis. In 1870 the syllabus of the experimental physics course consisted of sound, light, heat, mechanics, electricity and magnetism.

The Owens' authorities, and Roscoe in particular, had high expectations of physics at Manchester and in 1870 the bold decision was taken to appoint two professors, a senior and a junior, to replace the departing Jack, the nature of the appointments depending on the availability of candidates. Balfour Stewart, then Director of the Kew Observatory, was approached. After some reluctance, for he was perfectly happy in his existing post, he agreed to come as the senior professor. Stewart proposed that the physics course be split into two parts. The 'A' course would include statics and dynamics, forces and properties of matter; the 'B' course would include areas of special interest to Stewart including the "energies of nature and their laws of transmutation" and "cosmical physics" including "a sketch of astronomy, meteorology and terrestrial magnetism". He would assume the responsibilities of this course and the other professor, when appointed, would supervise the 'A' course. Stewart would also be director of the proposed physical laboratory (the college was still in Quay Street when he was appointed) and explained his views:

"The Physical Laboratory would not only be used in experimental [illustrations] of certain laws enunciated in the lectures but I think that some observational and also some experimental research ought always to be going on in order that the more advanced students should be brought into contact with nature. Then they ought to be taught the use of the various instruments and set to devise and work out experiments. They ought also to be taught the philosophy of experiment:

1. To pay attention to and evaluate all sources of error giving due weight to each and dismissing those that ought to be disregarded (thus it is a very common mistake to give inordinate importance to some utterly useless refinement).
2. To pay strict attention to [natural] phenomena as ... something new.
3. To reach the legitimate conclusion from an experiment, no more and no less." [15]

These were conditions rather than suggestions but the college raised no objection and Stewart was duly appointed. The second, junior, professorship went to Thomas H. Core, a graduate of Edinburgh, who had experience in various teaching jobs. The arrival of Core was timely because soon afterwards Balfour Stewart was severely injured in a major train crash; he was left disabled and unable to carry out his duties fully during a long convalescence. However, he slowly recovered, at least partially, and he and Core reorganised the course along the following lines:

1. An experimental course comprising a mechanics class taught by both Core and Stewart, a physics class with an experimental section and an energy class, both taught by Stewart.

2. A mathematical course comprising junior and senior sections, the former sometimes taught by Stewart and the latter by Core.
3. A practical physics course under the direction of Stewart.

There were 148 students in the physics class in 1871-2. Amongst the students in that year were two bright youngsters destined for distinction, J.H. Poynting and Arthur Schuster. Schuster, born in Frankfurt in 1851 of a well-to-do Jewish family, emigrated to Britain in 1870 with his family following the embodiment of Frankfurt into the German Empire after the Prussian-Austrian war. In the same year he embarked on spectroscopical research and then studied with Kirchhoff at Heidelberg where he obtained his Ph.D. in February 1873. In October of the same year he joined the Physical Laboratory at Owens as Honorary Demonstrator, this being the first session in which the college occupied the new building on Oxford Street, and immediately started to show his exceptional ability. At that time three rooms in the basement were provided for the use of the Physics Department. There was one other demonstrator, Francis Kingdon. Stewart, Core, Kingdon and Schuster comprised the entire staff. From time to time, as the number of physics students grew, further rooms were allocated. Though the accommodation provided was modest the physical laboratory was probably then the largest in England outside London.

When he was well enough Stewart continued his researches on terrestrial magnetism and related matters and a long series of papers followed (see Appendix 5) - the first major research effort in physics at Manchester. In 1874 the late E.R. Langworthy bequeathed £10,000 for the endowment of a chair in physics and in 1879 Stewart was named Langworthy Professor. As a teacher he was evidently more effective in the laboratory than in the lecture theatre. Schuster recalled:

"He was not a good lecturer and had difficulty keeping order in the lecture room . . In the laboratory he was an inspiring teacher and it would be no exaggeration to say that he was the godfather of much of our modern science." [16]

In similar vein, J.J. Thomson, who studied in the physical laboratory as well as under Reynolds, recalled the effectiveness of Stewart's laboratory teaching:

"We were allowed considerable latitude in the choice of experiments. We set up the apparatus for ourselves and spent as much time as we pleased in investigating any point of interest that turned up in the course of our work. This was much more interesting and more educational than the highly organised systems which are necessary when the classes are large. Balfour Stewart was enthusiastic about research and succeeded in imparting the same spirit to some of his pupils." [17]

During the 1870s and early 1880s Schuster developed into an outstanding physicist with a long string of publications to his name, mainly in the field of spectroscopy and the conduction of electricity through gases. In the latter area he showed that an electric current was passed through a gas by ions and that once the gas was ionised ("dissociated" was the term then used) a small potential was sufficient to maintain the discharge. He was also one of the first to indicate the path towards determining the ratio e/m for cathode rays using a magnetic field, a method which was ultimately to lead to the discovery of the electron. In the mid 1890s, following Röntgen's researches, he suggested that X-rays were transverse vibrations of the aether of small wavelength. In the next few years he spent a lot of time away from Manchester working in various laboratories. This included research at Göttingen under Weber, with Helmholtz in Berlin, and five years at the Cavendish Laboratory in Cambridge, working first with Clerk Maxwell and later with Lord Rayleigh with whom he collaborated in the determination of the Ohm in absolute measure. He was elected a Fellow of the Royal Society in 1879 at the age of 28. In 1881 a chair of Applied Mathematics was founded at Owens and Schuster was appointed. He remained at Manchester for the rest of his university career.

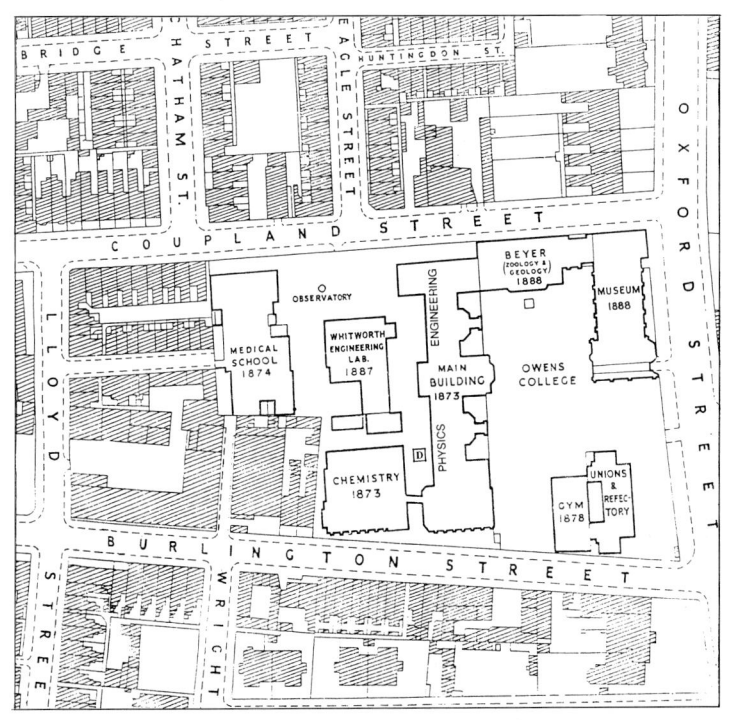

Plan showing the location of the Engineering and Physics Departments in the new Owens building, 1890. The Whitworth Laboratory and the dynamo house (marked 'D') were not completed until 1887 and 1888 respectively.

With Balfour Stewart as Langworthy Professor of Physics, Thomas Core as Professor of Physics and Schuster as Professor of Applied Mathematics the status of the Physics Department steadily increased and the number of papers produced grew year by year. The physics course in the mid 1880s had an increased electrical content compared with that of a few years previously. "Middle physics" included in the syllabus magnetism and electricity, units (theoretical and practical), voltaic cells, electrical machines, galvanometers and electrometers, the measurement of potential, current, resistance and capacity. A course in "observational physics" included terrestrial magnetism, reflecting Balfour Stewart's interest in the subject.

In 1887 Balfour Stewart died at the age of 59. Schuster was appointed without competition to the Langworthy Chair of Physics, and Applied Mathematics was left in the capable hands of Horace Lamb who had been appointed Professor of Mathematics in 1885. The eminent triumvirate of Schuster (physics), Reynolds (engineering) and Lamb (mathematics) were to be the backbone of the physical sciences at Owens for almost the next 20 years.

Professor Arthur Schuster aged about 55. He introduced electrical technology into teaching and research at Manchester

Thomas Core does not appear to have been interested in research and did not publish any papers. Schuster was both teacher and researcher; his interests were wide and not confined to what might be called "pure physics".

In the years 1868 to 1885 great advances had been made in all branches of electrical engineering in the outside world. They included improvements in dynamos, motors, telegraphy, electro-plating, cells and batteries, lighting, telephony, measurements, underground cables, microphony, arc furnaces, electric railways, public electricity supply, materials (e.g. the use of vulcanised rubber in cable insulation) and transformers. There was even a variable-speed a.c. motor on the market. Many successful companies were producing electrical goods in large quantities, including Siemens, Ferranti, Crompton, and Mather and Platt. Some first-rate scientists were turning their attention to electrical engineering and starting to make a name for themselves, amongst them John Hopkinson, a former Owens College student. It is hardly surprising therefore that Schuster, now well established in the Langworthy Chair and with the Physics Department making its way very successfully, should turn his attention to adding electrical engineering to the syllabus. This would require the acquisition of a suitably equipped laboratory to be fully effective. By the mid 1880s plans were afoot to create such a laboratory and to teach electrical engineering.

During the 1888-89 session the new dynamo house, situated in the courtyard behind the Physics Department (see plan on p. 39), was opened, and a course in electrical technology was introduced. The Owens College calendar of 1889-90 reported on the facilities available, and the course:

> The dynamo-house contains a gas engine of 7 h.p., two direct current and two alternate current dynamo machines specially constructed for purposes of instruction. Students wishing to devote themselves to Electrical Technology should have gone through a course of the ordinary physical measurements, such as is given in the first year's course of the Physical Laboratory. The special course of technical electricity will include the following:
> I. Preliminary: Magnetic behaviour of iron and steel; laws of electro-magnetic induction; distribution of currents in conductors; testing and calibration of electrical instruments; passage of currents through electrolytes; primary and secondary batteries; testing of cables and discovery of faults in a telegraph line.
> II: Investigation of a magnetic field; direct methods of winding armatures; alternate current dynamos and their properties; characteristic curves; loss of power due to heating; coupling of dynamo machines.
> III: Electric lighting; voltaic arc; properties of carbon filaments; photometry.
> IV: Telephone; microphone; induction balance; duplex telegraphy.

The fees for the whole session (5 days per week) were £17 - 17 - 0. (17 guineas). This was slightly more than the Engineering Department's fees of 16 guineas.

A search through the examination questions for the physics course in 1890 indicates that the subject was examined thoroughly. There were questions on:

Magnetic force and magnetic induction; experimental determination of a magnetic force; induced magnetism; magnetic fields induced by currents; coefficient of self induction; magnetic potential; magnetic moment; magnetic shell circuits; laws of electrolysis; heat generated by a wire; deposition of silver; self and mutual induction; effects produced by a magnet inside a coil; description of the Gramme armature of a dynamo machine and explanation of its action.

Questions on all the above were set by Professor Schuster. Professor Core also set an electrical paper with questions involving:

Faraday's ice-pail experiment; heat produced in discharging a 'jar'; experiments to show that an electric spark lasts only a very short time; how to get the greatest current from a number of interconnected cells; internal resistance of cells; tangent galvanometer; description of an experiment in which the energy of an electric current is converted into that of a rotating mass; description of an experiment in which the existence of a back current in a voltameter is proved.

Bearing in mind that electronics as a subject did not yet exist, the instruction given and the examination of it seems to have been thorough.

The new course in electrical technology proved successful. By 1895 the staff of the Physics Department had grown to five, namely the two professors (Schuster and Core) and three others, C.H. Lees (Lecturer), W. Gannon (Demonstrator and Assistant Lecturer) and A. Griffiths (Demonstrator and Assistant Lecturer). Each of the three junior staff was interested in electrical engineering and published papers in that area (see Appendix 5). Lees had studied electrical engineering at Central Technical College, London, under Professor Ayrton and in 1901 became a Council member of the Manchester section of the Institution of Electrical Engineers.

Whilst the Physics Department was steadily increasing in size the college in general was also growing, as reflected in the rise in the number of professors and the number of students as the years went by, as follows:

Year	Number of Professors	Total Number of Students
		(Figures in brackets = No. of women)
1855	6	52
1860	7	69
1865	9	113
1870	12	264
1875	20	545
1880	22	633
1885	24	745 (66)
1890	25	811 (61)
1895	28	992 (104)

In the spring of 1896 one of the three non-professorial members of staff of the Physics Department, William Gannon, left to take up an educational appointment in Staffordshire. To replace him Council resolved "That the Senate be authorised to appoint a Demonstrator and Assistant Lecturer in Physics with the status of a Junior Demonstrator and Assistant Lecturer at a salary of £100 per annum in the place of Mr Gannon." [18] To cope with the increasing demands of electrical technology Professor Schuster decided to appoint a specialist whose responsibility it would be to organise the electrical engineering courses offered by the Physics Department. In 1896 Robert Beattie, a 23 year-old Scottish graduate of Durham University, was appointed to fulfil the task. He was to remain in charge of electrical engineering matters in the Faculty of Science for the next 42 years.

References: Chapter 3

General:
Same as the general references listed in Chapter 2 (see page 24).

Specific to the text:
1. E. Fiddes (see general references), p.65.
2. Joseph Thompson (see general references), Chapter 14.
3. W.H. Chaloner (see general references), p. 8.
4. Ibid., p.18.
5. H.B. Charlton (see general references), p. 170.
6. Jack Allen. *The Life and Work of Osborne Reynolds.* First chapter of *Osborne Reynolds and Engineering Science Today.* Ed. J. D. Jackson and D. M. McDowell. Manchester, 1970. See pp. 2 - 3.
7. Ibid., pp. 15-16.
8. Minutes of the Trustees' Meetings for Educational Purposes, Owens College. **4**, p. 69. Manchester Central Reference Library.
9. J.J. Thomson. *Recollections and Reflections.* G. Bell and Sons Ltd., London, 1936.
10. Joseph Thompson (see general references, above), pp. 566 - 567.
11. See Jack Allen (above), p. 78.
12. J.D. Jackson. *Osborne Reynolds: scientist, engineer and pioneer.* Proc. Roy. Soc., Lond. A, **451**, pp. 49 - 86, 1995.
13. *Oxford English Dictionary*, current (full) edition. See under "Reynolds".
14. See Jack Allen (above), pp. 30 - 33.
15. Letter from Stewart to Roscoe dated 2nd June 1870. British Chemical Society Archives.
16. A. Schuster. *Biographical Fragments.* Macmillan, London, 1932.
17. See J.J. Thomson, *Recollections and Reflections*, above.
18. Minutes of the Council meeting of Owens College held on 15th May 1896. (John Rylands University of Manchester Library).

Chapter 4
1896-1907: The New Physical Laboratory and the end of the Schuster era.

The appointment of Robert Beattie in 1896 was a reflection of the increasing relevance of electrical technology at that time. The local importance of electrical engineering was such that youths from the Manchester environs were clamouring to be educated in the subject. Mather and Platt had been producing large electrical machines in Manchester for years (one of their notices from the 1887 Manchester Exhibition celebrating the Queen's Jubilee is shown on the opposite page) and Ferranti opened at Hollinwood in 1896. The British Westinghouse Electrical & Manufacturing Company (later to become the Metropolitan Vickers Electrical Company) was soon to set up in Trafford Park (1902), and there were many other local electrical firms. Electrical engineering was the 'up and coming' industry and the Physics Department at Owens clearly had a responsibility to ensure that the educational needs were met in an appropriate manner. Hitherto the teaching had been in the hands of physicists who were not specialist electrical engineers. Beattie's task would be to take a leading part in the teaching of the electrical technology and, preferably, do some research.

As Robert Beattie was to be in the forefront of electrical engineering at the university for over 40 years it is worthwhile taking a brief look at his background.[1] He was born at Fordoun, Kincardineshire, on 18th May 1873, the son of a farm overseer, also called Robert Beattie, and his wife Elizabeth (née Watt). The small house where he was born stood in the middle of open countryside with views of agricultural land all around. He was educated first at a local school and later, after the family moved to Johnston Mains, Laurencekirk when he was 12, at Robert Gordon's College, Aberdeen. In 1890 he entered the Durham College of Science, Newcastle-on-Tyne, as an open exhibitioner. The college was a constituent college of the University of Durham and all his subsequent degrees, (B.Sc., M.Sc. and D.Sc.) were conferred by the University of Durham. After the award of scholarships in 1891 and 1892 he obtained the Honours B.Sc. Degree in Experimental Physics in 1892. From 1892 to 1894 he undertook research at Durham College of Science and in 1894 was awarded an 1851 Exhibition Scholarship. From that year until 1896 he did further research in the same department and also in the electrical engineering laboratories at University College, London, under Professor J.A. Fleming who later (in 1904) invented the thermionic valve. He was then appointed Demonstrator and Assistant Lecturer in Physics at Manchester. The other staff of the Department in

1896-1907: The new physical laboratory and the end of the Schuster era

Advertisement for the Queen's Golden Jubilee Exhibition in Manchester, 1887

1896 when Beattie arrived comprised Arthur Schuster (Langworthy Professor), Thomas Core (Professor), C.H. Lees and Albert Griffiths. The latter two were Demonstrators and Assistant Lecturers, like Beattie. The academic staff therefore numbered five in all. At the same period the Engineering Department's staff also comprised five men; Osborne Reynolds (Professor), J.B. Millar and C.B. Dewhurst (Assistant Lecturers), and two Demonstrators, G.W. Wilson and W.H. Moorby.

The need for a new Physical Laboratory

At the time of Beattie's arrival the Physics Department was rapidly outgrowing its cramped quarters in the lower floor of the main building. It seemed clear to Schuster that nothing less than a new building would serve the requirements of physics if any expansion were to be possible and early in 1897 Schuster applied for Council's authorisation for the building of a new Physical Laboratory. Schuster's Memorandum to Council was about five pages in length but the requirement for the new building was so crucial to the future of electrical engineering at Owens that it has been decided to reproduce much of it here. Not only does the memorandum outline the reasons why a new building was necessary, but it also provides a useful insight into what was happening in some other physics departments at the time, and Schuster's ideas concerning the future.

The Owens College Memorandum prepared by Professor Arthur Schuster.

For some years past considerable difficulties have been felt in the Department of Physics owing to the insufficient accommodation provided for the Practical work in that subject. The following numbers will illustrate the gradual growth of the department in the last 20 years.

Session		
1875-76	Number of students attending the Physical Laboratory	13
1880-81	" " " " "	27
1885-86	" " " " "	62
1889-91	" " " " "	105
1895-96	" " " " "	183

The reason for this rapid growth is to be found in the fact that a knowledge of Physics has become indispensable to the Engineer and the Chemist, while the practical applications of Electricity are attracting an increasing number to the Physical Laboratory. Considerable importance is also being attached to the subject in the Preliminary training of medical men and nearly all English Universities have instituted examinations in Laboratory work at the earliest stage at which Physics is taken up as a subject of examination.

Before entering into the details of the special requirements of the College it may be well to give a short account of the great efforts which have of late years been made in other Colleges and Universities to erect buildings in which the teaching of practical physics may be satisfactorily carried on. Thirty years ago physical laboratories were practically unknown, a few rooms set apart for experimental work being considered sufficient to meet all requirements.

The first great impulse towards a better state of things in this country was given by the late Duke of Devonshire, who built the Cavendish Laboratory at Cambridge about the year 1873, at a cost which at the time was stated to be £20,000. The result has been that an important school of Physics has been established in Cambridge, and it is a decided advantage, to anyone who wishes to obtain a teaching appointment in this country, to have passed through that school. Abroad larger buildings even than that at Cambridge were erected at Strassburg and Berlin. In size and completeness we shall never be able to rival these in Manchester, but we cannot contemplate with indifference our inferiority to such Universities as Tubingen, Freiburg and Erlangen. Within the last few years Physical laboratories have been erected in all these places which offer considerably greater facilities than we can give our students here. I have special information with respect to the laboratory at Erlangen, which is a building standing by itself, 100 ft long and 62 ft wide, consisting of cellar and four stories, the large lecture room (50 ft by 45 ft) being an annex to the main buildings. The cost of the building was £8,400, that of the fittings was £2,200, the total cost being £10,600.

It cannot be said that the students in the Universities which I have named are either more numerous than ours, or require a more specially physical training, on the contrary it has not hitherto been the custom in Germany for intending electrical engineers to enter the University, while the practical branches of Physics form an important part of the teaching at Owens College. Some of the German Universities are beginning to move in the same direction and I am informed that at Cottingen it is intended to build a Laboratory in which special provision for electrical teaching will be made. If this movement becomes at all general we may look forward to a further increase in the efficiency of the Physical Laboratories in Germany.

The students using the Physical Laboratory at Owens College may be divided into four classes:

1. Those requiring elementary instruction; their numbers of late years has varied between 100 and 120.
2. Those taking up Physics as one of their University subjects, either for Honours or for the ordinary degree of B.Sc., their number averages about 18, and they are mostly preparing for the teaching profession.
3. Chemical Honours Students requiring instruction chiefly in those parts of Physics which have a chemical bearing such as Spectrum Analysis, or Sacharimetry. Their number varies from 12 to 20.
4. Those taking up electricity as a Profession. Their number of recent years has been about 20.

1. With regard to the elementary course every student, whether he is to be a Chemist, Engineer, or a Medical man, has to pass through it, for it is found that the teaching given in lectures alone cannot yield satisfactory results, unless it is supplemented by practical work taken into conjunction with the lectures. The room we are using for this purpose occupies an area of about 990 square feet (30 ft by 33 ft) and we have occasionally as many as 40 students working simultaneously in it. . . . In the new Laboratory of Erlangen a space of about 2,400 square feet is reserved for the purpose. At Cambridge a large hall has been added within the last two years to the Cavendish laboratory, and in it

the elementary work is being carried on. The provision made at University College, London for this purpose also largely exceeds ours, having an area of about 1,600 square feet (52 ft 6 in x 30ft).

2. With regard to the second of the above divisions, instruction at the Owens College is at present carried out in a series of rooms occupying a total area of about 1,600 square feet. Although the rooms are not very convenient, they answer their purposes fairly well, and would be sufficient provided the number of students should remain stationary; but it would be extremely disappointing if there were no marked increase within the next few years . . .

3. The third class of students mentioned above, viz. the Chemical Honours Students, at present have to carry on their work in the same room as the others, and some inconvenience arises as for much of the work the rooms have to be darkened. We are greatly in need of a room which could be specially fitted up for spectroscopic work.

4. It is however the electrical students who at present suffer most from want of space. Some years ago a Dynamo house was built, which has proved of great utility. But in the training of electricians attention has to be paid to a great many matters besides the mere handling of dynamo machinery, which can be learnt in the workshop. The electrical engineer ought, for instance, to be able to verify the instrument he is using, and he ought to be able to test, with regard to their magnetic properties, different samples of steel supplied to him.

We have no room where the necessary instruments can be fitted up; we must in consequence make temporary arrangements carrying the apparatus from room to room as occasion arises, this adds much to the labour of the attendants and detracts from the efficiency of the work. We are also in want of space to carry out certain tests for the electrical firms in the neighbourhood. The Professor of Physics has been approached within the last few weeks by a firm of steel manufacturers in Sheffield, with a view to our undertaking to test magnetic properties of certain kinds of steel manufactured by them. It would be greatly to the advantage of the College, if we could undertake such work, but with our present accommodation this is almost impossible.

The students we have so far trained have done very well in their subsequent career, and when our present Dynamo house was built there were very few institutions where similar instruction was given. But recently both University College and King's College, London, have built electrical Laboratories, and special attention is given to the subject in the Engineering Laboratory at Cambridge, the Professor of Engineering there being a distinguished electrician.

Thus while only a few years ago we compared well with other places, we are now being rapidly left behind, some of the public schools even fitting up efficient laboratories with extensive electrical appliances.

Attention has naturally been directed for several years past to the way in which the defects enumerated may be remedied. A few rooms might perhaps be found in the College buildings which could be added to the Physical Laboratory, but even this is doubtful. But it may be specially pointed out, that although this might afford temporary relief, in some ways it does not constitute a satisfactory solution of the difficulty. It is not so much the number of rooms at our disposal which is deficient as their size, their internal arrangement, and the means of communication between one part of the laboratory and the other. There are few if any institutions of importance to equal ours which are not possessed of a Physical Laboratory specially built as such, while most of the rooms we are now using were never intended for the purpose. The different parts of the Laboratory communicate only by means of a long corridor in the basement of the College, through which delicate instruments have to be carried in the midst of a number of students going from lecture to lecture, or to their lockers or to their common room. This corridor is a sort of thoroughfare between the College and both Medical School and Engineering Laboratories and from the nature of the case a part of the College over which it is difficult to exercise complete control.

There are at present only one attendant and one boy attached to the Physical Laboratory and it would be impossible for them to look after any additional rooms, while with more convenient arrangements, the same number of servants could deal with a larger establishment.

The Council must therefore face the question of entirely rebuilding the Physical Department, especially as the Professor of Physics has already once or twice drawn the attention to the fact that the lecture room accommodation is not sufficient. The number of students entering for the lectures is always now larger than that for which the room was originally intended. The seats had to be put closer together and movable desks had to be added and even then every available seat but one or two was occupied last year.

Our apparatus room is so full that two rooms have to be used in the attics to house some of the instruments, which consequently have to be carried down two stories along a narrow staircase once or twice a year . . .

It will be seen therefore that we are short of room in every part of the Laboratory and there seems to me to be no intermediate course possible. Either the College ought to erect an entirely new building, or it must be content to take only a second rank among the institutions in which Physics is taught. In a new building provision ought to be made for an increase in the number of students, such as may be reasonably expected in the next ten or twenty years; and the building should be so placed that a further extension would be possible.

A separate building such as would be sufficient for the purpose would occupy a space of about 270,000 cubic feet complete, and excluding the space occupied by roof, passages, staircases etc., it would leave 170,000 cubic feet for working rooms. The total space occupied by our present rooms exclusive of the two rooms in the attic, is 101,000 cubic feet, so that my suggested building would provide for an increase of about 70%. Calculating the expense of the building at 8d per cubic foot (the new College building at Newcastle, including a very good Physical Laboratory was put at a cost of 6½d per cubic foot), the cost of the building would be about £9,000. . .

Without any pretence at presenting an accurate estimate it may be roughly stated that a sum of £12,000 would enable the College to erect a building which would put the Physical Laboratory in the front rank among teaching institutions while a sum of £10,000 might possibly prove sufficient for the purpose. Until the latter sum has been secured, it would not be wise to attempt to build.

In the sections of the memorandum not reprinted here, Schuster presented further arguments and outlined some of his ideas regarding the equipment to be installed, of which electrical items were to form a prominent part. The document gives a fascinating insight into the hurly-burly of life in the basement of the old Waterhouse building a hundred years ago, with apparatus being carried around amidst the milling throng of students. The memorandum also showed that the need to collaborate with industry, thereby bringing in money, was as real then as it is now. Moreover, Schuster displayed considerable political astuteness in extolling the virtues of new buildings that had been erected in other universities and asking, in effect, whether Council wished the Physics Department to be in the "second rank" of physics-teaching institutions. He knew exactly which points would arouse the civic pride of Council members. Council was persuaded easily of the validity of Schuster's case. In the Minutes of the Council meeting held on 30th April 1897, beautifully hand-written in a ledger, it was recorded:

Received: A letter to the Principal from Mr J.P. Thomasson in reply to a memorandum drawn up by Professor Schuster and forwarded by the Principal as to the expediency of erecting a Physical Laboratory on a site of its own in the College.
Resolved: That Professor Schuster's memorandum be entered on the minutes.
Resolved: That the House and Finance Committee be requested in cooperation with Professor Schuster to take such steps as may be expedient and report thereon.

In the Council meeting of 15th October of the same year it was reported that gifts of £10,000 for building and £5,000 for maintenance had been received, and at the meeting of 29th October Mr J.W. Beaumont was appointed architect. Building work began soon afterwards.

The John Hopkinson Laboratories

At this point, as so often happens, fate took a hand in determining the course of events; in this case through the death in a climbing accident of the eminent electrical engineer John Hopkinson. He was one of a family of thirteen children of a very bright Lancashire family. Several of the siblings achieved outstanding careers including John's elder brother, Alfred, who became a King's Council, Member of Parliament and, in 1898, Principal of Owens College. Their father became an Alderman. A readable account of the family's early life is given in Alfred's autobiography, written in 1930. [2] John entered Owens College as a 16 year old in 1865 to take a 3-year science course. He excelled in his studies and went on to Trinity College, Cambridge. From there he went into the engineering industry and later set up as a consultant. During this period important papers on electrical engineering began to appear and he showed his legal skill in court by defending his own patent (for the 3-wire system of power transmission) successfully against infringement. As his practice expanded he was offered the post of Professor of Electrical Engineering at King's College, London, which he accepted in 1889. Both before and after that date he made numerous important contributions to the development of electrical engineering. The 'Hopkinson back-to-back' test, to name but one, was sufficient to make his name nationally known in the world of electrical engineering.

Hopkinson was an experienced mountaineer and often climbed in Switzerland. In the early hours of 27th August 1898, at the age of 49, he said goodbye to his wife at their hotel in Arolla and set out with his 18 year old son and his two daughters Alice and Lena, aged 23 and 19, to climb on the Petite Dent de Veisivi. They did not return and the next morning the four bodies were found roped together 700 ft below the summit. Their bodies are interred in the cemetery at Territet.[3]

The relevance of this tragic event to our story is that following John Hopkinson's death his family donated a large sum of money in his memory to Owens College, where he had studied. The letter from Alderman Hopkinson (John's father) in which the gift was offered was read before the University Council at its meeting of 2nd November 1898 and stated that he and Mrs Hopkinson and other near relatives of the late Dr John Hopkinson desired to erect a memorial to him in the wing of the new Physical Laboratory intended for Electro-Technology teaching and research, that the sum already offered for the purpose amounted to £1,500 including £400 from Mr Edward Hopkinson and £100 from Mr Charles Hopkinson and that the donors trusted that the name of Dr John Hopkinson would be associated with this portion of the laboratory. It was resolved by Council that other scientific friends of the late Dr John Hopkinson might desire to join in the memorial to him, and that Mr Mather* and Professor Schuster were asked to consider what steps might be advisable in the matter. Thus from Manchester's point of view, tragic though Hopkinson's death was, the money could not have arrived at a more opportune time, bearing in mind that the new Physical Laboratory was in the process of being built with its hoped-for provision for electrical engineering. In order to comply with Council's resolution regarding the possibility of additional donations, the Owens College Council member, eminent engineer and sometime Member of Parliament, William Mather, who was chairman of Mather and Platt, prepared a letter appealing for further funding, as follows:

Sir,

You will no doubt have heard that Mrs John Hopkinson, with her son and daughter, have provided funds for an extension of the Engineering laboratory of the University of Cambridge, in memory of the late Dr John Hopkinson. At the meeting of the Council held at Owens College held on November 2nd, a letter was read from Alderman Hopkinson, in which he expressed the desire that a memorial of his son should also be

---

* Mr Mather was knighted in 1902.

founded in connection with the Owens College, where he received his early scientific education. You may also have heard that a new Physical Laboratory is at the present moment being built at the Owens College, and it has been the intention of the Council of that College, to provide as far as their funds would allow, for a well equipped electro-technical laboratory in a separate wing of the new building. Alderman Hopkinson, his family and near relatives, have now placed the sum of £1,800 at the disposal of the council of the College, with the expression of their wish that the name of Dr John Hopkinson should be associated with the electro-technical wing. The sum given is sufficient to provide for the cost of building the dynamo house and engine room. Some of the friends of the late Dr Hopkinson have thought that others connected with him during his life time might be desirous of joining in a memorial of this kind which could be made more important if additional funds were provided to pay for equipment, and if possible the addition of a complete electro-technical laboratory; the whole wing to be called "The John Hopkinson Electro-technical Laboratory." The services rendered to electro-technical science by the late Dr Hopkinson seem to make a memorial of this kind specially appropriate. The additional sum required for the complete scheme would be about £1,200. Should you feel desirous of joining in this effort to commemorate the life and work of the late Dr John Hopkinson, whose premature death has created such widespread regret, we shall be obliged by you returning the enclosed form stating your intentions. About £400 has been already contributed by several gentlemen having sympathy with the object, including the signatories to this letter.

We enclose form of reply,
Yours faithfully,

W. Mather

The sum received from John Hopkinson's family exceeded the amount originally mentioned; at a meeting of Council on 18th January 1899 a letter from Dr Edward Hopkinson (John's brother) was read which stated that "the sum of £1,884 is enclosed". William Mather's appeal brought in extra funding and there were many gifts of equipment. For example, at its meeting of 7th November 1900, Council recorded the gifts of a 25h.p. multi-polar direct current motor from the British Westinghouse Electric and Manufacturing Co. Ltd. and of drive belting to equip the new laboratory from Messrs. F. Reddaway, and requested that letters of thanks be sent to the firms concerned. In addition, in 1898 "a few former students of the College" donated an oil painting of Dr Hopkinson painted by T.B. Kennington in 1891. From the early 1950s until the late 1980s it was missing, but after a search it was found in the early 1990s. It has been partially restored and now occupies a place of honour in the Manchester School of Engineering.

The opening of the new Physical Laboratory

In the planning of the new Physical Laboratory Schuster worked closely with the architect, J.W. Beaumont of Manchester, and Beaumont was sent to

1896-1907: The new physical laboratory and the end of the Schuster era

the continent to inspect the most modern of the German laboratories. The cornerstone was laid on 4th October, 1898. It is still un-eroded and clearly visible at the south-east corner of the building. At the ceremony Henry Simon, the engineer, made a speech in which he commented on Manchester's reputation as "one of the most glorious centres of industry and

Dr John Hopkinson

enterprise in the world" and added, "Why then should [the city] not provide a technical university in every way equal to the importance and standing which Manchester occupied in the world?" He went on to express his hopes that the efforts of practical men would be combined with those of the man of science. [4]

The New Physical Laboratory was opened on 29th June 1900. The guest of honour, Lord Rayleigh, praised the city and college for their efforts and expressed high hopes for the future. Schuster, in his speech, reiterated his view that Owens was founded to serve the community and serve science. He said his laboratory would train young men and women to serve Manchester enterprise and to educate researchers to advance the frontiers of knowledge. [5]

On the wall of the ground floor corridor in the west side of the building a large plaque was set inscribed as follows:

THE JOHN HOPKINSON LABORATORIES
This portion of the building
was erected and equipped in the year 1899
in memory of
DR. JOHN HOPKINSON, F.R.S.
by his parents, relatives and friends

A further memorial was installed in the form of a bas-relief sculpture of John Hopkinson, set high into the east wall of the dynamo house. It is still in the room at the time of writing (1997) though some years ago it was moved to the north wall. The room is now used by the Psychology Department as a student common room.

A complete description of the building and its facilities was given in two books, of 1900 [6] and 1906 [7], and will not be repeated here. However, the electrical engineering facilities in the John Hopkinson wing, all located on the ground floor, are worthy of mention. They comprised a dynamo house, photometer room, accumulator room, standardising room, alternating current laboratory and an electrochemical laboratory. The description is as follows:

Dynamo house

The building, 56 ft long by 27ft wide, was fitted with 17 examples of d.c. and a.c. generators and motors of outputs ranging up to 25 h.p. driven by belts from a line of shafting in the room immediately underneath. Power was supplied to the shafting at one end by a high speed gas engine donated by Mr W.J. Crossley, M.P. coupled through a clutch to the shaft and at the other end by a 25 h.p. d.c. motor presented by the British Westinghouse Co. and driven off the public supply. Arrangements were made so that one half or the whole of the main shaft could be driven. Current could be sent to any part of the building though a distribution system. Amongst the larger power units were a pair of similar d.c. shunt machines each of 18 h.p. enabling the 'Hopkinson' and many other tests to be carried out. A supply of current at low voltage for the electro-chemical

1896-1907: The new physical laboratory and the end of the Schuster era

laboratory was supplied by a 15 kW "Castle" dynamo of the double-commutator type capable of delivering 500 amperes at 30 volts or 1000 amperes at 15 volts depending on the connections made. Two machines presented by Mather and Platt were particularly useful for educational purposes, one being an 18 kW single-phase alternator with a rotating armature, the sections of which could be grouped in different ways to vary the current output over a wide range; the other was a 12 kW rotating-field polyphase alternator having an armature ring-wound with six independent phases which were connected to six separate pairs of terminals, thus permitting a large number of different combinations to be formed, and the machine to be used as a single-, two- or three-phaser. Also present were a pair of similar single-phase Wilde machines representing one of the earliest types of commercially successful alternator. They were presented by Dr Henry Wilde, F.R.S. for the purpose of illustrating the property of synchronous running originally discovered by him. There were a number of smaller machines, all of which could be belted to the main drive shaft as necessary, and various single- and polyphase transformers and other necessary equipment. A room in the basement in the same wing as the dynamo house was used as a photometric gallery.

Accumulator room

This contained two batteries each of 34 cells, presented by the Chloride Electric Storage Co. and the Tudor Accumulator Co. respectively and having respective capacities of 300 and 135 a.h. corresponding to a maximum discharge rate of 100A. By inserting plugs appropriately into a board any required voltage up to the available limit of 120v could be supplied to any part of the building. [A similar arrangement operated in the electrical engineering department until the 1980s.]

Standardising room

This contained appropriate equipment for the accurate calibration of direct and alternating current ammeters, voltmeters and wattmeters, for high-voltage testing and for general standardisation work.

Electrical and magnetic testing laboratory

The room was equipped with apparatus for the purpose of giving students a working knowledge of the testing of magnetic materials and commercial measuring instruments; galvanometers of different kinds, portable insulation-testing equipment, instruments for permeability and hysteresis testing; potentiometer, resistance and electro-motive standards, Kelvin double bridge and other resistance-testing sets and a Kelvin balance and many other instruments.

Alternating current laboratory

In addition to a large stock of alternating current measuring instruments the laboratory contained a Duddell oscillograph, secohmmeter, capacity and inductance standards, etc. The room also contained a small motor-generator as a local supply of single-phase current of variable frequency but single-phase as well as 2-phase and 3-phase currents of fixed frequency from the dynamo house were also available.

Electro-chemical laboratory

This was set up because of the rapid commercial development that was taking place in the field. It was housed in a 36 ft by 38 ft room with provision for what were described as physico-chemical measurements and in the preparation of chemical substances by electrolysis, and equipment was available for research work.

Electrical Engineering at Manchester University

The New Physical Laboratory at Owens College, completed in 1900

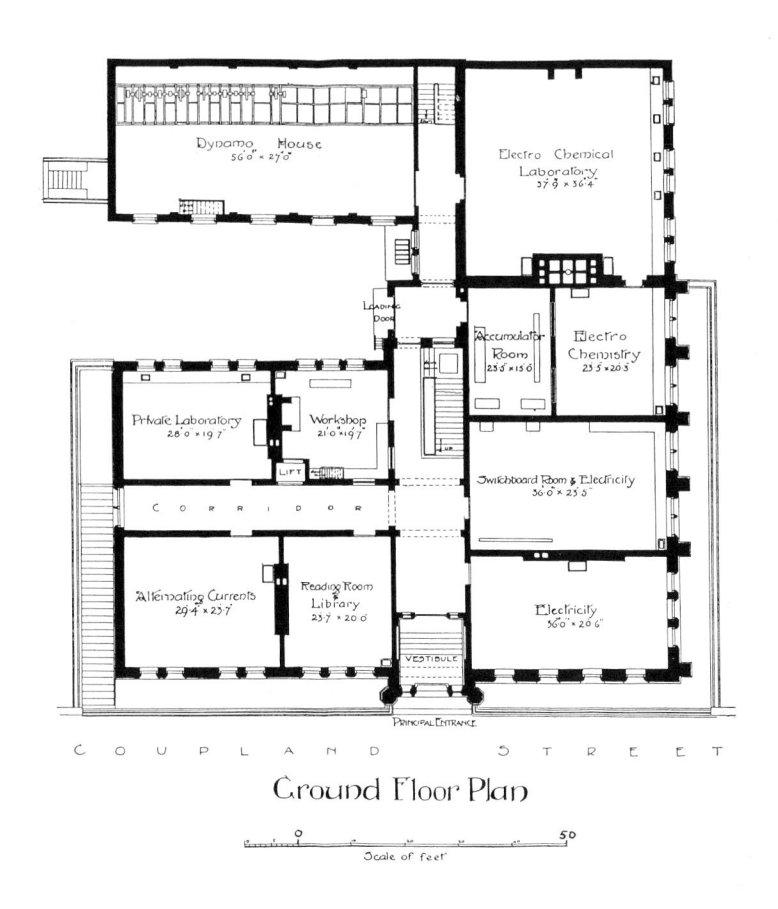

Plan of the ground floor, which housed the Electro-technical Laboratories

1896-1907: The new physical laboratory and the end of the Schuster era

Although the electro-chemical laboratory was in one sense an electrical laboratory, by nature it was more akin to chemistry or metallurgy than physics. It was therefore not part of the 'electro-technology' section of the Physics Department which Robert Beattie ran, and it operated as a section of the main department rather than as part of the 'electrical wing'. In 1900 Dr R.S. Hutton, an Owens chemistry graduate of 1897, was appointed Demonstrator and Lecturer in Electro-chemistry at the same salary as Robert Beattie (£100 p.a.) and thereafter the electro-chemical laboratory was in his charge. Hutton remained in Manchester until 1908 and in 1912 married Professor Schuster's daughter, Sybil. He had a long and distinguished career as a metallurgist both in industry and at Cambridge University and lived to the age of 93. In 1964 he wrote a book of recollections, parts of which give an interesting insight into life in the Manchester Physics Department in the first decade of the present century.[8]

The new physical laboratory, fully equipped and with its new electro-technic and electro-chemical laboratories, was generally reckoned at the time to be the best in Britain and was said to rank fourth in the world in size, behind John Hopkins, Darmstadt and Strassburg [9].

Robert Beattie's promotion

In 1900 Robert Beattie was promoted from Assistant Lecturer and Demonstrator to Lecturer and Demonstrator, at which status he was to remain until 1912. In September 1899 he had been awarded his M.Sc. and in June 1903 he obtained his D.Sc., both of Durham. From the time of his arrival in Manchester he had been engaged in research in his specialist subject of instrumentation and for 20 years from 1896 published a series of papers on the subject (see Appendix 6). Professor Schuster was clearly well satisfied with his achievements for at its meeting of 2nd July 1902 Council resolved "That the stipend of Mr Robert Beattie, Lecturer in Electro-technics, be increased to £150 as from 29th September next, the arrangement as to the share of fees to continue as at present."

Owens College becomes the Victoria University of Manchester

It will be recalled from the previous chapter that in 1873 Owens College became part of a federal university with two other constituent colleges, University College, Liverpool and Yorkshire College, Leeds, with provision for other colleges to join, though none did. The arrangement was always an uneasy one; so far as Owens was concerned the system had been a compromise arrangement when, in the early 1880s, attempts to make Owens a university in its own right failed. In the latter years of the 19th century moves to abandon the federal arrangement gathered pace but some feared that splitting the link with the other two colleges might weaken Manchester's status, though it was by far the strongest of the three and was the federal university's administrative centre. Schuster took a leading role in the independence movement. In a letter to the *Manchester Guardian* he decried the timidity of those who wished the federal arrangement to continue. Schuster's arguments were supported by the College's Professor of Philosophy, Samuel Alexander. In his *Plea for an Independent University in Manchester* Alexander stressed the "growth of that sense of organic connection of a university with its city and district." He added that a Manchester University "may well aspire to represent worthily the intellectual energy of a city which is illustrated by the names of Dalton and Joule." [10]

The movement for independence won the day and on 15th July 1903 the College was reconstituted as the Victoria University of Manchester. The changes became law in 1904 and the legal status of the University has remained essentially the same to the present time. Schuster was elected the first Dean of the Faculty of Science.

The fact that Manchester was at last a university in its own right made little difference to the day-to-day affairs of the physics department. Research was taking place actively in several different fields, including Beattie's own area of electrical instrumentation. A good all-round honours course was being offered including the option of specialisation in Electrical Technology supervised by Beattie or in Electro-Chemistry supervised by Hutton. Mathematics was taken by all students, while the general outline of the physics syllabus was as follows:

First year: Heat; light; sound; magnetism and electricity. The laboratory work included mechanics, specific gravity and laws of pendulum vibration.
Second year: Theoretical physics; physical optics; magnetism and electricity; kinetic theory of gases. Laboratory work included measurements in magnetism and electricity and light.
Third year: Theoretical physics. The laboratory work consisted of an original research project, a scheme which had been introduced by Schuster in the 1880s.

Arthur Schuster and his staff member Charles Lees wrote two text books on practical physics which undoubtedly would have been based on the Manchester physics course at the time and will therefore reflect the content of the general course physics in more detail than is given above. [11] In addition to the standard physics course, specialist courses were available in electro-technology and electro-chemistry. Details of Beattie's electro-technology courses were as follows:

Electro-technology course (Elementary version):
Magnetism of iron; theory of series, shunt and compound dynamos; motors,

accumulators; incandescent lamps; arc lamps and their mechanism; alternate current dynamos; transformers.

A special course in electro-technology was also offered for those wishing to take up a career in this area. It comprised:

Calibration of instruments; investigation of laws of induction; magnetisation of iron; determination of the characteristics and efficiencies of series, shunt and compound dynamos; experimental verification of laws; coupling of dynamos; accumulators - charging and testing; photometry of arc and incandescent lamps; characteristics and coupling of alternate and multiphase current dynamos; motors; transformers; transmission of power.

An electro-technology examination paper from June 1904 is given in Appendix 7.

Events in the Engineering Department

At this point, it is perhaps worthwhile to look briefly at what was happening in the Engineering Department in the same period. Although the Physics Department had moved into its large new building, Engineering was still in its quarters in the basement of the main Waterhouse building and in the Whitworth Laboratories in the back quadrangle; their turn to move to larger premises would come in 1908. The Departmental staff in 1900, apart from Osborne Reynolds, consisted of J.B. Millar and C.B. Dewhurst, each of whom was a Demonstrator and Assistant Lecturer, and two Demonstrators, G.W. Wilson and J.H. Grindley. The course continued to be a good all-round one in civil and mechanical engineering with an excellent range of laboratory facilities but there was no electrical content in the course so it will not be discussed further here. Over the years Osborne Reynolds had built a reputation as one of the finest engineers in the world, not just in one field of expertise but in several. Sadly, in the early years of the present century his mental powers were beginning to fail though he was only 60. His paper *On the Sub-Mechanics of the Universe* communicated to the Royal Society in 1902 was subjected to five months of criticism by the referees before it was published and, it is generally agreed, was very difficult to follow. Sir J.J. Thomson described it as "the most obscure of all his writings, as at this time his mind was beginning to fail", and Sir Horace Lamb took the view that:

"Unfortunately, illness had already begun gravely to impair his powers of expression, and the memoir, as it stands, is affected by omissions and discontinuities which render it unusually difficult to follow. No one who has studied the work of Reynolds can doubt that it embodies ideas of great value, as well as of striking originality; but it is to be feared that their significance will hardly be appreciated until some future investigator, treading a parallel path, recognises them with the true sympathy of genius, and puts them in their proper light." [12]

Sadly, Reynolds' decline was becoming obvious; the once-great mind was no longer what it was. In 1905 he had to bow to the inevitable and, at the age of 62, he resigned. He remained as Honorary Professor and retained his seat on the University Senate. The university calendar of 1905-06 said that he would take part in advising post-graduate students on research. However, because of his deteriorating condition this task was not a success.* In 1905 a fund was opened to commission a painting of Reynolds by the artist John Collier. It was soon completed and was presented to the university by Dr Ward, Master of Peterhouse, on 18th November 1905. It currently hangs in the Simon Laboratories.

Stanley Dunkerley, who had graduated at Owens in mathematics in 1889 and engineering in 1890, was appointed in 1905 to replace Reynolds. In 1905-06 the staff comprised the newly-appointed Dunkerley as Professor and Director of the Whitworth Laboratories, Reynolds in his honorary capacity, J.B. Millar and C.B. Dewhurst (Lecturers), A.H. Gibson (an engineering graduate of 1903 who, in 1920, would become professor and head of the department) as Demonstrator and Assistant Lecturer, and Fred Pickford (an engineering graduate of 1904) as Demonstrator. Millar, who had joined Osborne Reynolds in 1869, was in his final session (1905-06) as a staff member. In the 1906-07 and 1907-08 calendars Dunkerley was the Professor of Engineering and Director of the Whitworth Laboratories and Reynolds was named Beyer Research Professor. But in the staff list of 1908-09 Reynolds' name did not appear at all. By then he had retired, in rapidly declining health, to Watchet in Somerset.

Life in the Physics Department and the retirement of Schuster

Returning to events in the Physics Department, 1906 saw the 25th anniversary of Arthur Schuster's elevation to a professorship and the occasion was duly marked with celebrations and the publication of a 142-page book, *The Physical Laboratories of the University of Manchester*. Schuster was a popular figure, not only in Manchester but nationally and internationally. His health had never been very robust but he travelled widely and was known in scientific circles throughout the world. Though his physics department was now famous the number of staff was still only small. Thomas Core retired in 1905 and in 1906 the staff consisted of (with their year of appointment), Schuster (1881 [applied mathematics] and 1888 [physics]), Charles Lees (1891), Robert Beattie (1896),

---

* In 1903 Osborne Reynolds asked the newly-graduated A.H. Gibson, who had expressed a wish to do research, what type of work he wished to follow. Gibson said he would like to do research in hydraulics, whereupon Reynolds said, "That's impossible - I've done it all!" [13] As many readers will know, Gibson subsequently distinguished himself in hydraulics research.

1896-1907: The new physical laboratory and the end of the Schuster era

Two views of the dynamo house, early 1900s

R.S. Hutton (1900), G.C. Simpson (1905), Herbert Stansfield (1906) and Sidney Russ (1906). Thus the place was small enough to have very much a family atmosphere. Schuster had instituted fortnightly colloquia in the old building in 1894 in which accounts were given of current research. All staff and the more senior students were invited and the intimate character of the meetings was enhanced by the presence of Mrs Schuster who provided the tea and took an active part in the social life of the staff and students. She organised many activities for the University, for instance the foundation of a staff bicycle club which held regular outings into the neighbouring parts of Cheshire and Derbyshire, and she took an interest in the sports played by women students. [14]

Schuster had a lively wit and could produce a timely 'put-down' when necessary. In the early 1900s Marie Stopes, the famous author of *Married Love*, was a member of the Botany staff at Manchester University, being an expert on palaeobotany. She was rather egotistical and when she once told Schuster that she could not help making a new discovery every day, he responded by asking whether, if the new discovery of today showed that the one of yesterday was wrong, did she count it as one discovery or two? [15] Schuster always fiercely defended Manchester University against unwarranted criticism, especially remarks of a snobbish kind. A Manchester industrialist once asked Schuster to allow his son, who had just come down from Cambridge, to take a course in electrical engineering, but condescendingly gave the impression that he thought it was rather a come-down for the boy to study in Manchester after Cambridge, at which Schuster bristled and remarked: "I don't know who your son has associated with at Cambridge, whether with Dukes' sons or Earls' sons, but he will find in Manchester a lot of students a great deal better than himself." Later he wondered whether he should have restrained himself in the interests of preserving friendly relationships with local industry, but the industrialist evidently had also had second thoughts about what he had said because the next day he sent Schuster a brace of pheasants as a peace offering. [16]

As the Physics Department moved through the first decade of the 20th century Schuster made the decision that the time had come for him to retire. Much of his best experimental work of the 1880s and 1890s had been done with the help of his private research assistant, Arthur Stanton, who had graduated from Edinburgh University in 1881. But in the 1890s Stanton became a victim of drug addiction and in 1898 he died.[17] After that, most of Schuster's research was of a theoretical nature until the arrival from Erlangen in 1906 of Hans Geiger as his research assistant in the laboratory. Schuster felt, no doubt correctly, that by that stage of his life his best work was behind him. Moreover, he had no financial worries. His family, all of whom had come over from Germany to Britain with him, were talented in many areas of expertise and all did well for themselves financially. Schuster decided that if a brilliant young physicist could be found to replace him he would retire and let the younger man take over. Schuster knew of Ernest Rutherford, a New Zealander who was currently professor of physics at McGill University, Montreal and had a high opinion of his potential. Schuster let it be known that he would stand aside, at the age of 56, if Rutherford, then 35, could be persuaded to come to Manchester. In a letter to Rutherford, Schuster said: "Manchester is not at all a bad place", and there would be "plenty of time for research work, and the laboratory is beginning to be known. [18]

Rutherford was interested. He had already been approached by Yale and by King's College, London. He declined Yale on the grounds that he wished to return to Britain (he had graduated from and researched at Cambridge in the 1890s under the Owens graduate J.J. Thomson) and he felt that a move to King's College would involve a sacrifice of research facilities. The situation at Manchester was more promising. In a letter to Schuster, Rutherford said he was "inclined to consider very favourably the suggestion of becoming a candidate for the position you propose. The fine laboratory you have built up is a great attraction to me." [19]

Schuster wanted Rutherford in preference to anyone else because he was convinced his successor should be a relatively young experimentalist who would dominate the new areas opened up by the discoveries of X-rays and radioactivity. Rutherford had been very active in these areas at McGill, and Schuster had been impressed by Rutherford's masterly Bakerian lecture in 1903 on the radioactive changes in the radium, thorium and actinium families. In 1906 Rutherford turned down an invitation for the post of secretary of the Smithsonian Institution on the grounds that he wanted to return to Britain. Schuster was very proud of what he had achieved in Manchester and on 7th October 1906 remarked in a letter to Rutherford:

"I am so strongly attached to the place I could not bear to leave my position except to someone who will keep up its reputation and increase it. There is no-one to whom I would leave it with greater freedom from anxiety than yourself." [20]

Rutherford needed no further persuasion. He accepted the Manchester post and arrived on 6th June 1907 to take up the Langworthy Chair of Physics. At the same time Schuster resigned his professorship at the age of 56 and was awarded the title of Honorary Professor. So ended a proud era in the history of the university, but another was about to begin.

Electrical Engineering at Manchester University

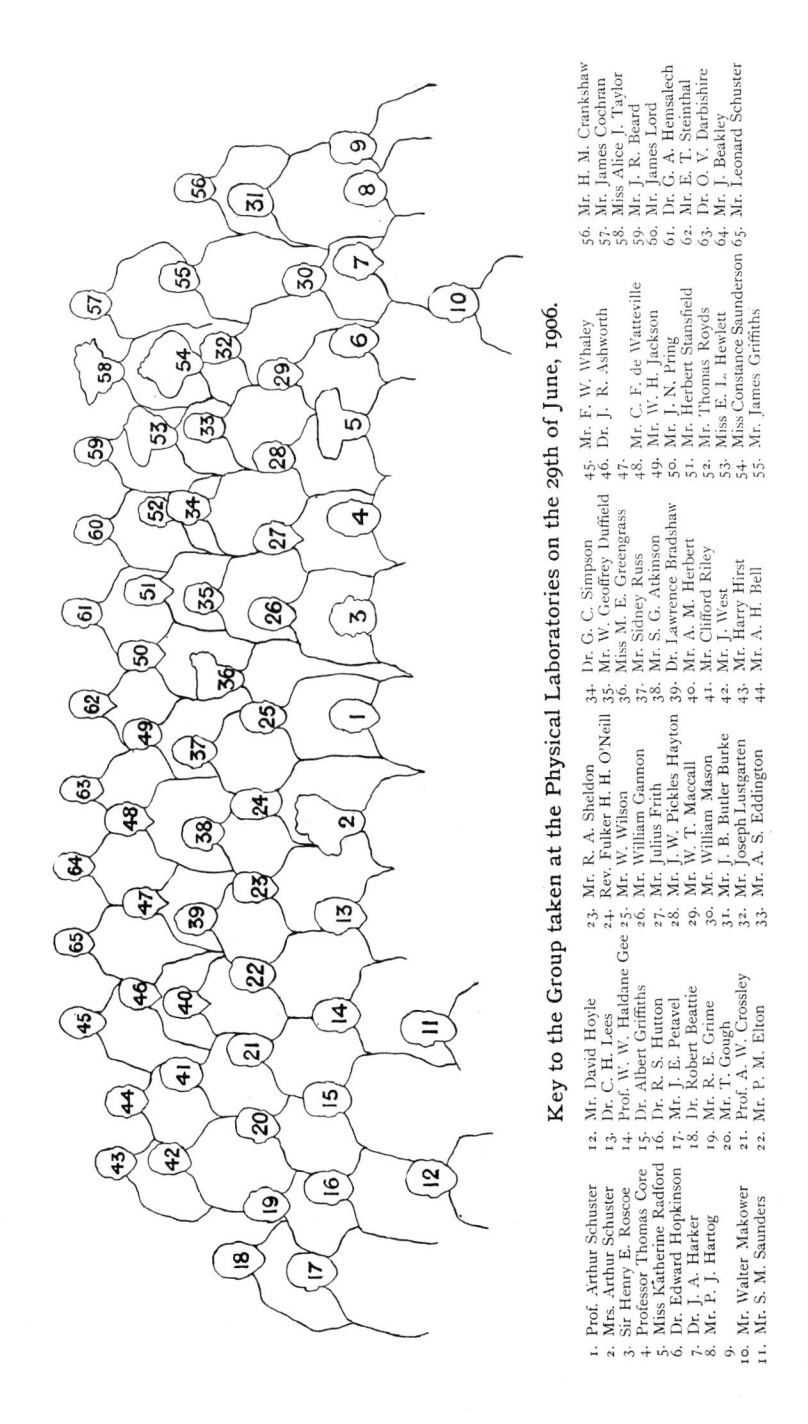

1896-1907: The new physical laboratory and the end of the Schuster era

Key to the Group taken at the Physical Laboratories on the 29th of June, 1906.

1. Prof. Arthur Schuster
2. Mrs. Arthur Schuster
3. Sir Henry E. Roscoe
4. Professor Thomas Core
5. Miss Katherine Radford
6. Dr. Edward Hopkinson
7. Dr. J. A. Harker
8. Mr. P. J. Hartog
9.
10. Mr. Walter Makower
11. Mr. S. M. Saunders
12. Mr. David Hoyle
13. Dr. C. H. Lees
14. Prof. W. W. Haldane Gee
15. Dr. Albert Griffiths
16. Dr. R. S. Hutton
17. Mr. J. E. Petavel
18. Dr. Robert Beattie
19. Mr. R. E. Grime
20. Mr. T. Gough
21. Prof. A. W. Crossley
22. Mr. P. M. Elton
23. Mr. R. A. Sheldon
24. Rev. Fulker H. H. O'Neill
25. Mr. W. Wilson
26. Mr. William Gannon
27. Mr. Julius Frith
28. Mr. J. W. Pickles Hayton
29. Mr. W. T. Maccall
30. Mr. William Mason
31. Mr. J. B. Butler Burke
32. Mr. Joseph Lustgarten
33. Mr. A. S. Eddington
34. Dr. G. C. Simpson
35. Mr. W. Geoffrey Duffield
36. Miss M. E. Greengrass
37. Mr. Sidney Russ
38. Mr. S. G. Atkinson
39. Dr. Lawrence Bradshaw
40. Mr. A. M. Herbert
41. Mr. Clifford Riley
42. Mr. J. West
43. Mr. Harry Hirst
44. Mr. A. H. Bell
45. Mr. F. W. Whaley
46. Dr. J. R. Ashworth
47.
48. Mr. C. F. de Watteville
49. Mr. W. H. Jackson
50. Mr. J. N. Pring
51. Mr. Herbert Stansfield
52. Mr. Thomas Royds
53. Miss E. I. Hewlett
54. Miss Constance Saunderson
55. Mr. James Griffiths
56. Mr. H. M. Crankshaw
57. Mr. James Cochran
58. Miss Alice J. Taylor
59. Mr. J. R. Beard
60. Mr. James Lord
61. Dr. G. A. Hemsalech
62. Mr. E. T. Steinthal
63. Dr. O. V. Darbishire
64. Mr. J. Beakley
65. Mr. Leonard Schuster

References: Chapter 4

1. I am greatly indebted to Mrs Janet Paterson (Professor Beattie's daughter) for allowing me to read the family history of the Beattie family and to use portions of it in the preparation of this book. Without her help and that of her husband, Dr J.A. Paterson, many of the details of Professor Beattie's life and work would have been impossible to ascertain.
2. Sir Alfred Hopkinson. *Penultima.* Martin Hopkinson Ltd., London, 1930.
3. Percy Dunsheath. *A History of Electrical Engineering.* Faber and Faber, London, 1962; also James Grieg: *John Hopkinson - Electrical Engineer.* A Science Museum booklet. H.M.S.O., 1970.
4. *Manchester Guardian*, 5th October, 1898, p. 10.
5. Ibid., 30th June, 1900, p.11.
6. P.J. Hartog. *The Owens College, Manchester.* J.E. Cornish, Manchester, 1900.
7. *The Physical Laboratories of the University of Manchester.* Manchester University Press, 1906.
8. R.S. Hutton. *Recollections of a Technologist.* Pitman, London, 1964.
9. *Physical Laboratories* (see ref. 7.), p. 2; *The New Physical Laboratories of the Owens College Manchester.* Nature, **58**, 1898, p. 621.
10. *Manchester Guardian*, 7th June 1902, p. 6; 9th June 1902, p. 9; 11th June 1902, p. 12;. 10th July 1901, p. 12;
11. A. Schuster and C.H. Lees: *An Intermediate Course of Practical Physics.* Macmillan, London, 1896; and *Advanced Exercises in Practical Physics.* Cambridge University Press, 1901.
12. Jack Allen. *The Life and Work of Osborne Reynolds.* First chapter of *Osborne Reynolds and Engineering Science Today.* Ed. J. D. Jackson and D. M. McDowell. Manchester, 1970. See page 57.
13. Part of recollections of the Electrical Engineering Department tape-recorded for the author by Professor Eric Laithwaite. The story was told to Eric Laithwaite by Harold Gerrard, long-serving member of the Electrical Engineering Department's staff, who had been told it by A. H. Gibson.
14. R.S. Hutton (see ref. 8), p. 35
15. Ibid., pp. 105-106.
16. A. Schuster. *Biographical Fragments.* pp. 248-257. Macmillan, London, 1932.
17. Hutton (see ref. 8), p. 33.
18. Letter from A. Schuster to E. Rutherford, 7th September 1906. Rutherford Papers, Cambridge.
19. Letter from Rutherford to Schuster, 26th September 1906. Schuster Papers, Royal Society of London. See also J.B. Birks, *Rutherford at Manchester*, New York, 1963.
20. Letter from Schuster to Rutherford, 7th October 1906. Rutherford Papers, Cambridge; also Birks (see ref. 19), pp. 52, 65.

Chapter 5
1907-1912: Electro-technics in the early Rutherford years

When Rutherford arrived in Manchester he was already known worldwide as one of the most promising physicists of his day. Born in Spring Grove, New Zealand in 1871, he was the fourth of a family of 12 children of a wheelwright and flaxmiller. Winning scholarships to Nelson College and Canterbury College, Christchurch, his earliest research projects were on the magnetisation of iron by high-frequency discharges and magnetic viscosity. He was admitted to the Cavendish Laboratory and Trinity College, Cambridge, in 1895 on a scholarship, and succeeded in transmitting wireless signals over the then record distance of half a mile. Under the leadership of the ex-Owens physicist J.J. Thomson, Rutherford then discovered the three types of uranium radiation. In 1898 he was appointed Professor of Physics at McGill University, Canada, and in collaboration with Frederick Soddy he formulated the theory of atomic disintegration to account for the tremendous heat energy radiated by uranium.

Rutherford's research work in Manchester

Although Rutherford in 1907 was already the author of many scientific papers and two books, *Radioactivity* (1904) and *Radioactive Transformations* (1906), his reputation had been established by the quality, originality and importance of his publications rather than their number. It is not at all surprising that Schuster saw the 35 year old Rutherford as just the man that Manchester needed and, as matters turned out, he was not to be disappointed. Following his arrival in June 1907, Rutherford soon set about building up a new research team. During his Manchester years (1907-1919) he assembled a glittering array of young physicists whose names were to become famous through their discoveries. Apart from Hans Geiger, whom Rutherford inherited from Schuster, they included Andrade, Bohr, Chadwick, Darwin, Marsden, Moseley, Nuttall and Robinson.

It is not the intention here to give a full account of Rutherford's work in Manchester; firstly, because it is fully documented elsewhere [1], and secondly, because it is not of direct concern to the development of electrical engineering in the Physics Department at that time, other than that the general air of liveliness of Rutherford's team must have acted to some extent as an inspiration to those working in fields unrelated to atomic physics. Here, however, is a very brief summary of Rutherford's Manchester work. He established that alpha particles were doubly ionised helium ions by counting the number given off with a counter devised by one of his team, Geiger. This led to the revolutionary concept of the atom as a sort of miniature universe in which the mass is concentrated in the nucleus surrounded by planetary electrons. His assistant, the young Danish physicist Niels Bohr, applied the quantum theory (1913) to Rutherford's work and established the concept of what became known as the Rutherford-Bohr atom. Later, in a series of experiments carried out between 1915 and 1919 (see Chapter 7), Rutherford discovered that alpha-ray bombardments induced atomic transformation in atmospheric nitrogen, liberating hydrogen nuclei. [2] For his brilliant early work in Manchester he was awarded the Nobel Prize for Chemistry in 1908, a fact currently commemorated by a small blue plaque placed in the wall at the entrance to Coupland Street.

Events and personalities

Whilst Rutherford and his team were establishing their research work in atomic physics, Beattie was unspectacularly researching in his chosen subject of electrical measurements, but he was always wary of visitors. This suspicion dated from early in 1904 when a visitor from France, a Monsieur Grassot, was shown Beattie's latest invention, an ingenious method of adapting a galvanometer to measure magnetic flux by removing its control spring. The visitor watched and listened with interest as Beattie explained his device and then returned to France. Shortly afterwards Grassot published a paper on a method of measuring flux which, to all intents and purposes, was Beattie's device, and took out a patent in all countries. It soon became known as the Grassot Fluxmeter, a fact which did not please Beattie. The story of Beattie's displeasure at losing the credit for his idea in this unfortunate way is well established in the folklore of the Electrical Engineering Department, having been heard from several independent sources, all originating from Beattie himself. [3]

The other leading member of the physics staff who, like Beattie, was not involved in Rutherford's work was Robert Hutton who was in charge of the electro-chemical laboratory. For six years Hutton did some of his work in conjunction with Joseph Petavel who had come to Manchester in 1902, when he was 28, as a Harling Research Fellow. He helped Hutton with his work on an electric furnace and their work was described in a joint I.E.E. paper in 1902.[4] Hutton has given an amusing account of some of their collaboration:

"Co-operative research with Petavel was not always easy, for he kept rather abnormal hours and often arrived about mid-day and stayed until midnight or later. He told me that his landlady thought he was a night-watchman, until he started to work with me, when she suggested that he had been promoted to the post of night-stoker, owing to the dirty nature of electric furnace work. Sometimes when I felt too tired to stay the whole course, he worked on alone, and on one occasion came round to my rooms and woke me up, reporting that he had scribbled down some results but was sure he would not be able to read them in the morning, so we must draft our report forthwith." [5]

Despite his foibles Petavel was a very bright man whose contribution to the design of high-pressure apparatus exhibited much originality, little work having previously been done at high pressures, especially in conjunction with high temperatures. He was also an expert on the subject of explosions, which would make his expertise of particular value during the 1914-18 war. The esteem in which he was held may be judged by the fact that when in 1908 Stanley Dunkerley, who in 1905 had been appointed Professor of Engineering in succession to Osborne Reynolds, had to resign owing to ill-health, Petavel was appointed Professor of Engineering in his place. He stayed until 1919 when he left to become Director of the National Physical Laboratory, and was knighted in 1920.

The Physics Department in those days was a lively place, inhabited mainly by the students and the youthful research team that Rutherford was assembling. But at the other end of the age range the Department was occasionally visited by Dr Henry Wilde, the old Manchester inventor whose date of birth was 1833, in the reign of George IV. He claimed to have been the inventor of the commercial dynamo and there is much evidence to support his claim. [6] Wilde's paper on his commercial dynamo was read before the Royal Society by Michael Faraday in 1866. [7] (Papers had to be read by a Fellow, which Wilde was not, at that time). Wilde seems to have been rather unfortunate in the overlapping of his inventions with those of others and never took kindly to credit not being given where, in his opinion, it was due. He was of a litigious disposition, not at all averse to going to court to establish the priority of his inventions, a trait which caused his estrangement with a lot of people. However, he had always been on good terms with Schuster and liked to visit the Physics Department occasionally to see him and to look round the laboratories. Hutton recalls one such visit:

"He had a great respect for Schuster, who treated him very courteously. I myself once experienced his critical attitude when, with the object of trying to interest him, I gave him a large piece of the metal calcium which had just arrived; it was described in the literature as a yellow metal, but turned out to be white when reasonably pure. Wilde took this away with gratitude, but returned the next day and accused me of misleading him for, on cutting the metal under paraffin, he found it had a greyish and not a whitish colour!" [8]

Although Schuster had resigned the Langworthy Chair in 1907 to make way for Rutherford, as Honorary Professor he was often to be found in the

Department. Over the years Schuster had established numerous contacts at home and abroad and even after his retirement was influential in attracting funds to the department, thereby strengthening Rutherford's team. He also provided funds out of his own pocket to establish a Readership in Mathematical Physics, of which one of the holders was Charles G. Darwin, and a Readership in Meteorology which G.I. Taylor held. In addition, he continued to finance the research assistantship under Rutherford of his protégé, Hans Geiger.

The bearded Schuster must have presented a patriarchal image to the young students and research students of the Physics Department at that time but he had always been and remained a popular and much-respected figure. In 1908, when Schuster heard that the Manchester electrical engineer Henry Royce was collaborating with the Hon. C.S. Rolls in designing and making a superior motor car, he immediately ordered one, and the first Rolls-Royce produced at the Derby factory (Registration Number R 548) was received by Schuster on 1st September 1908. Hutton recalls that his wife (Schuster's daughter, Sybil, whom he married in 1912) was one of the family drivers of the vehicle as soon as she was legally entitled to drive at the age of 17.[9] After Schuster's retirement he continued to live in Manchester until 1913 when he was elected Physical Secretary to the Royal Society and left to live in Berkshire in order to be nearer London. He died in 1934.

Events in the Engineering Department

Whilst Rutherford was rapidly making his mark in Physics, the Engineering Department was expanding and looking to the future. It is interesting to take a brief look at what was happening there, as although at the time the electro-technics elements of teaching and research were a part of Physics, it would not be long before that link was almost severed and instead the electro-technics group would be linked with Engineering. Stanley Dunkerley had been appointed Professor and Director of the Whitworth Laboratories in 1905 in succession to Osborne Reynolds but in 1908 he too had to resign owing to ill health after only three years in the post, and he died in 1912. As previously stated, Dunkerley's successor as professor was J.E. Petavel who had recently been engaged in research in the Physics Department.

The Engineering Department had been housed in the old Waterhouse building since 1873, and the addition of the Whitworth Laboratory in the back quadrangle in 1887 had provided much improved facilities. But in the early 1900s, with the increase in student numbers, accommodation was proving inadequate and proposals were put forward for the construction of a new engineering building. The university calendar of 1907-08 reported on the plan for the siting of a new building to house the Engineering Department and reported that the Corporation of Manchester, at the request of the university, had arranged for the closing of a portion of Eagle Street, between Coupland Street and Huntingdon Street, and had sold that portion of the street to the university. In the meantime, until the new building was ready, some rooms in the Faculty of Arts would be used for classes. The calendar stated that building had not yet begun. The following year's calendar, 1908-09, noted that work on the new building was proceeding and that the boiler house, engine house and workshop were already finished, but the new laboratories, lecture rooms and drawing offices would not be ready until the following year. The new laboratories were indeed completed in 1909 and retained the name of the Whitworth Laboratories. They were officially opened on 15th July of that year by the leading engineer Sir Alexander Kennedy, senior partner in the firm Kennedy & Donkin, and Emeritus Professor of Engineering at University College, London. At the time of the opening there was still some work to be done on the building but it was soon completed and everything was ready in time for the new session of 1909-1910.

In the 1909-10 session the staff of the department, in addition to the newly appointed J.E. Petavel, comprised C.B. Dewhurst (Senior Lecturer), A.H. Gibson (Demonstrator and Lecturer in Hydraulics), C.H. Lander (Senior Instructor in Drawing and Assistant Lecturer), Andrew Robertson (Demonstrator), and Robert Colton (Junior Instructor in Drawing and Demonstrator). Gibson had graduated from the department as recently as 1903 but had already gained his D.Sc. The following year he resigned on his election to the Chair of Engineering at the University of St. Andrews. The 1909-10 calendar recorded that temporary assistance had been given "in the drawing department" by C.M. Mason. Mason, who had graduated in the Engineering Department in 1908, was destined to become a long-serving member of staff known to generations of students until his retirement in 1955. By 1911-12 the number of engineering staff had risen to seven: Petavel, Dewhurst, Pickford, Colton, Cook, Lander and Robertson. With the exception of Petavel and Dewhurst who were aged 37 and about 40 respectively, it was a young group, the last four of those named on this list having graduated from the department in 1905 and Pickford in 1904.

The university annual statement in 1910 reported that: "Pursuant to the arrangements made with the Corporation of Manchester and of special clauses in their private act of Parliament of 1908, Coupland Street has been closed to traffic and great advantage has been found in the diminution of noise in the rooms abutting on that street." This change would be of benefit to the Physics Department, the Engineering Department and, when it was founded, the Electro-technics Department. The street closures had in effect meant that Eagle Street ceased to exist. (see plans on pages 39 and 98).

Some physics graduates of the period

Returning to events in the Physics Department, several notable people graduated in the period covered in this chapter. In 1907 Stephen Butterworth graduated with a first, specialising in electro-technics; he became an authority on filter amplifiers. 1909 saw the graduation of Ernest Marsden (1st class) who became a key member of Rutherford's team and whose name would become familiar to all students of atomic physics. In the same year Harold Gerrard graduated with a second class. (Second class degrees in physics were not subdivided into II(1) and II(2) until 1922). He was soon to become a founder member of the Electro-technics Department and to be a long-serving stalwart, not retiring fully until 1959. In 1910 Henry Cotton (better known as Harry Cotton) graduated with a first. He became Professor of Electrical Engineering at Nottingham and was known to generations of students through his many text books, for example *Electrical Technology* and *Applied Electricity*, the contents which, it is generally agreed, were derived from Beattie's lecture notes. He died in 1985 at the age of 96.

The first-class honours physics graduates of 1911 included James Chadwick, Joseph Higham, Harold Robinson, J.M. Nuttall, H.P. Walmsley and John Wood. All except Higham were recruited into Rutherford's research group and made names for themselves. Chadwick enjoyed a particularly distinguished career in nuclear physics for which he was awarded numerous medals of learned societies. He was knighted in 1945 and created a Companion of Honour in 1970. J.M. (Jimmy) Nuttall, known for the Geiger-Nuttall relationship of atomic physics, became a long-serving staff member of the Physics Department, retiring in 1956. The "odd man out" amongst the 1911 physics graduates was Joseph Higham (known to all his friends as Joe), who threw in his lot with electro-technics rather than atomic physics. In 1920 he joined the staff of the Electro-technics Department and remained a loyal and much-respected staff member until 1952 when he was forced to retire through illness. He and Harold Gerrard enjoyed the privilege of being onlookers at two major scientific achievements; firstly, Rutherford's brilliant early Manchester researches in atomic physics, and secondly, the equally brilliant post World War II work of Williams and Kilburn in designing and building the world's first successful stored-program computer. Rarely can one person, let alone two, have been close witnesses to two

1907-1912: Electro-technics in the early Rutherford years

N.Eccles   S.Kinoshita   R.Rossi   W.Kay   G.N.Antonoff   E.Marsden   W.C.Lautsberry
F.W.Whaley   H.C.Greenwood   W.Wilson   W.Borodowsky   Miss. M.White   E.J.Evans   H.Geiger   T.Tuomikoski
S.Russ   H.Stansfield   H.Bateman   Prof. Schuster   Prof. Rutherford   R.Beattie   J.N.Pring   W.Makower
R.E.Slade   　　　　　　　　　　　　　　　　　　　　　　　　　　　W.A. Harwood

Physics group - 1909

　　　　　　　　　J.M.Nuttall   W.Kay   T. Taylor   J.Chadwick
C.G.Darwin   R.Wilson   D.C.H.Florance   Miss M.White   -   H.Robinson   -   -   T.Tuomikoski
H.Geiger   W.Makower   Prof. Schuster   Prof. Rutherford   R.Beattie   H.Stansfield   E.J.Evans
R.Rossi   H.Moseley   J.N.Pring   H. Gerrard   E.Marsden

Physics group - 1912

73

such outstanding achievements. The year 1912 saw the graduation, again first class, of another man destined to achieve much; he was A.B. Wood, who became Director of the Royal Naval Scientific Service; in effect, the Royal Navy's chief scientist. His name will reappear in Chapter 7.

Extension of the electro-technical laboratories

The opening of the John Hopkinson Electro-technical Laboratories in 1900 heralded the start of a new era for teaching and research in the subject and the facilities available were then as good as in any university laboratory in Britain. There were now several electrical engineering firms in the Manchester area; this fact, combined with the facilities that Manchester offered and the reputation it enjoyed for teaching the subject, ensured that the electro-technics section of the Physics Department was a sought-after place at which to study. Consequently student numbers steadily increased following the opening of the John Hopkinson laboratories. The result was that as the first decade of the century drew to its close the age-old problem of lack of accommodation once again emerged. A major contributory factor to the problem was that the enrolment at that time included not only students working for degrees but also those studying for certificates.

Professor Rutherford presented a case for the provision of improved accommodation to the university authorities who seem to have readily accepted the arguments. The annual statement signed in July 1911 by the Vice-chancellor, Professor Alfred Hopkinson, carried a paragraph on the proposed extensions. Under the heading *New Buildings* a sub-section headed *New Laboratories for Electro-technics and Physics* reported:

"The number of students receiving instruction in Electro-technics has largely increased during the last five years. Better accommodation is required for this department, which is in charge of Dr Beattie. The rooms devoted to the subject are scattered, and owing to this, serious inconvenience is felt in the Physical laboratories. At the same time the accommodation for the very important researches carried on by Professor Rutherford and his Staff, and the numerous Research Fellows and students, many of whom have come from a distance, was found to be quite inadequate, and the conditions under which the work was carried on prejudicial to its development. Inconvenience was also felt from the fact that there was only one lecture room in the Physical laboratory, and another of smaller size was urgently needed. The Council accordingly determined to erect a new building on a site adjoining the old Physical laboratory. The new building will contain a lecture room to accommodate about 80 students, and proper accommodation, in addition to the John Hopkinson laboratories, for the Department of Electro-technics, as well as some additional accommodation for Physics.

1907-1912: Electro-technics in the early Rutherford years

Immediately adjoining this a small building is to be erected, part of which will be let to the glass blower and to the instrument maker who regularly do work for the University, and whose workshops, it is desirable, should be in close proximity to it. The cost of the new buildings, including the two workshops, for which rent will be paid, will be £7,370 apart from equipment. The Council has no funds available for the purpose, but it was felt to be essential to carry it out, as very valuable work would otherwise have been crippled. Towards the equipment of the new laboratories Mr J.F. Cheetham and Mr Hermann Woolley have each generously given £250, and Mr Neville Clegg £200. It is expected that the new buildings will be completed in the course of the present year, and that part at least will be in use at the beginning of the session."

The Vice-chancellor, Alfred Hopkinson, in approving the annual statement, must have felt a glow of satisfaction in noting the success of the Electro-technical Laboratories, dedicated to the memory of his brother, John. Another indication of the increasing status of electro-technics was the doubling of its staff in 1911, from one (Robert Beattie) to two by the appointment at a salary £125 p.a. of Harold Gerrard, who had graduated in Physics in 1909, as Junior Demonstrator in Electrotechnics, both men operating in the Physics Department. Gerrard, from Lowton, near Leigh in Lancashire, was born in May 1888 and had attended Leigh Grammar School. At that time the physics staff consisted of Rutherford (Professor), C.G. Darwin (Reader in mathematical physics), Beattie (Lecturer and Demonstrator in electro-technics), H. Stansfield, E.J. Evans and H.G.J. Moseley (Lecturers and Demonstrators), Hans Geiger (Lecturer and Research Assistant), J.N. Pring (Lecturer and Demonstrator in electro-chemistry) and Gerrard, the newly appointed electro-technics staff member. It will be noticed that Beattie, as one of the most senior of the physics staff, sat next to Rutherford in both of the group photographs shown on page 73.

John Pring, who had graduated with first-class honours in chemistry from Manchester in 1904, had been appointed to take charge of the electro-chemical laboratory following the resignation of Hutton, who had taken up a position in a silver-plate manufacturing company in Sheffield. In his memoirs Hutton said that his departure was due to the poor pay associated with his lecturing job, an ever-present complaint amongst university staff. He said that after three or four years of postgraduate work, much of it at his own expense, he had been appointed Lecturer in charge of the Electro-chemical Laboratory in 1900 at a salary of £100 per annum (the same as Beattie) and that after eight years he was receiving £250 per annum which, he said, was as highly paid as any other non-professorial member of the Manchester University staff at the time, with little prospect of increase. It is extremely likely that Beattie, who had been on the staff four years longer than Hutton and was three and a half years older, was receiving the same. Hutton remarked in his memoirs that a movement for improvements in salaries was afoot at the time he left. A delegation of "Memorialists of the Junior Staff" appealed to the University for increased stipends, and soon after Hutton departed some improvement took place.[10]

The Edward Stocks Massey bequest

The increasing importance of electro-technics had been recognised for many years, firstly with the building of the John Hopkinson wing in 1900 and now by the extensions which, by the end of 1911, were rapidly nearing completion. Robert Beattie, having been appointed in 1896, had 15 years of solid achievement in teaching and research behind him and, as we have seen, was one of the most senior staff members of the Physics Department. Moreover he was still publishing papers regularly, three of which appeared in 1911-12 co-authored with his new staff member, Harold Gerrard (see Appendix 6). It is not surprising therefore that thoughts were turning in the university hierarchy to setting up Beattie's electro-technics group as an independent department with Beattie as its professor. However, as always, money to found a chair was a problem. Fortunately, not for the first time in the history of the University, an unexpected bequest came to the rescue. One remembers crucial legacies of an earlier era; that of John Owens to found the College, Charles Beyer and Edward Langworthy for chairs in engineering and physics respectively, Joseph Whitworth for the new engineering laboratories of 1887, and more recently, the gift from John Hopkinson's friends and relatives for the new electrical wing. Now it was to be the turn of electro-technics once again to benefit through the will of a Burnley brewer, Edward Stocks Massey.

Edward Stocks Massey J.P. was a member of a well-known and respected Burnley family. Brief details of his life are as follows [11]:

Born in Brierfield in 1849 of a reasonably well-off family (his father was an alderman and twice Mayor of Burnley in the 1870s) he was educated privately and became so interested in cotton spinning that his father turned over the management of the Victoria Mill in Trafalgar Street to him. He did this successfully and introduced many of the most recent cotton-spinning appliances. The family also owned Massey's Brewery and Alderman Massey took his two sons, Charles Massey and Edward Stocks Massey, into partnership. Three years later Alderman Massey died and the businesses were left in the hands of the two sons, but the breweries were doing so well that they sold the cotton mill. Again Edward was influential in introducing the latest technology into the brewery and the business expanded, Edward becoming managing director and his brother having a seat on the board until his death. The company owned a large number of public houses in Burnley and the surrounding area. In 1889 the business was formed into a limited company. Edward later became chairman.

Edward was a great lover of art and music and at one time was a leading member of the Burnley Choral Society and was a regular attender at Hallé's concerts. He also made many gifts and was liberal in his support of local churches and schools. In 1887, at the age of 38, he married a London girl, Eleanor Harrison. Thereafter he resided in a large residence, Bamford Hall, near Rochdale, and little more was seen of him in Burnley. He died following a stroke at his home on 27th December 1909, at the age of 60. His wife survived him and lived until 1921. There were no children.

There is nothing in the above narrative that is at all remarkable. But what *was* unusual were not only the provisions in his will, but his announcement some years before his death of what some of those provisions would be. The announcement came in a letter to the Mayor of Burnley dated 29th March 1904. It stated that:

1) After death duties the estate would exceed £125,000.
2) The majority of the money would be given to the town of Burnley.
3) The bequest would be subject to the condition that in the event of any licence or licences of any of Massey's public houses being withheld by the police or the magistrates (or both) within the parliamentary borough of Burnley, for any reason whatsoever, then the full value of such property as declared by valuation should be deducted from the town of Burnley and such deductions would be duly dealt with as directed in the will.
4) Any person, committee or corporation, body or institution disputing the validity of the will or any codicil of it shall forfeit any benefits under the will or codicil and any money intended for such beneficiary "shall be forfeited".
5) Under the will a wide discretion as to the disposal of the bequest to the town would be left to the borough council, who would be required to confirm their decision as to the way the bequest is to be applied by two resolutions:
   (a) agreed by a majority of three-fourths of the Council, and
   (b) confirmed after an interval of not more than two years and not less than one year by a resolution of a like majority of the Council.

It is clear from the last clause that he did not wish the Council to be hasty in its decision-making concerning the disposition of the money. The letter pointed out, darkly, that the refusal of the licence of the Wheat Sheaf Inn, Burnley, had lost £1,700 (the certified value of the property), and that "this loss might have been avoided if the frequenters of the house, who were inhabitants of Burnley, had been more careful as to their conduct."

Edward Stocks Massey's latest will at the time of his death had been drawn up and signed only 10 months earlier, on 23rd April, 1909 [12]. It was an 11-page document, containing the provisions foretold in the above letter, and detailed bequests. As he had stated in his letter to the mayor, the bulk of his estate was to go to the town of Burnley, under the conditions stipulated above, but only after the death of his wife, who was generously provided for. Various personal bequests (watches, etc.) would be made available immediately. The bequest to Burnley would prove of great value over the years; scholarships would be set up, amongst other initiatives, from which young people are still benefiting to the present day. Also (and this is the relevance to our story) a bequest was to be made to Manchester University. The relevant clauses as far as the university was concerned were: *

Page 7: . . . . . UPON THE TRUSTS hereinafter declared (that is to say) UPON TRUST to raise and pay thereout unto the University of Manchester the sum of Six thousand eight hundred pounds and such further additional sum (if any) as hereinafter mentioned to be applicable for the general purposes of such Institution and such sum of Six thousand eight hundred pounds and further additional sum (if any) to be paid free of duty and which duty is to be paid out of the residue of the residuary trust funds and income aforesaid . . . . .

Pages 8 to 9: . . . . PROVIDED ALWAYS AND I HEREBY DECLARE that in case the license of any licensed premises comprised in the list contained in the paper Writing consisting of two pages and headed "List of Licensed Premises belonging to Masseys Burnley Brewery Limited" and which has been dated at the foot the day of the date and this my Will and has been signed by me at the foot of each page and hereinafter referred to as "the said List" shall at any time after the date of this my Will and before the day of my decease and while the premises to which such license shall relate shall continue to belong to the said Company be forfeited or refused to be renewed by the Licensing Justices or other Licensing authority then there shall be raised and paid to the University of Manchester and in addition to the said sum of six thousand eight hundred pounds hereinbefore bequeathed to such University a further sum equal to the value as hereinafter defined of the said licensed premises the license or licenses of which shall be so forfeited or refused to be renewed as aforesaid and as free from duty AND I DECLARE that the further sum now being bequeathed is the further or additional sum referred to in the original bequest to the said University AND I DECLARE that the respective values of the licensed premises respectively comprised in the said List shall for the purpose of the foregoing provisions be deemed to be the respective amounts appearing as the values thereof respectively in the said List and which amounts are the values placed thereon by Mr Charles Parsons in the year One thousand eight hundred and ninety six AND I EXPRESSLY DECLARE that any compensation which may be allowed on any forfeiture of a license or refusal to renew a license shall not be taken into account AND I DECLARE that in case any question shall arise in reference to the licensed premises to which the provision hereinbefore contained shall relate or as to the value thereof for the purpose of the foregoing provisions or otherwise as to the further sum (if any) to be paid to the said University the decision of my Trustees or Trustee shall be absolute and conclusive . . . . .

The donation of the money from the Edward Stocks Massey bequest coincided exactly with the period when the University was thinking of setting up "Electro technics" as a department in its own right. The terms of

---

* They are quoted here in full because the convoluted legal language has a certain charm.

the bequest allowed the university to use the money as it wished and it was decided that, as there was a pressing need for the establishment of a chair in electro-technics, the money should be used for that purpose. As the foregoing sections will have shown, the bequest was to be £6,800, to be boosted by additional sums should any of the Massey public houses in and around Burnley lose their licences through the disorderly conduct of the patrons or any other cause. When the final details were worked out, the University benefited to the total of £10,386. God moves in a mysterious way, and we can but marvel at his ingenuity in turning the transgressions of Burnley's citizens to Manchester University's advantage.

References: Chapter 5.

1. J.B. Birks. *Rutherford at Manchester*, p. 47. New York, 1963.
2. *Chambers Biographical Dictionary*. Chambers, London, 1990.
3. The story of Beattie, Grassot and the fluxmeter has been heard from several people over many years. Grassot's paper is: M.E. Grassot, *Fluxmètre*. J.Phys., Series 4, **3**, p. 696, Sept. 1904 (in French). See also Beattie's paper: *An electric quantometer*. Electrician **50**, pp. 383-385, 1902.
4. R.S. Hutton and J.E. Petavel. Journal IEE, Vol. 32.
5. R.S. Hutton. *Recollections of a Technologist*, pp 44-45. Pitman, London, 1963.
6. A. Schuster. *Biographical Fragments*, pp. 248-257. Macmillan, London, 1932.
7. Henry Wilde, Proc. Royal Soc., **15**, p. 107, 1866.
8. Hutton (see above), p. 105.
9. Ibid, p. 105.
10. Ibid, p. 37.
11. Obituary of Edward Stocks Massey and a copy of his letter to the Mayor of Burnley, both published in *The Burnley Express*, 29th December, 1909; *Men of Burnley*, article in *The Burnley Express*, 12th October 1963; *Stocks Massey - the man behind the money*, article in *The Burnley Express*, 20th March 1981; Copy of a page from the Minutes of the General Purposes Committee of Burnley Borough Council, 24th January, 1910;
12. Edward Stocks Massey's will, dated 23rd April, 1909.

In connection with my Edward Stocks Massey enquiries I wish to thank the following people for their co-operation and helpfulness and for kindly providing photocopies of the above documents:

Mr K.G. Spencer, local historian, of Burnley, for responding to my initial enquiries and suggesting whom to contact to obtain the necessary information (all duly obtained, as above and below).

Mr Brian Whittle, Chief Executive Officer and Town Clerk to Burnley Borough Council (for providing a photocopy of the will).

Mr J.D. Hodgkinson, District Librarian, Lancashire County Council (for providing photocopies of the newspaper cuttings).

Chapter 6
1912-1914: The founding of the Electro-technics department

When the university decided to utilise the timely arrival of more than £10,000 from the Edward Stocks Massey bequest to found a chair of electro-technics, the way was open for the creation of the new department. But the opening of the new laboratories, the creation of the chair for Robert Beattie and the founding of the department did not occur simultaneously. The first of these three events was the opening of the new extension to the physical laboratories, much of it intended for the use of electro-technics, which took place with due ceremony on Friday evening, 1st March, 1912, a day better remembered for the fact that earlier in the day suffragettes went on the rampage in London, smashing windows in West End shops and at number 10 Downing Street. The opening took the form of a conversazione and a special brochure was produced, the cover of which is shown on the opposite page. The proceedings began at 8.15 p.m. with a reception on the first floor of the physical laboratory, with experiments and exhibits on continuous view throughout the laboratories. At 9.15 in the large lecture theatre of the main physical laboratory building the Honorary Degree of D.Sc. was conferred on Mr S.Z. de Ferranti, who at the time was President of the Institution of Electrical Engineers. This was followed by an Address given by Professor Arthur Schuster, again in the large lecture theatre, immediately after which Professor Schuster formally declared the new laboratories open. Refreshments were available all evening in the main physics building and in the extensions.

The exhibits and demonstrations on display represented the work of the whole of the Physics Department, not just the electro-technics section of it, and there were also exhibits from the Manchester Municipal School of Technology (the early forerunner of UMIST), other universities and some commercial manufacturers of scientific equipment. A detailed account of the exhibits was given in the brochure. The following is a brief summary:

Room 1.   Refreshments.
Room 2.   Radioactive minerals and radioactive preparations. New radioactive minerals.
Room 3.   Colour photography - a display of various processes; gyroscopes; an experiment with heat insulation; replica of Mother of Pearl; effect of radium emanation on bacteria.
Room 4.   Radioactive exhibits - a number of working experiments were on display. Magnetic storms (due to tramway currents).

1912-1914: The founding of the Electro-technics Department

University of Manchester.

Physical and Electrotechnical Laboratories.

EXPERIMENTS AND EXHIBITS

AT A

Conversazione

HELD ON THE OCCASION OF THE

OPENING OF THE EXTENSIONS

BY

Prof. ARTHUR SCHUSTER, Ph.D., D.Sc., LL.D., F.R.S.

ON

FRIDAY EVENING, MARCH 1st, 1912.

Manchester:
The Manchester Courier Ltd., Cannon Street.

Front cover of the brochure for the opening of the extensions

Room 5. Radioactive exhibits. More experiments illustrating alpha-ray tubes containing emanation; the range of alpha-rays; the condensation of radium emanation by liquid air; production of helium by radium; a method of separating helium; radioactive shadows; experiment to show the presence of a new substance in the thorium emanation (thorium A) of mean life one-fifth of a second; phosphorescent effects produced by actinium and thorium emanations; experiments with scintillations.

Room 6. Exhibition of apparatus by the Cambridge Scientific Instrument Co. Ltd. (More than 22 exhibits, some of them shown working.)

Room 7. Large lecture theatre.

Room 8. Optical apparatus, etc. A number of different exhibits including gratings, etc; photographs showing the effect of high pressure on arc spectra.

Room 9. Changes in spectrum lines produced by a magnetic field (Zeeman effect).

Passageway. Model showing wave motion.

Room 10. Exhibition of liquid crystals; experiments with polarised light; sound waves made visible (Toepler's experiment); fluorescence of iodine vapour.

Room 11. Meteorological exhibits: balloon ascents; kite ascents; electrified clouds; large Wehnelt discharge tube showing passage of heavy current through a gas at low pressure.

Room 12. Exhibits from the Manchester Municipal School of Technology: Brownian movement and related exhibits; optical exhibit - various optical instruments used for special purposes.

Room 13. High-frequency experiments (Tesla coil) with wireless lighting [lighting of a bulb not connected to an external circuit]; stationary electric waves.

Room 14. Counting of atoms and matter (Dr Geiger's counters demonstrated); a high-pressure pump, capable of yielding 37,000 p.s.i.

Room 15. Exhibition of gas discharge-tubes - a large discharge tube excited by 2,000 volt direct-current dynamo.

Electrochemical laboratory:

Room 16. Exhibition of refractory silica ware; electric furnaces for use with high gaseous pressure; electro-deposition of zinc; migration of colloids in an electrical field; ionisation produced by glowing carbon in vacua; vessels for use at high temperatures (silica ware [melting point 1700°C] and magnesia ware [melting point 2400°C]); silica trees grown by osmosis; experiments with water voltameter; demonstration of a flour explosion.

Room 17. Dynamo room - machinery in motion.

Experiments and exhibits in the new buildings:

Room 18. Apparatus constructed and in course of construction by Mr Charles W. Cook, Scientific Instrument and Apparatus Maker to the University. (A long list of items was available for inspection).

Room 19. The jumping ring, illustrating electromagnetic repulsion in an alternating magnetic field; the spinning egg, showing the action of a rotating magnetic field on a conductor; more exhibits from the Manchester Municipal School of Technology: apparatus for testing india-rubber by the hysteresis method; photographs of electric discharges up to 130,000 volts; insulating material used in high-tension work; pin and suspension porcelain insulators; terminal and switch insulators; photographs of electric discharges from a Wimshurst machine; portable photometer for measuring the illumination of streets.

Room 20. The 'Blitzafel' - a 20,000 volt discharge on a glass surface.

Room 21. The telewriter, for transmitting writing by electricity; models of Ferranti meters; a wireless telegraph station exhibit.

Room 22. Exhibit of apparatus by Mr Otto Baumbach, glass-blower to the University. Also a demonstration of glass blowing.

Room 23. Refreshments.

Electro-technics lecture room. Experiments on vortex motion, etc., due to the late Professor Osborne Reynolds, illustrating his theory of the sub-mechanics of the Universe.

It will be noticed from this list that the current work on atomic physics was well represented. Robert Beattie and Harold Gerrard had various demonstrations working and ran their generators in the dynamo house. The unguarded belt drives that were currently used on the electrical machines, and were also used throughout industry at that time, would nowadays be regarded as a Health and Safety Inspector's nightmare. Forty-seven years after the opening ceremony Harold Gerrard, recalling that evening, said that many of the ladies present at the conversazione wore full flowing skirts or dresses, in the fashion of the day, and it was a matter of great concern to Beattie and himself to ensure that they did not get too near the machines, otherwise something disastrous might have happened. [1]

A sad event had taken place a week before the opening of the laboratories with the death on 21st February 1912 at the age of 69 of the brilliant and much-respected Osborne Reynolds, the 'father' of engineering at Manchester University. The decline in physical health and mental ability that had forced his retirement in 1905 had continued in the subsequent years. He died at his home in Watchet, Somerset. The cause of his death was given as "aphasia; progressive softening of the brain", which was a diagnosis often used at that time. [2] In modern medical terminology it might be the equivalent of Alzheimer's disease, but there are alternative possibilities of a similarly degenerative kind. [3] Clearly the brochure for the opening of the new laboratories must only have been printed shortly beforehand, as Reynolds was referred to as "the late Professor Osborne Reynolds".

Electrical Engineering at Manchester University

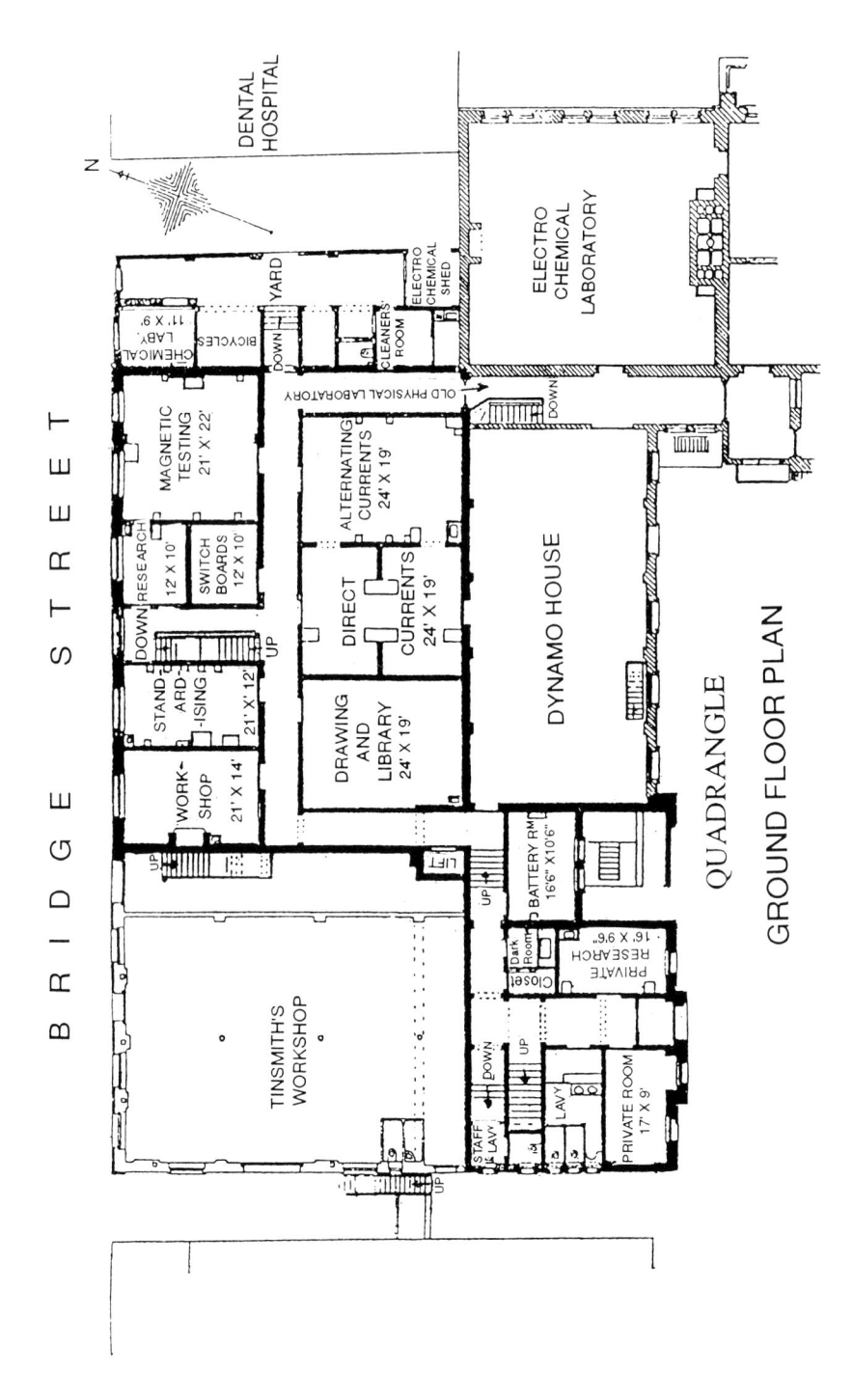

New Extensions: Ground floor plan

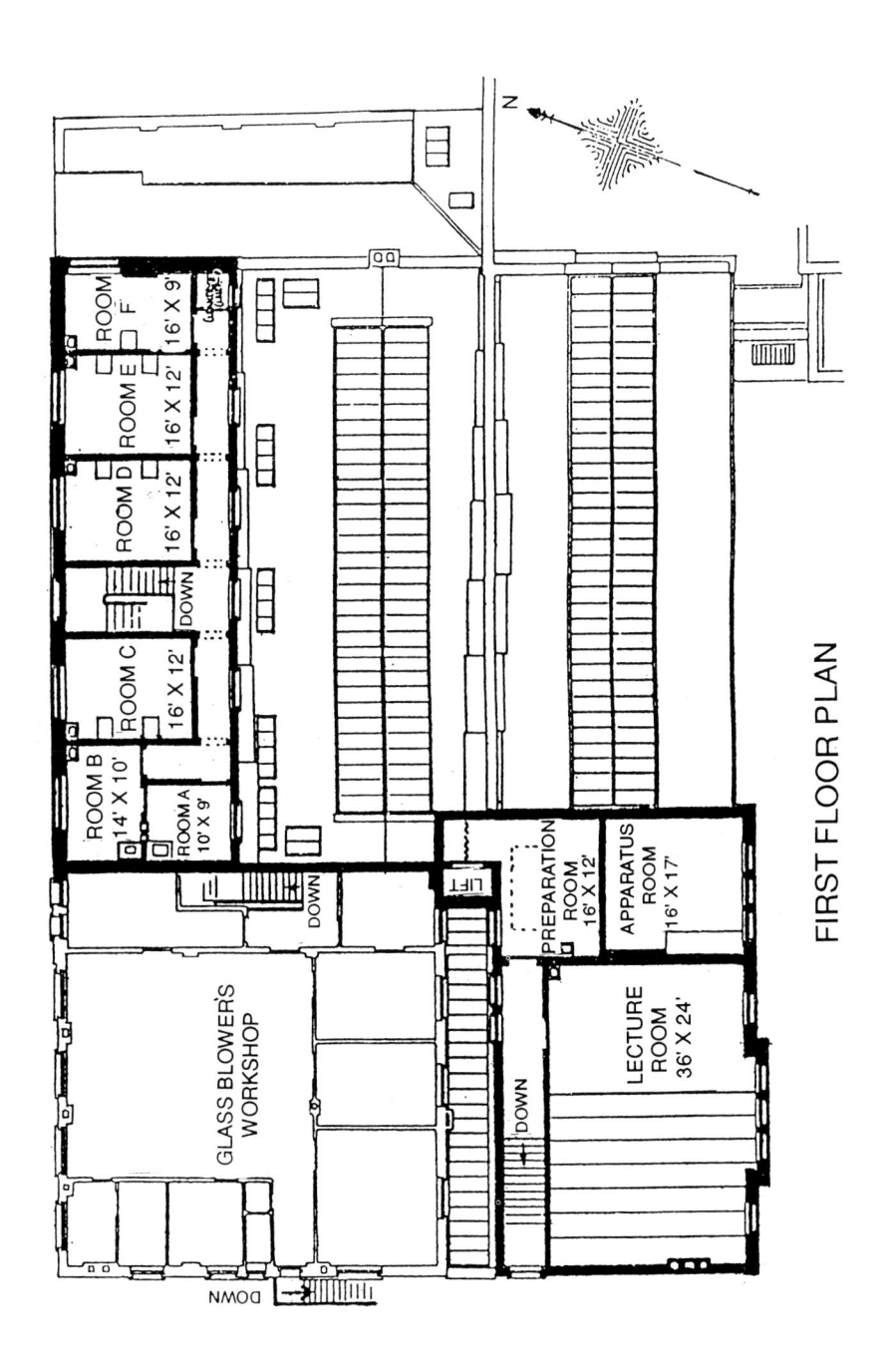

1912-1914: The founding of the Electro-technics Department

New Extensions: First floor plan

Facilities available in the new Electro-technics Department

These were outlined in a separate brochure available when the extensions were opened [4] and a booklet produced by the Engineering Department two years later, which covered civil, mechanical and electrical engineering. [5] The following account, taken from these sources, should be read in conjunction with the plans on pages 84 and 85:

"The electrotechnical laboratories occupy the whole of the ground floor of the north wing. The three principal rooms on the ground floor of this wing abut upon the Dynamo House on one side and open into the central corridor on the other. This part of the wing is only one storey high, and, like the Dynamo House, is lighted from the top by a lantern roof. Of the three rooms referred to, that in the centre communicates directly with the Dynamo House, of which it may be regarded as a constituent part, being intended chiefly for elementary tests on direct current machines. The room is divided into several bays by brick walls, with the object of increasing both the bench space and the wall area available for attaching instruments.

On the right, the Direct Current Room opens into a room containing two small motor-alternator sets, an oscillograph and other alternating current apparatus. This room serves the purpose of an Alternating Current Laboratory, where students can acquire a working knowledge of the properties of alternating currents before proceeding to make tests on the larger machines in the Dynamo House. The room on the left of the Direct Current Room fulfils the double function of a Drawing Room and a Departmental Library.

The remaining rooms on this floor are situated on the north side of the central corridor. They comprise a small workshop, a Standardising Room, a Research Room, a Switchboard Room, and a room fitted up for Magnetic Testing and other introductory work of a general nature. In the basement are a store room and a photometric room.

The central corridor of the north wing leads on the right to the main Physics Laboratory, and on the left to the west wing of the extension. Here on the ground floor are to be found a Battery Room, Photographic Dark Room, Private Research Room and Private Room, together with general and staff lavatories. Above, on the first floor, which is served by a hand lift, are the Lecture, Preparation and Apparatus Rooms. The Lecture Room, with a seating accommodation for 100, is specially fitted up with a view to the teaching of Electrotechnics, the lecture table having all necessary appliances for demonstration work. To facilitate the carrying out of experiments on alternating currents, a switchboard enables a motor-alternator to be controlled from the lecture room itself, and large scale electrical instruments attached to the wall can be put in circuit with any piece of apparatus on the lecture table, thus enabling the course of an experiment or test to be followed by a large class.

A noteworthy feature of the new buildings is the system of bare wires run on insulators, which has been adapted throughout for experimental circuits. This system has proved so satisfactory in the main Laboratory that it has been employed wherever possible in the present extension. From the battery, which is of 600 ampere-hour capacity with a maximum discharge rate of 300 amperes, heavy bare copper conductors run along a subway beneath the main corridor to the switchboard room in the north wing. From this, by means of plug boards, current can be distributed over the whole building. The heavy currents required for use in the Dynamo House are, however, taken direct from the battery to a special plug board in the Dynamo House, and a similar arrangement has been adopted for the Lecture Room. All the alternating current machines are connected up to a special plug board occupying a central position in the Direct Current Room, and from this connection is made to the principal battery board, so that alternating current is available anywhere throughout the buildings. The new battery, it should be remarked, is additional to the smaller battery already installed in the main Physics Department, but can be used in conjunction with it if required. It is also conveniently situated for supplying current to the Engineering and Electrochemical Departments.

Although the Dynamo House had been built and equipped in 1900, it contained a number of items of much newer equipment. These included a Rosenberg dynamo, the gift of Messrs. Mather & Platt, a Siemens-Schuckert single-phase motor, two motor-alternator sets for experiments on synchronising and parallel running, a booster-balancer set, an interpole motor specially fitted up for experimental work, and a number of induction motors and transformers.

Robert Beattie promoted to Professor

The next stage in the development of electro-technics as a department came in May 1912, ten weeks after the opening ceremony, with the news that Robert Beattie was to be promoted to the status of the Edward Stocks Massey Professor of Electro-technics. A letter to Dr Beattie from the Registrar, dated May 10th 1912, read [6]:

Dear Sir,

I beg to inform you that at a meeting of the Council held on Wednesday, you were appointed Professor of Electro-Technics and Director of the Electro-Technical Laboratory, at a stipend of £500 per annum, as from the 29th September, 1912: the appointment to be in lieu of your present appointment.

Yours faithfully,
Edward Fiddes,
Registrar

The stipend of £500 would be equivalent to nearly £30,000 at 1997 values. [7] The formal letter of appointment was backed up by a personal note written three days later (on Beattie's 39th birthday) from the Vice-chancellor, Alfred Hopkinson, which is reproduced on the opposite page. [8] The text reads:

Dear Beattie,

You will be receiving from the Registrar to-day formal intimation of your appointment as Professor of Electro-Technics but I wish also to add my formal congratulations and of welcome to you as a member of the Senate. I hope the department will continue its growth and long flourish under your direction.

     Yours sincerely

     Alfred Hopkinson

The continuing growth of electro-technics and the promotion of Beattie must have continued to please Hopkinson, bearing in mind the part his brother had unwittingly played, posthumously, in the creation of the electro-technical laboratories. In the following year, 1913, Hopkinson resigned and was succeeded as Vice-chancellor by Frederick Weiss, the Professor of Botany in the University, on a short term basis, to be succeeded in turn by Henry Miers in 1915. Alfred Hopkinson had served Manchester long and well, having been Professor of Law from 1875, Principal of Owens College from 1898, in which capacity he steered through the transition of the college into the Victoria University of Manchester in 1903, and then continuing as the university's first Vice-chancellor until 1913.

With the new electro-technics extensions opened and Beattie now a professor, the question arises whether electro-technics now constituted a department in its own right. The 1912-1913 university calendar listed Robert Beattie as Edward Stocks Massey Professor of Electro-technics and Director of the Electro-technical Laboratories but his name was still listed amongst the physics staff. Similarly, Harold Gerrard was listed amongst the physics staff as Junior Demonstrator in Electro-technics. But Alfred Hopkinson's letter suggests that he, as Vice-chancellor, regarded Electro-technics as a Department. Any ambiguity on what some might regard as an academic point disappeared in 1913 when the university calendar listed Electro-technics as a Department in its own right with a staff of two; Robert Beattie as professor and Harold Gerrard as his junior demonstrator. They were assisted by Arthur White, a technician trained by William Kay in the Physics Department.

Robert Beattie's marriage

During his years as a lecturer, Robert Beattie lived in lodgings in Manchester and at the time of his promotion to professor in 1912 he was still, at 39, unmarried. His family had moved from Laurencekirk to

1912-1914: The founding of the Electro-technics Department

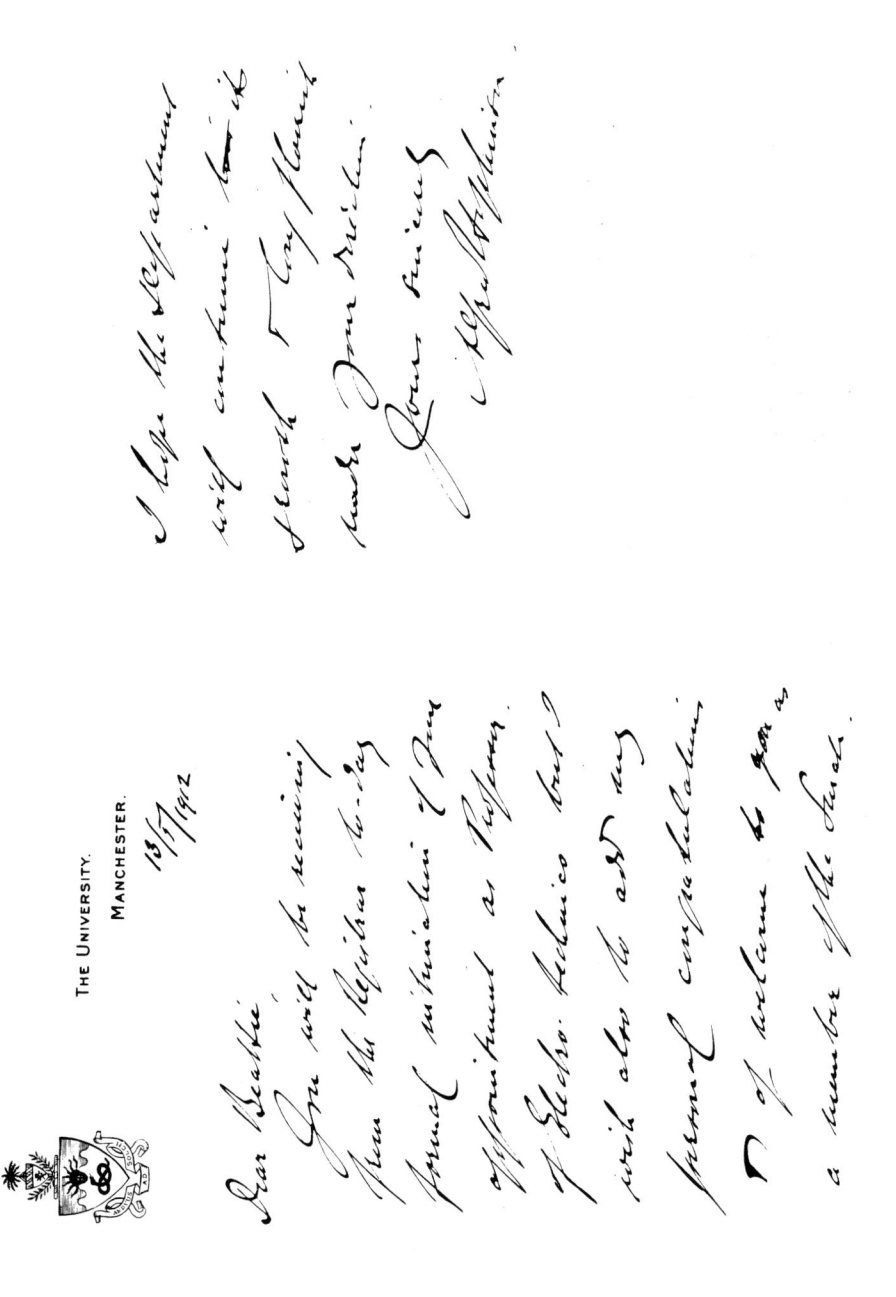

*Letter from the Vice-chancellor, Alfred Hopkinson, to Robert Beattie.
May 1912*

Friockheim, Angus, in 1903 and soon afterwards he met Janet Kyd, one of the daughters of a local master shoemaker. She and Robert became engaged around 1906 and eventually, on 13th March, 1913 when Robert was nearly 40 and she 36, they were married. One can only speculate why the engagement was so long but a possible reason is Janet's sense of family responsibility. Her father had died in 1908 at the age of 72 and in 1912 an elder sister had died aged 38 of tuberculosis. Afterwards her mother left Friockheim to live in Montrose with a married daughter and her husband. Janet was a teacher whose salary was probably needed whilst the family was together. The long engagement may have been due to her unwillingness to leave home until the family broke up. Alternatively, or perhaps as a contributory factor, Robert might have been waiting for promotion before marrying, bearing in mind the relatively low pay he received as a lecturer. The marriage took place at 27, Balmain Street, Montrose, which was the home of the married sister and her husband. An indication of Robert Beattie's modesty so soon after his promotion to a chair is that on the marriage certificate he stated his occupation as 'Teacher in a School', the same as Janet gave for hers. Until the time of the marriage he had been living at 66, Claremont Road, Moss Side but as soon as they were married the couple set up home at "White Knowle", White Knowle Road, Buxton, whence he used to travel in to the university each day by train. [9]

The undergraduate course

When the new Department of Electro-technics was founded a major change of policy came about, regarding undergraduate courses. Hitherto, when electro-technics was a part of physics, students who wished to specialise in electrical engineering graduated in physics, having taken a number of physics subjects in conjunction with some electrical subjects. From the founding of the new department this link was changed; instead electro-technics would be part of engineering from the point of undergraduate degrees. Thus, engineering students wishing to specialise in electrical engineering would take a number of civil and mechanical engineering subjects in conjunction with some electrical subjects, and their degree would be in Engineering, not Physics. This arrangement continued until as late as 1968. During the whole of this period the word "Electrical" was not mentioned on their degree certificates; it was just "Engineering". In practice the Electrical Engineering staff decided what class of degree their students would receive at the end of the three years of their course, and the Engineering staff did the same with their civil and mechanical students, but all the names, civil, electrical and mechanical, were then combined and published on a common "engineering" results list. From a practical point of view this did not matter when the "electrical" students applied for jobs, as any reference they received from the university staff would state that they had specialised in electrical engineering.

However, in spite of this major policy change Physics students continued to be able to take an electro-technics option. The procedure in their case was that after taking a basic Physics course up to Part I Honours standard they could take Electrical Engineering* papers amongst the options in the latter part of their course. Up to 1921 this was made possible by a regulation to the effect that students who had passed Part I in Physics could *either*:

(a) Present a thesis or dissertation on some special branch of Physics; *or*
(b) Present themselves for examination in some special department of Physics or other subject approved by the Faculty.

From 1921 onwards the regulation was formalised into the following:

Candidates who have passed the Part I Examination must subsequently attend satisfactorily an advanced course or courses in one of the following:
(a) Pure Physics; (b) Electro-Technics; (c) Physical Chemistry.

This wording of the regulations for the Physics course continued unchanged into the Second World War and in a related form right through to 1963 when the Physics and Electronic Engineering course was founded. Consequently, Physics students have always been able to include electrical subjects in their choice if they so wished.

The lectures and laboratory teaching which the students received from Robert Beattie and Harold Gerrard after the change was much the same as before. Their job was to teach the "electrical engineering" part of the course, and it made little difference whether the students were doing physics and mathematics in the rest of their course, as they did before the change, or engineering and mathematics, as the Engineering students did after 1912, provided the lecturers kept an eye on the syllabuses of the non-electrical part of the course in order to ensure the whole course remained well balanced.

A further important point concerning the courses at that time was that most of the science departments, including Engineering and Physics, offered "Certificate of Proficiency" courses as well as degree courses, a system that had operated since the foundation of Owens College and would continue, at any rate in Engineering, until 1948. The existence of the certificate courses, which were obviously of lower standard than the degree courses, must have added considerably to the work load of the staff, but from the University's point of view they provided a valuable source of income.

---

* Although the Department was officially called "Electro-technics", it and its courses were often referred to as "Electrical Engineering" in university publications such as the Calendar, in which the phrase "Electro-technics", when it appeared, was sometimes hyphenated and sometimes not.

Details of the courses in Electrical Engineering in 1914 are taken from the university calendar for 1914-15 and are as follows:

Course for Certificate in Engineering

The Engineering Certificates carry no title to a degree, but successful completion of the three years' course will furnish evidence of thorough scientific training for the engineering profession in its various branches. The courses, however, are not intended to supersede the practical training which can only be obtained in the office of a Civil Engineer or the workshop of a Mechanical or Electrical Engineer.

Some of the principal Engineers in Manchester and the neighbourhood have signified their willingness to receive certificated Engineering students of the University into their works, either as premium apprentices for a short term or as ordinary apprentices without premium; and to confer upon them such privileges as they may from time to time be found qualified to avail themselves of.

The certificate course consists of:

a) Two courses of lectures, each of three hours a week for a session, from: Surveying, Graphic Statics and Theory of Structures, Strength and Testing of Materials, Kinematics and Dynamics of Machines, Heat Engines, Hydraulics.

b) Not less than three courses of lectures, each of three hours a week for a session, in Pure and Applied Mathematics or Experimental Mechanics.

c) Two courses in the Engineering Laboratory, each of one day a week for a session.

d) Two courses of Drawing and Design, each of one day a week for a session, including special courses of lectures on Design and Drawing.

e) Two courses, each of not less than two hours a week, to be chosen from: Physics; Geology; Electrical Measurements and Theory of Electrical Machines; Electrotechnics; Metallurgy and Fuel; Chemistry; Mining; French or German.

At the end of the third year students are examined on these subjects. The award of the Certificate is dependent on the marks obtained in the examinations and on the work done in the department during the three years.

Degree course in Electrical Engineering

As mentioned earlier, the degree was designated "Engineering" and a considerable portion of the course consisted of civil and mechanical engineering subjects, supplemented by the usual mathematics and physics lectures.

Only the electrical engineering part of the course will be quoted here.

I. *Introductory General Course*

Monday and Friday from 11.30 a.m. to 12.30 p.m. for half the session, commencing during the Lent Term.

In this class an elementary descriptive account of Applied Electricity is given suitable for Civil and Mechanical Engineering students who do not wish to specialise in Electro-technics, and to serve as a first-year course for Electrical Engineering students. Students taking this course must possess a knowledge of the elements of Magnetism and Electricity or attend Physics II B at the same time.

1912-1914: The founding of the Electro-technics Department

II. **Final B.Sc. Course**
Monday, Wednesday and Friday from 12.30 to 1.30 p.m. throughout the session:
(1) *Michaelmas term*: Direct current machinery: magnetisation of iron. The direct-current dynamo: armature winding; field-magnet design; armature reaction and commutation; operation losses and performance characteristics. The direct-current motor: performance characteristics; starters; speed regulation; series-parallel control; efficiency. Motor generators. Balancers. Boosters. Coupling and operation of generators. Testing and rating of direct-current machinery.
(2) *Lent term*: Alternating-current machinery. Single-phase alternators: calculation of e.m.f.; wave form; elementary theory of alternating-current flow; calculation and measurement of power; performance characteristics. Single-phase synchronous motors. Synchronising and parallel running. Single-phase transformers. Polyphase generators. The induction motor. The series alternating-current motor.
(3) *Summer term*: Arc lamps and their mechanism. Incandescent lamps. Photometry. Electric lighting. Storage batteries.
Text books. Franklin & Esty's *Elements of Electrical Engineering*; Hay's *Continuous Current Engineering*; Hay's *Principles of Alternating-Current Working*. Also for reference: Hawkins & Wallis' *The Dynamo*; Thompson's *Dynamo Electric Machinery*; Crocker & Wheeler's *Management of Electrical Machinery*; Parr's *Electrical Measuring Instruments*.

III. **Honours Course**. Tuesday, Thursday and Saturday, from 9.30 to 10.30 throughout the session.
(1) *Michaelmas term*: Higher direct-current work. Theory of armature winding. Theory of commutation. Design of direct-current dynamos and motors. Special types of direct-current machines.
(2) *Lent term*: Higher single-phase work. The single-phase alternator: theory of armature reaction and regulation; predetermination of pressure drop; Influence of wave form. Calculations by symbolic method. Graphical theory of the synchronous motor. Synchronisation and parallel running. Theory and design of transformers.
(3) *Summer term*: Higher polyphase work. The polyphase alternator and its design. Polyphase transformers. Theory and design of induction motors. The asynchronous generator. Compounding and compensating alternators. Commutator alternating-current motors. Theory of phase meters and alternating-current energy meters.
Reference books: In addition to those recommended for (I), Hobart's *Dynamo Design*; Parshall & Hobart's *Electric Generators*; Hobart's *Electric Motors*; Steinmetz's *Theoretical Elements of Electrical Engineering*; McAllister's *Alternating-Current Motors*; Hay's *Alternating Currents*.

IV. *Generation Transmission and Distribution of Electrical Energy.*
Three hours per week during the Summer term, at times to be arranged.
Systems of Distribution. Generating and substations. Switchboards. Regulating and safety devices. Central station economics. Calculation of size of conductors. Arrangement of cables and feeders. Overhead construction. Electric traction. Central station testing.
This course is supplementary to Electrotechnics III and may be taken concurrently with it.
Reference books: Dawson's *Electric Traction Pocket Book*; Parshall & Hobart's *Electrical Railway Engineering*; Andrew's *Electricity Control*; Wordingham's *Central Electrical Stations*; Bell's *Electric Power Transmission*; Wilson's *Electric Traction*.

Electrical Drawing and Design. At times to be arranged.

The course is intended for third-year Engineering students who are attending Electrotechnics III, and others attending this class who have the requisite knowledge of mechanical drawing. Opportunity is given of gaining familiarity with the general principles of Design as applied to typical examples of direct- and alternating-current machinery. A Drawing Room has been fitted up for the purpose, and, if found necessary, special lectures, supplementing those on Design in Electrotechnics III, are given.

Laboratory work

In connection with the laboratory work a systematic course of practical work has been arranged, comprising: Testing of conductors, insulators and magnetic materials. Investigation of commercial measuring instruments, meters, incandescent and arc lamps, etc. Regulation and efficiency tests of direct-current generators, motors, balancers and boosters. Alternating current measurements. Oscillographic study of wave forms. Performance tests of single and polyphase alternators, synchronous motors, transformers, induction motors, etc. Synchronising of alternators. High-tension experiments.

Students are expected to have completed a course of practical electrical measurements such as that given in the first and second year courses in the physical laboratories before embarking on the above work.

The laboratories are open each weekday from 9.30 a.m. to 4.30 p.m. and on Saturdays from 9.30 a.m. to 12.30 p.m.

It is not easy to judge this electrical engineering syllabus of 1914 from the viewpoint of the late 1990s, but one feels it might have been in some respects a little old-fashioned even by the standards of 1914. Only seven weeks after the opening of the John Hopkinson extensions in 1912 the value of radio had been demonstrated in dramatic fashion when the sinking liner *Titanic* summoned help by radio and as a result the Cunard cargo-liner *Carpathia* steamed to the scene and was able to pick up over 700 survivors. Earlier, in 1910, radio played a major part in the arrest of the murderer Dr Crippen. Yet in 1914 radio was not mentioned in the electrical engineering syllabus, nor were telegraphy or telephony. Perhaps these subjects were mentioned in passing somewhere in the various courses, but they do not seem to have formed a substantial part of any of them. Whether such devices as the diode (invented in 1904) or triode (1906) were described is impossible to know with certainty, but there is no evidence from the examination papers of the period that they were, nor that the then infant subject of electronics received any type of attention. On the other hand, coverage of instrumentation, machines, power transmission and distribution and similar related topics was extremely thorough, both in the lectures and laboratory work. All the above courses, degree and certificate, were given by Robert Beattie and Harold Gerrard who also supervised all the laboratory work. Between them they clearly had a substantial work load. The content

1912-1914: The founding of the Electro-technics Department

of the syllabuses reflects their special interests of instrumentation and machines.

With the founding of the new Department of Electro-technics, albeit with a staff of only two, there was reason for optimism. But it had been operating as an independent unit for only two years when, on 4th August 1914, Britain was plunged into war, and all plans for the future had to be suspended.

References: Chapter 6

1. Anecdote told by Harold Gerrrard at his retirement dinner in December, 1959.
2. Certified copy of Osborne Reynolds' Death Certificate.
3. The opinion of a medical practitioner who has been consulted on the matter.
4. *An Account of the Physical and Electrotechnical Laboratories and of the New Extensions.* The Manchester Courier Ltd., Manchester, 1912.
5. *University of Manchester Faculty of Science: Engineering Department.* University Press: Sherratt & Hughes, Publishers to the University of Manchester, 1914.
6. Copy of a letter supplied by Mrs J.E. Paterson, Professor Beattie's daughter.*
7. *Whitaker's Almanack, 1997.* Section on "The purchasing power of the £, year by year", p. 599.
8. Copy of a letter supplied by Mrs J.E. Paterson.*
9. Professor Beattie's family details supplied by Mrs J.E. Paterson.*

* The kind help of Mrs Paterson and her husband, Dr J.A. Paterson, is acknowledged with thanks.

Chapter 7

1914-1918: The First World War

Following the outbreak of war in August 1914 the normal work of the university was severely disrupted. Although compulsory conscription did not come into effect until March 1916, large numbers of staff members, students and other university personnel volunteered for active service when the war started. At best they were to be absent for four years; at worst they would be injured, taken prisoner or killed. Consequently the number of students passing through the various departments fell to a tiny trickle, a state of affairs which would continue for the duration of the war.

Special wartime arrangements

In December 1914 the University obtained from the Privy Council power to waive the requirement of the statutes as to the period of study after matriculation and before graduation. Further, the University amended the conditions in such a way as would prevent, as far as possible, any member of the University being placed at a disadvantage through going on active service, undertaking other approved duties in connection with national defence during the war, or from internment as civilian prisoners in any enemy country. Ordinances were passed empowering the Senate to accept a certain period of War Service in lieu of a portion of the three years' course, and to waive the full requirements of the existing Ordinances in any Faculty or Department, except that of Medicine, as to the passing of examinations or as to the order in which examinations were to be taken.

The tenure of scholarships and Local Authority Awards was also postponed in the case of the many students who were engaged in national defence and approved war service. Power was also granted by the Court to the Senate to confer degrees *in absentia* on duly-qualified students who were actually engaged at the time in war service. [1]

In short, with Britain fighting for its life, the University took what sensible measures it could to smooth the path for all participants by the temporary relaxation of its many rules.

Effect of the war on student life

Following the outbreak of war, students who had not yet joined up found life very different from what they had been used to in peacetime. The University's branch of the Officer Training Corps (O.T.C.) was very active, as would be expected. From the start of the war to the end of July 1915, 550 cadets were granted commissions, of which 210 were non-members of the university but members of the Training Corps. In 1915 the University decreed that until further notice no lectures would be given on Wednesdays and Saturdays so that students in the O.T.C. could take part in military training, and also stated that students unable to join the O.T.C. would be required to undertake on those days some form of work of civic or national usefulness so that all students should feel these days were set apart for work for their country.

In March 1915 the Registrar published a list of present and past members of the university who were serving with His Majesty's Forces in the war. [2] Subsequently updated lists were published periodically. The lists gave the nature of their service and particulars of the Battalion and Regiment to which they were attached.

Students in the Engineering Department (including Electro-technics)

It is perhaps not surprising that engineering students, with their special interest in mechanical and electrical equipment, would be in great demand by the Armed Forces. In those days, even if war had not broken out, the number of students entering each year would only have been in the region of 20. As a large proportion of the Engineering Department's students enlisted when the war started there were few remaining other than some from overseas and those who were medically unfit. Sadly, it was not long before heavy casualties began to take their toll. In the Engineering Department's student records of the war years the handwritten words "killed in action" appear all too frequently against the names of students who had temporarily left their course to fight in the war. Of the 21 students who entered the Department in 1912 (civil, mechanical and electrical combined), several enlisted before the completion of their course. Of these, four had already been killed by the summer of 1915, which in normal circumstances would have been their date of graduation. One student who should have graduated in 1916 had been killed before that date. The decline in the number of students graduating, year by year, is illustrated by the fact that there were 10 Honours Engineering graduates in 1915; in 1916 the number had fallen to four, in 1917 there were no honours graduates at all, and in 1918 three, all of whom were foreign. As hardly any students entered during the war it would be well into the 1920s before student numbers returned to normal.

In the university as a whole, by the summer of 1916 the number of students who had withdrawn from their courses to serve with the armed forces or to engage in other approved war service had risen to over 1300. The number of students reported killed or "missing" already numbered 90. [3]

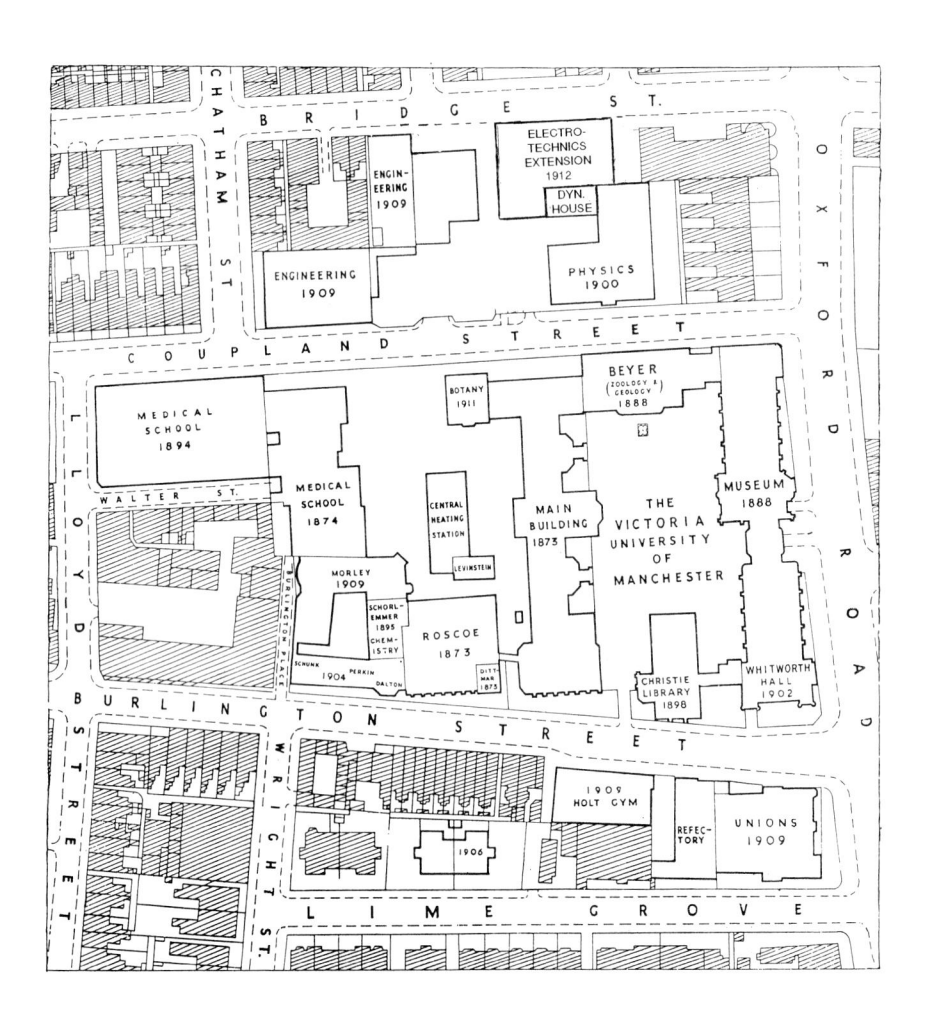

Plan of the university at the start of the war, 1914

1914-1918: The First World War

Contribution to the war effort by the science departments

Most of the ordinary research work of the various science departments was abandoned once the war started and the work of the staff members who remained in Manchester (many had joined up or were on approved war service elsewhere) was turned to special services, both advisory and experimental, in connection with the war.[4] In the Engineering Department (civil and mechanical) much of the work centred on the construction of experimental apparatus for research on the materials used in aeroplane construction, and on appliances used in anti-submarine warfare. Many thousands of tests were done in the laboratories for the Navy, Army, and Air Board. In the Chemistry Department the work centred on the design and testing of high explosives and other munitions; samples of all such materials manufactured in the Manchester area were tested in the Chemistry Department's laboratories. In Physics and Electro-technics a lot of work was done for the Admiralty and is described in the next section.

The war work of Rutherford, Beattie and Gerrard

It will be recalled that the Electro-technics Department was founded with a staff of two; Robert Beattie as Professor and Harold Gerrard as Junior Demonstrator. In 1914, just before the war started, Gerrard was promoted to Senior Demonstrator, a rank he retained during the war years. As soon as the war began the best scientific brains in the country were pressed into action on behalf of the war effort. Professor Rutherford (then 'Sir Ernest' as he had been knighted in 1914) was invited to join the Board of Inventions and Research for the Admiralty, and was to asked to devise effective methods of detecting the presence of submarines. This assignment, involving work very different from what he had been doing, was of vital importance to the country, for even in the early days of the war shipping was suffering heavy losses at the hands of enemy submarines. Two anti-submarine committees were set up, a Lancashire committee and a Clyde committee. Rutherford was joined on the Lancashire committee by his colleague Robert Beattie, Edward Hopkinson of Mather and Platt (the brother of the late John and of Alfred (former vice-chancellor of the University)), Dr E.W. Marchant of Liverpool University, Professor Miles Walker of the Electrical Engineering Department of the Manchester Municipal School of Technology, and a number of others. The chairman was John Taylor of Mather and Platt and the vice-chairman A.P.M. Fleming of British Westinghouse. Setting most of his research aside, Rutherford entered into the task of devising ways of detecting the presence of submarines with extraordinary energy. The full story of anti-submarine warfare in the First

World War is well told elsewhere [5,6,7] and this account will be limited in the main to the part played by Rutherford, Beattie, Gerrard, and one of Rutherford's physics graduates of 1912, Albert B. Wood. The work of these four people seems to have been crucial in fighting the submarine menace and its importance cannot be overstated.

Rutherford started by considering the ways in which a submarine could be detected under water: the sounds it makes, the heat it gives off, the electromagnetic disturbance created by a large metal object moving in the earth's magnetic field, and its ordinary visual characteristics. All of this might seem very obvious but the fact is that the first principles of submarine detection needed to be laid down; no serious thought had been given to the subject hitherto, even though submarines had been in existence since about 1900. The methods identified by Rutherford were 'passive', relying on the receipt of information sent out in one form or other by the submarine. 'Active' methods, whereby a signal was projected towards the submarine and the echoes coming back were received and interpreted, were still in the future.

Rutherford had a large testing tank built in his laboratory and started experiments. It did not take him long to decide that of the four possible methods of submarine detection outlined above the only viable one was that of listening to the sound the submarine makes, a finding that has remained valid until recent times. Consequently he concentrated his efforts in this area, using devices called hydrophones (basically a microphone attached to a sound-receiving diaphragm suitable for use under water), and devices called Broca tubes, named after their inventor (a French scientist). These consisted essentially of aneroid barometer capsules, single and multiple, with a stethoscope attached. Experiments were also made with carbon granule microphones, single-contact microphones and magnetophones (telephone earpieces). Much of the wartime anti-submarine effort would involve the use of hydrophones, on board ship or submarine (it was important for our own submarines to be able to detect enemy ships), towed in capsules, or underwater in groups connected by wires to the shore. This last method was valuable in protecting estuaries and harbours. Harold Gerrard assisted in these experiments in the earlier part of 1915, and Rutherford also called up the assistance of his former student A.B. Wood who, after graduation, had gone to Liverpool University as an Oliver Lodge Fellow and Lecturer in Physics under Lodge himself. Though Wood continued his duties at Liverpool he often visited Manchester to help Rutherford's experimental work. Gerrard and Wood continued their water-tank investigation of listening devices under Rutherford's direction until the autumn of 1915.

1914-1918: The First World War

A tragic casualty

Whilst Gerrard and Wood were busily engaged in their important experiments, other former Rutherford students and staff had gone their various ways. Geiger, who was German, had returned to Germany before the war started, so was technically one of the enemy. Chadwick was on a visit to Germany when the war began and was interned there for the duration. D.H. Florance, a member of Rutherford's teaching staff, had joined up, almost apologising for so doing in a letter to Rutherford in which he stated: "I felt sure you would have no objection to my offering my services to the common cause. I was anxious to carry out the work you left me to do and so I delayed as long as possible, but I felt the call for the strong and vigorous was so urgent that I could no longer hold out and consequently applied for a commission". Florance must have set Rutherford's mind wondering whether bright young scientists could be better used, when he said in one of his letters: "We need scientific men in the artillery, and some of the officers who do not happen to know that the exterior angle of a triangle equals the sum of of the two interior and opposite angles find themselves tied up." [8] In July 1915, A.S. Eve* wrote to Rutherford suggesting that H.G.J. Moseley, a member of Rutherford's teaching staff who had joined up when the war started and who was already internationally renowned for his brilliant work on X-ray spectra, would be more valuable "if he were set to solve some scientific problem" rather than acting as a Brigade Signals Officer in the Royal Engineers, though the problem was that Moseley "would naturally be very much incensed" if he were simply brought home for his own security. Rutherford reacted immediately by writing to R.T. Glazebrook, an eminent scientist who was Director of the National Physical Laboratory at the time, to see whether suitable scientific work could be found. But the request was overtaken by events for in August 1915 Moseley was killed by a sniper's bullet whilst serving in the Dardanelles. So ended the life of one of Rutherford's brightest stars at the age of 28. In an obituary in *Nature* Rutherford followed his scientific tribute by stating:

"Scientific men in this country have viewed with mingled feelings of pride and apprehension the enlistment in the new armies of so many of our promising young men of science - with pride for their ready and ungrudging response to their country's call, and with apprehension of irreparable losses to science . . . It is a national tragedy that our military organisation at the start was so inelastic as to be unable, with a few exceptions, to utilise the offers of services of our scientific men

---

* A.S. Eve was British by birth but spent most of his career as a distinguished physicist in Canada. In the war he was in the British Forces and was Director of Research at the Admiralty Experimental Station at Harwich from 1917-18. In 1937 he wrote a biography of Rutherford.

except as combatants in the front line. Our regret for the untimely death of Moseley is all the more poignant because we recognise that his services would have been far more useful to his country in one of the numerous fields of scientific enquiry rendered necessary by the war than by the exposure to the chances of a Turkish bullet."

In the context of the times, when women were giving white feathers to non-uniformed young men in the streets and people like Schuster were subjected to opprobrium because of their Germanic names, these were brave words.

Gerrard and Wood are sent to Hawkcraig

The only naval establishment where any anti-submarine work was being done at that time was at the Admiralty Experimental Station at Hawkcraig, a rocky promontory on the north side of the Forth and adjoining the little seaside village of Aberdour (Fifeshire), about five miles east of Rosyth. It was in the charge of Commander C.P. Ryan. Rutherford was sent to visit the establishment on 15th and 16th September 1915 and was accompanied by Sir Richard Paget*, Sir Richard Threlfall and (on the 16th) the Duke of Buccleuch. Rutherford was impressed by the work that Ryan, a good, practically-minded man, was doing on a miserly budget with no scientific support. Immediately after the visit Rutherford submitted his report which gave hearty approval to Ryan's work and urged that it be supported. He pointed out that no submarines had been made available for the scientific visitors to listen to, as "all were on active service". He recommended that "a ship be placed immediately at Commander Ryan's disposal in order to carry out practical experiments on listening to submarines and other ships", and complained that Ryan was "only allowed to spend £1 on any one experiment". Rutherford suggested that, instead of this niggardliness, Commander Ryan should be given "all necessary financial and scientific assistance" and, if and when he required it, "the services of two or three trained physicists should be placed at his disposal". It is with regard to this last recommendation that Harold Gerrard and Albert Wood once again come into our story, for Ryan immediately took up the offer of scientific assistance and they were to be the two scientists sent to help.

On 7th October 1915 Wood received a curt note from Rutherford. "I want to see you on Saturday morning, preferably between 11 and 12, about an important matter". Wood was duly told what Rutherford had in mind and three days later received another scribbled note: "Matters of which I spoke

---

* Sir Richard Paget, barrister, scientist, musician and a pioneer of the artificial production of speech, was Assistant Secretary to the Admiralty Board of Invention and Research, 1915-18.

are moving rapidly and your services may be wanted possibly within a week. Consult Wilberforce [the Professor of Physics at Liverpool] and make tentative arrangements for your departure. The Admiralty will arrange pay - but I am sure it will be quite satisfactory. I hope to see Gerrard on Monday in connection with the same matter. Give my regards to Prof. Wilberforce and tell him I am sorry to interfere with his Department but hope he can make suitable arrangements to take over." In an addendum Rutherford wrote: "If you are appointed you had better give up university salary for the time you are away." The note shows the forceful side of Rutherford's nature. Neither Wood (at Liverpool) nor Gerrard (Beattie's demonstrator) were under his jurisdiction, but the needs of the war came first. Rutherford needed them, so he took them.

Rutherford produced a further, much more detailed, technical report at the end of September 1915, in which he reviewed the whole subject of submarine detection and outlined the methods he thought would be successful. This report was used as the central basis of all anti-submarine work for the remainder of the war. It has already been stated that hydrophones seemed to be the most effective means of detection in the prevailing state of knowledge, but the problem remained of identifying the direction from which any sound was coming and, if possible, assessing the distance. The methods Rutherford considered were: (1) to use the instinctive directional sense of the human ear; (2) to use devices such as trumpet-shaped receivers in the water; (3) to use interference methods; and (4) to compare the sounds heard on the opposite sides of the listening ship. Rutherford and his assistants had already tried (1) in his experimental tank with little success; Ryan had found (2) and (3) were unpromising, so efforts were to be concentrated on (4). Rutherford's report contained much-needed scientific information relating to the speed of sound in water, the amplitude of motion, the pressure of sound waves and how these factors affected the response of a steel diaphragm in water.

At Rutherford's direction, Gerrard and Wood left Manchester for Hawkcraig on 17th November 1915, taking with them F.W. Pye (of Liverpool University) as a mechanical assistant. They were taken by ship from Edinburgh to Hawkcraig where they met Commander Ryan and his small band of helpers. They discovered that, prior to Rutherford's visit two months previously, neither Ryan nor anyone else at the naval outpost had ever heard of Rutherford, such was the state of scientific knowledge in the navy. However, work soon started in earnest, and many sea trials on underwater detection were made by Gerrard and Wood from H.M. Drifter *Hiedra* which had been placed at the disposal of the two young scientists as a result of Rutherford's report to the Admiralty. Wood and Gerrard were each paid £1 per day for their services whilst attached to the Royal Navy; considerably more than they received from their university posts.

The experiments which Gerrard and Wood did were carried out not under Ryan but under the personal direction of Rutherford himself through letters containing specific instructions, about 40 of which were communicated during the next few months. Letters from Rutherford contained statements such as "I am getting several other pieces of apparatus constructed, and we shall keep you very busy after this testing", and "I am sorry to hear the weather has gone wrong and that you have not got any further with the experiments." But Gerrard and Wood were clearly inventing equipment of their own for in another letter Rutherford wrote: "Your little device may be quite useful if it works well in practice. I would certainly show it to Ryan."

The naval establishment, in which everyone had a specific rank, was clearly not used to dealing with young civilians placed in their midst; they did not fit into the pecking order. Following the death of Moseley, Rutherford was anxious that no more of his bright young men should be killed and requested that their position be formalised. In February 1916 the two men received "Admiralty Indispensibility Certificates" and also Navy Armlets and were advised that "You are exempted from the provisions of the Military Service Act, under which the Government have power to grant exemptions to those in the employ of H.M. Government. Should you be approached by the Local Recruiting Authorities, it will be sufficient to exhibit the Indispensibility Certificates."

Gerrard's account of his work

The following brief summary of the work is taken from Harold Gerrard's account written at the end of the war: [9]

"On November 16th, 1915 I was attached to Hawkcraig Experimental Station, and worked with Mr A.B. Wood under the direction of Sir E. Rutherford and Captain (then Commander) C.P. Ryan until Professor Bragg arrived on May 8th, 1916.* During this period, experimental facilities were in a very primitive state and a considerable amount of time was taken up in equipping the ship Hiedra, the laboratory and the workshop. The experimental work was entirely acoustical, and as the subject was new, a good deal of work had necessarily to be done on lines which did not lead to any practical devices, but which served to eliminate the least promising directions of attack, and laid the foundations of future progress. Nevertheless, two devices were developed on the Station which have since been largely used in the Service, namely listening gear on submarines, and the P.D.H. [Portable Directional Hydrophone] Mark I.

---

* Professor W.H. Bragg, who had recently moved from Leeds University to University College, London, was appointed Resident Director of Research at Hawkcraig in May 1916.

1914-1918: The First World War

On many occasions Commander Ryan was assisted in his experiments on towing bodies, under-water signalling, and the attachment of hydrophones to the hulls both of surface ships and submarines.

Some tests were carried out which led to a modification of the Hervey-Gardner listening plates for bell signals which were fitted on many warships.

After the arrival of Professor Bragg at Hawkcraig, on May 8th 1916, I worked under his direction and in collaboration with Mr F.L. Hopwood. The Morris-Sykes twin-plate directional hydrophone arrived at Hawkcraig about this time, and a good deal of work was done in the direction of simplifying this apparatus and investigating its properties. It was finally given up in favour of the single plate type, which was equally good as regards sensitivity and directional properties, was less liable to failure and simpler to construct.

As the bi-directional instrument possessed ambiguity of direction, attempts were made to convert it into a unidirectional instrument, and it was discovered that an obstacle placed near one side gave it this property. This led ultimately to the baffle as now used on the Mark I P.D.H. A good deal of experimental investigation was carried out on the subject of baffles, with a view to giving a clear account of their action, and to improve by their means the directional properties of the hydrophone.

Many tests were made with a water syren constructed by Mr A.Q. Carnegie, and with a magnetophone constructed by Mr S.G. Brown. Experiments on binaural listening were carried out with many forms of listeners, and a trombone type of compensator was constructed. This method of direction finding was laid on one side, however, as it was considered that the directional hydrophone was more suitable, and there was not time to develop the two systems side by side.

A directional hydrophone was mounted on submarine B.3, so as to enable the submarine to get accurate bearing of any sound heard. Towards the end of 1916, and at the beginning of 1917, much time was spent in testing and adjusting the Mark I P.D.H.s which were being constructed for service use."

Life at Hawkcraig had its lighter moments. Most of Ryan's staff were R.N.V.R. officers, one of whom was an Irishman, Lt. Herbert Hamilton Harty, none other than the man who would later, as Sir Hamilton Harty, be one of the most famous conductors of the Hallé Orchestra. He and his wife (Agnes Nicholls, a top-class singer) lived at Hawkcraig in a flat above Wood. One of Ryan's projects involved the use of pairs of hydrophones fitted to submarines, one port and one starboard, which would indicate to an intelligent trained operator the approximate bearing of another ship. In selecting the hydrophone pairs for this purpose he made use of the musician's particular skills. Wood wrote "It seems a little bizarre to recall Lt. Hamilton Harty sitting amongst a pile of hydrophones using a little hammer to tap the steel diaphragms and arranging them in pairs as 'port low' and 'starboard high' pitch." Others in the company included Lt. Brett who was a professional violinist and Lt. Rose, a London theatre manager. Wood recalled that Hamilton Harty, Agnes Nicholls, Brett and Rose once gave a concert in the village hall in Aberdour to entertain the staff and villagers.

According to Wood's account, it was "first class London talent - front seats 2s. 6d!"

Other stories involved Ryan's two dogs which were great favourites of everyone. They were a white cairn terrier and a red water spaniel. The terrier was a 'wanderer', and Ryan told Wood that it had once been shot and buried by a gamekeeper but had dug itself out and returned home to Hawkcraig. The water spaniel was stone deaf and on one occasion witnessed by Wood it stood over the lighted fuse of a charge due to explode when blasting rock for a new jetty at Hawkcraig. They threw stones at it to try to move it and shouted, without avail, but luckily the fuse failed and all was well. This same dog was in the habit of stealing Ryan's lunch from the table in his private hut, until one day the beef-steak was wired up to a high-voltage circuit. The terrified dog was seen galloping out of the hut at high speed without the steak and the misdemeanour was not repeated! [10]

Sir Richard Paget, who was often at Hawkcraig and took part in the experimental work, was a lively man of many talents. Wood recalls one experience:

"Early in the year 1916 Sir Richard Paget, on being informed of our failure to obtain a reliable frequency analysis of a submarine's propeller noise due to resonances in hydrophone diaphragms and microphones, suggested listening direct with the two ears under water. Meeting with little enthusiasm from Gerrard and myself, he went out with a sailor in a small boat and a submarine circled round at a moderate distance. Sir Richard, who had a 'skull-note' of G sharp when he tapped his head, leaned over the side of the boat, with the sailor sitting on his legs, put his head under the water, came up, tapped his head and ran up the scale to the required note, which the sailor duly wrote down. Then down again to fish up another note, and so on. I am not sure that the notes he obtained in this way were very reliable - but at any rate he didn't develop pneumonia!"

No electronic amplifiers were available in those days, until Gerrard and Wood were fortunate to receive a three-electrode valve designed by H.J. Round of Marconi. It was about two inches in diameter and five inches high with a glass 'pip' at the top about one inch long. The pip contained asbestos and was used, in conjunction with a small heating coil (a few turns of wire outside the glass pip in series with the valve filament), to control the degree of 'softness' of the vacuum. When in correct adjustment the valve was a very efficient amplifier and at its best was as good as two or three 'hard' valves which were supplied towards the end of the war (in 1917 and 1918).

There were many famous visitors to Hawkcraig whilst Gerrard and Wood were there. Rutherford and Sir Richard Paget came often to check on progress; Admiral and Lady Beatty (who lived at Aberdour House) often came; the work being done was important to Beatty, whose Battle Squadron

1914-1918: The First World War

was based at Rosyth. Wood and Gerrard had often watched and listened to Admiral Beatty's Battle Squadron as they went out to the North Sea from Rosyth but there was an air of great despondency on 1st and 2nd June 1916 when the stragglers of the Fleet limped back from the Battle of Jutland. Though the outcome of the battle was seen as a disaster at the time, the German fleet had been severely mauled and it was later concluded that the result was perhaps not as bad as was first thought. Shortly afterwards King George V was at Aberdour visiting Admiral Beatty, and the Prime Minister, Herbert Asquith, was at Rosyth. Admiral Jellicoe was a frequent visitor.

In the May of 1916 the scientific unit at Hawkcraig was expanded by the addition of more scientists and Professor W.H. Bragg was sent up as their Resident Director of Research. But relationships between him and Cdr Ryan were uneasy from the start. Jealousy and mistrust abounded. Ryan had built up Hawkcraig himself, and the sudden influx of high-ranking civilians over whom he had no control rankled. He thought the scientists were dismissive of his work and that their approach to problems was too 'academic'. The hostility was compounded by the technical inferiority of the British fleet that had been demonstrated in dramatic fashion at Jutland. The Navy felt that the scientists had not invented speedily or effectively enough. On one occasion the skipper of *Hiedra* responded to Bragg's request to perform a certain manoeuvre which was evidently contrary to Ryan's wishes and when the ship returned to Hawkcraig Ryan punished the skipper with 14 days' confinement for disobeying his orders. Bragg apologised to Ryan and said it was his fault that the skipper had acted as he did, but Ryan insisted the punishment should stand. On another occasion, in Bragg's absence, a length of hydrophone cable which had been laid at his request for use as a telephone line was removed by Ryan's staff without explanation to the scientists except by saying it was "by order".

The Admiralty mistrusted not only the scientists but also Britain's foreign allies. On the morning when a delegation of French scientists and officers was about to visit Hawkcraig, Bragg received an urgent signal from the Admiralty: "On no account show them anything." Relationships between Bragg and Ryan deteriorated further and the differences were taken to Arthur Balfour who had been appointed First Lord of the Admiralty following the sacking of Churchill after the failure of the Dardanelles expedition. It was suggested that Ryan be moved to another job but Bragg, wanting to see fair play, said that Ryan had been there first so should stay. The upshot was that it was decided that the scientific group should be moved to Parkeston Quay, Harwich, close to Commodore (later Admiral) Tyrwhitt's base from which his vessels waged war in the North Sea, where a new scientific unit would be formed. One of the great advantages of Parkeston was that ships, large and small, and submarines were coming and going all the time so there was ample opportunity to test for their location by hydrophones of various types and by magnetic loops.

Wood and Gerrard moved to Parkeston Quay at the end of 1916 to be followed by the whole scientific unit under Bragg's direction. The following is Gerrard's account of his work during the remainder of the war: [11]

"At the end of 1916, I was transferred to Parkeston Quay, but before there I attended, with Mr Hopwood, some trials conducted by Commander Middleton at Portsmouth, in which it was demonstrated that six M.L.s, equipped with Mark I P.D.H.s could track down and surround a submarine when once they had heard it. The efficiency of Mark I as a submarine detector was thus definitely established.

During the first month or two of 1917 at Parkeston Quay, a good deal of time was spent in designing the equipment of the laboratories, and this, combined with the unfinished state of the laboratories and the lack of ships for experimental work, made it difficult to make progress.

As a result of an urgent request from the Admiralty for a simple means of signalling from submarine to surface craft, I suggested a pneumatic riveter hammering out Morse signals on the side of a boat. Tests were made with various sizes, and it was judged to be a useful means of short range signalling. The development of this and other means of signalling was afterwards handed over to Mr Young.

Further tests were carried out with the water syren fitted on a destroyer.

As there seemed to be an opening for a towed device which would give an indication on contact, I suggested the device now known as the "Magnetic Trailer", and sufficient work was done to prove that it could be used to give a signal, or explode a charge incorporated with the magnet, on coming into contact with a submarine. The same device could also be adapted as a means of firing mines or paravanes.

Some assistance was rendered to the hydrophone officer at Portsmouth in connection with the fitting of Mark I P.D.H.s and M.L.s and drifters.

As a result of a suggestion made by Mr Sinnot of the Post Office, it was discovered that submarines could be heard by means of a coil on a ship connected to a valve amplifier and telephones. A good deal of work was done on this subject, leading up to the V.F. method of detection with electrodes developed by Mr Young and later to the C.F. method. The subject of detection from moving ships was entrusted to Mr Young and Mr Wood and I carried on experiments with coils on the bottom of the sea, with a view to using such coils for the defence of estuaries and narrow straits. We also suggested that such coils could be used in defensive mining schemes in place of magnetophone mines, and the idea was taken up by Captain Fraser, C.D.M.

It was discovered, however, that the direct current induced by the magnetic field of a ship was more easily detected than the V.F. effect, and numerous observations, both visual and photographic, were made with a coil on the bottom of the sea at Harwich. This method of detection promised to be very useful in the case of estuaries and narrow straits, but the range was unknown. Accordingly, after some successful preliminary experiments on the Firth of Forth, I proceeded to Wemyss Bay to establish an experimental station there with large loops laid in very deep water in the Clyde Estuary.

1914-1918: The First World War

It was at once observed that loops of large area were subject to severe disturbances, partly due to a tramway system some miles away, and partly to other causes, and that these disturbances interfered seriously with the legitimate use of the loops. The disturbances were studied as thoroughly as possible with the incomplete apparatus available at that time, but before any means of eliminating them had been devised, I was recalled to Parkeston Quay at the end of February, 1918.

I was sent to Portsmouth to study S.A. Marks I and II with a view to carrying on with the experimental work in place of Dr Rankins, but other arrangements were made and I returned to Parkeston Quay. "Ebro II" was fitted out with S.A. Mark I and some trials were made with the gear, but the experiments were soon dropped.

Some tests were made on the use of a valve and crystal rectifier in conjunction with a hydrophone, to examine the possibility of devising an automatic warning. It was shown that the French Army valve set was not suitable for use with hydrophones as the maximum possible output was too low.

The remainder of the time has been spent on electrode work of various kinds. As a result of a suggestion of mine, Dr Newbery was consulted as to the most suitable electrode, and he recommended the silver-silver-chloride type which has become the standard service pattern. [*]

The possibility of using electrode systems in conjunction with mining schemes, both automatic and controlled, has been explored, and the disturbances arising from movements of the sea investigated.

The disturbances affecting long cables in the sea terminating in electrodes have been investigated over long periods, and E.M.F's associated with tidal motion have been shown definitely to exist. This work is now in progress, but is being seriously interrrupted by damage to the cables during the prevalent gales."

The above report by Harold Gerrard had been written in response to a notice to all Parkeston staff on Armistice Day, 11th November 1918, by the new Director, A.S. Eve, who had taken over from Bragg, requiring "before 25th November", information on (a) date of joining, (b) an outline of the work carried out by that member of the staff, (c) a list of reports and patents, and (d) recommendations for future developments. He concluded, in referring to the future, "the policy of the Admiralty is not yet known". In his report Gerrard listed the following patents:

Magnetic Trailer	-	Lieut Partridge and H. Gerrard.
Silver Electrode	-	E. Newbery and H. Gerrard.
Electrode Mine	-	Colonel A.S. Eve, Captain A.G. Ionides and H. Gerrard.
Electrode & Acoustic Mine	-	A.B. Wood and H. Gerrard.
Non-polarising Sea Cell	-	A.B. Wood and H. Gerrard.

Gerrard added: "In the early days the names of physicists were not associated with any patents taken out, so that my name is not associated with any patents concerning the Mark I P.D.H. and baffle, and loop detectors, although a considerable amount of the work in connection with them was done by me."

---

* Dr E. Newbery was a Temporary Lecturer in Physics at Manchester University, having been appointed because several of the lecturing staff were away on active service or other war work.

Gerrard included at the end of his report a list of future projects that would prove useful in the anti-submarine war, as requested by Eve, but it would be pointless to list them as virtually all work was suspended almost from the moment the war ended. Wood said that he heard the war was over when he was boarding a ship to go out to one of the gateships at Harwich. Wood said: "When the captain told me of the startling news he said I should be wasting my time 'going on with what I was going on with'."

It should perhaps be mentioned here that "echo-detection" was only invented as the war was drawing to a close, even though the idea of sending out a "supersonic" (now called ultrasonic) signal and detecting the reflected echo, which would not only give a bearing but a range as well, had been suggested following the sinking of the *Titanic* in 1912. The development of the system was carried out at Parkeston. It was shrouded in great secrecy for many years and given the code name ASDICS, meaning "Anti-Submarine Division-ics", the "ics" meaning "pertaining to" as in statics, dynamics, electronics, physics, etc. Wood recalled that the cutting of the first quartz crystals for Asdic supplies was done by Farmer & Brindley, a firm of tombstone makers in Lambeth accustomed to cutting marble slabs. War has a way of seeking out and utilising the best available expertise!

In the course of their work Gerrard and Wood frequently went out in submarines to carry out their tests. The work was hazardous; communications were primitive and the possibility of collisions was never far away. Years later Gerrard confided that he "narrowly missed a watery grave when two submarines nearly collided under water".[12] Wood recalled that submarine crews were given better food than ordinary naval personnel, with good reason, in his opinion, so one of the few benefits of working in submarines at sea was that a good meal was guaranteed. He remarked that it would be difficult to imagine a more peaceful spot for the enjoyment of a quiet lunch than in a submarine sitting on the bottom of the North Sea!

Considerable space has been devoted above to Harold Gerrard's part in the fight against the enemy. The reason is that (a) his work was important and contributed significantly to the war effort, and (b) he comprised one half of the staff of the Electro-technics Department at the time and it can therefore be argued that what he did constituted half of the Department's work during the war. As most of this book is concerned with what the Department was doing, it seems only right to record his work reasonably fully, even though most of it was done outside the department. At that time winning the war was of far greater importance than continuing normal research work or teaching the tiny handful of students who were still in the Department. The latter task was left in the capable hands of Robert Beattie.

1914-1918: The First World War

Robert Beattie's work for the war effort and the Department

Beattie, who was 41 when the war started, retained his Manchester base during the war but, like Rutherford, he made frequent visits to check the work being done in establishments such as Hawkcraig and Parkeston Quay and to advise. His role is less well documented than that of Rutherford or Gerrard, but as he served with Rutherford on the Lancashire Anti-Submarine Committee it seems clear that he worked in collaboration with him during the hydrophone tests in the water tank at Manchester and in formulating some of the instructions sent to scientists such as Wood and Gerrard in their research outposts. This view is supported by the University calendar of 1917-18 which records that " . . . In his capacity as a member of the Board of Inventions for the Admiralty, Professor Sir Ernest Rutherford has made special investigations in the physics laboratory on problems connected with the Board. In this he has been assisted by Professor Beattie, the laboratory staff and other assistants." The same calendar also records, under the heading 'Electro-Technics': "Investigations on electromagnetic problems connected with the war are being conducted under Professor Beattie." The following year's calendar (1918-19) published in the summer of 1918 recorded, again under the heading 'Electro-Technics': "Investigations on electro-magnetic problems connected with the war continue to be conducted under Professor Beattie and Dr Newbery on behalf of the Admiralty and the Ministry of Munitions." In the course of this work he made many visits to research establishments. He was present at Gareloch when a British submarine, K. 13, sank there during commissioning trials on 29th January 1917 with the loss of 32 of the 80 men on board. The accident deeply upset him and it was one of the few events of the war he ever spoke of to his family in later years. [13, 14]

Apart from his war work, Beattie's main function, as the sole member of staff still present in the department, was to "hold the fort" during the war and to single-handedly do all the necessary teaching and administration. He seems to have found time to have done a little research as well, for in 1916 he published a paper, "A permeameter for straight bars". It prompted a surprising amount of correspondence in the Journal in which it was published (*The Electrician*) including an exchange with Silvanus P. Thompson, the well-known writer of text books on electrical subjects, who at the time was a Professor of Applied Physics at the City & Guilds College in London. Beattie was not one to suffer fools gladly and he believed in plain speaking, as is shown by a rather tetchy exchange with another correspondent who had used rather pretentious phraseology when commenting on Beattie's paper in the columns of *The Electrician*. Beattie responded somewhat sarcastically: "It may be a retrograde step to call a force a force, rather than a space rate of change of energy, but it certainly seems to me that by doing so there is a distinct gain in clearness of treatment and simplicity of statement."

Two events of personal importance to Robert Beattie and his wife took place during the war, one happy and one sad. In April 1915 a son, Robert Kyd, was born (Kyd, it may be remembered, was Mrs Beattie's maiden name); he will reappear in our story in chapters 9 and 10. And in August 1918 a daughter, Janet, was born but she did not survive. Mrs Beattie had travelled to Scotland for the birth of each child. After the war another daughter, Janet Elizabeth, was born in Buxton. Her help in passing on information about her father for this book is much appreciated.

Rutherford's 'other work' during the war

It will have been noticed from the earlier parts of this chapter that Rutherford was very heavily involved during most of the war in his work as a key member of the Anti-Submarine Committee of the Board of Inventions. Moreover, as most of his staff had departed owing to the war, he was left to soldier on in his department as best he could with a much smaller number of staff than normal. Nevertheless, Rutherford found time to do a little research work; he was assisted only by his faithful Laboratory Steward, William Kay. It was as a result of this little bit of part-time research, done almost as a hobby in the few spare moments at his disposal, that Rutherford "split the atom".

The story of this particular research work is well documented, for example in Wilson's biography of Rutherford [15], so only the barest outline will be given here. In 1914 one of Rutherford's research students, Ernest Marsden, had noticed an anomaly concerning the scattering of alpha-particles. Under certain conditions of absorption in air, and especially if he examined "H-particles" (the rapidly moving nuclei of hydrogen atoms), emerging obliquely, Marsden found too many of the H-particles appearing; far more than the impact of alpha-particles could account for. The work was written up by Marsden as "The passing of alpha particles though hydrogen" and published in the *Philosophical Magazine* in two articles, one in 1914 and the other in 1915. But he and Rutherford were uneasy about what was actually happening. After making this investigation Marsden had moved to New Zealand to take up his first university post and in 1915, in the middle of his anti-submarine work, Rutherford wrote to Marsden asking if he would "mind" the professor continuing the experiments himself. Marsden did not mind, in fact he was delighted, so Rutherford started a careful series of experiments which he continued for three years as and when time permitted, sometimes with the results noted by himself and sometimes by William Kay. The particular observer's name was recorded in Rutherford's notebooks at every stage of the experiments.

It did not take Rutherford long to realise that he had hit upon something very important indeed; by 1917 he was able to write:

"From the results so far obtained it is difficult to avoid the conclusion that the long-range atoms arising from collision of alpha-particles with nitrogen are not nitrogen atoms, but probably atoms of hydrogen . . . If this be the case we must conclude that the nitrogen atom is disintegrated under the intense forces developed in a close collision with a swift alpha-particle and that the hydrogen atom which is liberated formed a constituent part of the nitrogen nucleus."

Thus, the alpha-particles were 'splitting' nitrogen nuclei. By October 1917 Rutherford was fully aware that he was making "new" particles and that these particles were atoms of an element that had not originally been present in the apparatus. He therefore knew that he was splitting atoms, but he still had to show what they were being split into. More experiments followed which showed that if he shot alpha-particles, which are helium nuclei, into nitrogen gas, hydrogen nuclei came out at the far end. This meant that the alpha-particles were splitting the nitrogen atoms and knocking hydrogen nuclei, the positive particles of the atomic nucleus, out of the nitrogen nuclei. After more careful work to eliminate the possibility of spurious effects, followed by a lot of tidying-up experiments, his papers on the subject were published in 1919. The first two were concerned with his work on hydrogen; the next on oxygen and nitrogen. The fourth, entitled "An anomalous effect in nitrogen", incorporated the work described in the previous ones, and cautiously indicated that a major discovery might have been made.

By being the first man to disintegrate the nucleus of the atom, Rutherford ensured his place as one of the greatest of all scientists. To have done it almost on his own when beset by other urgent work and the country fighting for its existence is truly remarkable.

War casualties

As the war dragged on the casualties suffered by university staff and students steadily mounted. The university calendars published their names and rank year by year. Moseley, a gifted staff member, had been an early casualty, as described earlier. Another was H.G.S. Delépine, son of the University's Professor of Public Health. He graduated in engineering in 1910 and was appointed Assistant Lecturer and Demonstrator in Engineering in 1913; he died of wounds at Ypres in April 1915, four months before

Moseley. Both were Second Lieutenants. Two young Lecturers in Law were killed before the end of 1916, and as the war dragged on the same sad story continued. Harold Gerrard's younger brother, Percy, was killed at Le Cateau on 10th October 1918, a month before the war ended. By the time Armistice was declared, of the 3500 or so members of the university who had served in the war, 500 were dead. Their names are in a *Roll of Honour* published by the University in 1920 and they are also commemorated on a war memorial to the right of the entrance to the main university building on the west side of the quadrangle. The war had eventually been won, but at a terrible price in lives.

References: Chapter 7

1. These changes to the regulations are quoted in the university calendars of 1914-15 to 1920-21.
2. Vice-chancellor's annual statement, summer 1915.
3. Vice-chancellor's annual statement, summer, 1916.
4. Vice-chancellor's annual statements, 1915-1919.
5. Willem Hackmann. *Seek and Strike: Sonar, anti-submarine warfare and the Royal Navy, 1914-54.* H.M.S.O., 1984.
6. David Wilson. *Rutherford - Simple Genius.* Hodder & Stoughton, London, 1983; Chapter 12.
7. Special issue of the Journal of the Royal Naval Scientific Service to commemorate the life and work of Albert B. Wood, O.B.E., D.Sc. July 1965.
8. David Wilson's biography of Rutherford (see Ref. 6); Page 343.
9. Report written by Harold Gerrard on his work during the First World War, dated 18th November 1918, written in response to a request from Col. A.S. Eve.
10. See A.B. Wood's account of the Hawkcraig period in reference 7.
11. Information contained in Harold Gerrard's report; see reference 9.
12. Interview with Harold Gerrard in the *Manchester University Engineering Society Magazine*, Summer Term, 1949.
13. Information from the Beattie family history has been kindly passed on by Mrs Janet Paterson, whose help is acknowledged.
14. Information passed on by Mrs Paterson (see Ref. 10).
    Edwyn Gray. *A Damned Un-English Weapon.* Seeley, Service &. Co., 1971. (gives an account of the circumstances surrounding the sinking of the K.13).
15. Wilson's biography of Rutherford (see ref. 6); Chapter 13.

Much of the information about Harold Gerrard's part in the war, including some personal details, a copy of his report of 1918 and a copy of the A.B. Wood commemorative issue, was kindly supplied in 1991 by his sons and daughter, Messrs P.M. and T.O. Gerrard and Mrs Ann Burton, whose help is acknowledged.

Chapter 8

1918-1930: Consolidation in the post-war years

At the end of the war, which had so ravaged the youth of the country and had seriously curtailed ordinary academic work, the university was left with the task of picking up the pieces and getting back to normal. For the infant department of Electro-technics, whose work had barely begun when the war started and which had survived under the sole care of Robert Beattie whilst Harold Gerrard was away on war duty, this meant a period of quiet consolidation during the next few years. Many of the young scientists and engineers who had fought in the war or who had made significant contributions through their scientific war work had changed their outlook on life since the pre-war days and no longer wished to go back to their old jobs. Consequently the next year or two would see many staff changes in universities. A.B. Wood, for example, did not return to his former post at Liverpool University but stayed with the Admiralty where he enjoyed a distinguished scientific career. Harold Gerrard, however, was content with his lot in the Electro-technics Department and within weeks of the armistice returned to his pre-war post of Senior Demonstrator. So in 1919 the two-man Department attempted to resume where it had left off five years earlier.

The new Ph.D. degree

Few major initiatives had been put into effect within the university during the war years; there had been a holding operation in which the administration had been ticking over until such time as peace returned. But a notable exception was the decision, in 1918, to institute a new degree of Ph.D. Previously the only science degrees available were B.Sc., M.Sc. and D.Sc. If a Ph.D. was wanted it had to be obtained abroad. Heidelberg, for instance, had offered such a degree for many years and it had been customary for some bright British graduates to be sent to Germany to work for one. It is not surprising that the war led people to wonder why our scientists should be sent to a vanquished enemy nation to work for a Ph.D. when we could introduce a similar degree of our own. The arguments were summarised in the Manchester University Calendar* of 1918-1919:

"In consequence of representations made by the Universities of Canada and the United States, the question of making fuller provision for post-graduate study and research in the first instance from students in America and the overseas dominions, was

---

* There was no proper calendar in the 1918-19 session "owing to the shortage of paper and labour" but a pamphlet was produced containing the above information on the new Ph.D., and also staff lists, examination results, new and amended ordinances and regulations, etc.

considered at conferences first of the Northern Universities, and then of the British Universities held in the Spring of 1917. Subsequently, and on the lines generally adopted at those conferences, the Court, on the recommendation of the Senate and Council, instituted a new Doctorate of Philosophy (Ph.D.) attainable in any Faculty by graduates of approved Universities after a course of training in advanced study and research extending over not less than two years. The ordinances and regulations of this new degree have been adopted, and steps are being taken to bring them into operation.

It is hoped that this degree will encourage research within the university not only on the part of foreign and colonial graduates who would under pre-war conditions have migrated to German and Austrian Universities, but also on the part of our own graduates, and will lead to the establishment of flourishing schools of advanced work in several of the Departments."

This was a major step which, over the intervening years, has undoubtedly been successful. The introduction of the new degree may also have elevated the status of the D.Sc. as a by-product. Prior to the introduction of the Ph.D., many very able graduates obtained their D.Sc. within seven or eight years of the award of their first degree. Within a few years of the introduction of the Ph.D. a much more lengthy period appeared to be required in most cases before the D.Sc. was awarded.

In most departments the new Ph.D. degree was welcomed, but Robert Beattie did not approve of it. He had obtained his D.Sc. (Durham) in 1903 at the age of 30, and he considered that the normal progression for the best students should be B.Sc., M.Sc. and D.Sc. This might seem a reactionary view but it has to be remembered that the Ph.D., being new in those days and therefore something of an unknown quantity, was not viewed in the same light as the well-established Ph.D. of later years, and Beattie's view was shared by many others. Some regarded it as a German import. Whatever the rights and wrongs of Beattie's opinion of the degree, the fact remained that he would not accept the enrolment for the Ph.D. of any of the staff or graduates of his department. Consequently the only postgraduate degrees awarded in the Electro-technics Department from the introduction of the Ph.D. degree in 1918 until Beattie's retirement 20 years later were M.Sc.s. Almost inevitably this meant that some bright students, looking for a Ph.D., would be lost to other departments or universities where study for the new degree was welcomed. The first Ph.D. awarded to a post-graduate electro-technics student at Manchester did not come until 1942 (after Beattie's retirement) when A.E. Chester, who had gained a first-class honours B.Sc. in 1937 and an M.Sc. in 1938, was awarded one.

Harold Gerrard's promotion

When Harold Gerrard returned to the Electro-technics Department at the end of 1918 as Senior Demonstrator he was 30 years old. In the following autumn he was promoted to Senior Lecturer; a significant promotion. At the time of his promotion he had only three publications to his name, all written jointly with Beattie. However, he had done well in the war as Rutherford's protégé. Much of the work he did must have impressed Rutherford, and would have been published had it not been classified as "Secret". Gerrard remained at the grade of Senior Lecturer until his retirement as a full-time member of staff in 1953.

Important professorial changes

1919 and 1920 were marked by major staff changes. After 12 happy and fruitful years in Manchester Sir Ernest Rutherford announced that he was leaving in order to succeed Sir Joseph (J.J.) Thomson as Cavendish Professor of Physics at Cambridge. Thomson, at 61, had recently been installed as Master of Trinity College and had been persuaded as a result to relinquish the Cavendish Professorship from March 1919. Manchester had made strenuous efforts to retain its star professor whose salary, £1,250, was one of the highest of any professor in Britain.[1] There was much corresponding and behind-the-scenes lobbying before Rutherford agreed to move to Cambridge. In his letter of resignation to the Vice-chancellor, Henry A. Miers, Rutherford wrote that it was with "feelings of great regret that I sever my connection with a university with which I have been proud to be connected for the past 12 years" and added that it was "only after much hesitation and pressure that I felt it my duty to take up a more difficult task elsewhere". In a formal reply to the Council some months later Rutherford referred to Manchester as "the most progressive of our universities".[2] Rutherford left in the autumn of 1919. There is no doubt that he enjoyed his Manchester years. Though he still had major work to do and discoveries to make at Cambridge University, it would not be unjust to that institution to state that his greatest work of all was done in Manchester. Years later he wrote to Geiger in reminiscent vein: "They were happy days in Manchester and we wrought better than we knew."[3]

Rutherford was not the only one to depart in the aftermath of the war. Joseph Petavel resigned as Professor of Engineering in 1919 to become Director of the National Physical Laboratory. He was knighted the next year. In 1920 Horace Lamb retired as Professor of Mathematics, having held the post for 35 years. He was knighted in 1931. C.A. Edwards, the Professor of Metallurgy, also resigned in 1920 and was succeeded the following year by F.C. Thompson. Thus, within a period of a few months, there was a change of leadership in four science departments. W.L. Bragg, already a Nobel Laureate (jointly with his father, W.H., under whom Gerrard had served at

Parkeston Quay) was appointed to the Langworthy Chair of Physics and Sydney Chapman to the Mathematics chair.

The future of engineering in the university

The resignation of Petavel from the Engineering chair precipitated a debate within the university on the future of engineering in the Faculties of Science and Technology, a matter that would re-surface in the early 1940s and again in the late 1980s. At its meeting held on 16th October 1919 the Senate resolved:

As to the resignation of Professor Petavel and consequent arrangements: It was reported that Professor Petavel had consented to devote part of his time to the University during the present term and, if necessary, to pay some visits to the University next term.
Resolved:
(1) That, whilst postponing a more formal resolution, the members of the Senate desire to express to Professor Petavel their cordial congratulations on his appointment as Director of the National Physical Laboratory. They are pleased to know that he will continue for the present to give his services for the University.
(2) That the Council be recommended to appoint Mr Henry Baker, B.Sc., to give part-time assistance in the Engineering Department for the present session.
(3) That the Council be asked to consider the whole question of Engineering teaching and research in the University, including its relation to the different Faculties, and that the Senate thinks it is desirable that the situation should be considered by a Joint Committee of Senate and Council which should, amongst other steps, consult engineering experts from outside the University.
That a Committee of six members, including one representative of the College of Technology, be appointed to co-operate with the Council in the matter.

Usually university committees work and report at snail's pace but on this occasion the matter under discussion seems to have been resolved in double-quick time by university standards. At its meeting of 4th March 1920 Senate reported:

(ii) (a) The following report of the Committee on Engineering passed by the Council:-
(1) That two Schools of Engineering be maintained as at present, one in the Faculty of Science and the other in the Faculty of Technology.
(2) That the School of Engineering in the Faculty of Science should be a general School of Engineering, directed, if possible, by the type of Physicist-Engineer represented by those who have previously occupied the Chair in the University.
(3) That teaching should be maintained in two schools which will necessarily be of a somewhat different character; but, in the judgment of the committee, it is essential to secure close co-operation between the two, especially in respect of post-graduate work; for this purpose the more specialised work carried on in the College of Technology should play a most important part.
(4) That a Professor of Engineering in the Faculty of Science be appointed.

(b) The arrangements for the organisation of the Department and the appointment of a Professor.

1918-1930: Consolidation in the post-war years

Resolved: That a committee consisting of the Vice-Chancellor, Professors Bragg, Chapman, Dickie, Edwards, Lamb, Pyman, Stoney, Tout, be appointed to report on the reference from the Council.

These pronouncements from the Senate indicate clearly that the question of the appointment of a successor to Professor Petavel had been held in abeyance until the views of the committee looking at the structure of engineering in the university were made known. With the committee's recommendations now revealed, which were essentially that matters should continue as they were before, the way was open for the appointment of a new Professor of Engineering. It did not take long to find someone. The new man was Arnold Hartley Gibson, a Manchester Engineering graduate of 1903 who had already served a brief spell on the staff of the department in earlier days and had been Professor of Engineering at St. Andrews for the previous ten years. He was appointed to the post from the start of the 1920-21 session and was to remain in it until 1949, during which period he enjoyed a distinguished research career in the field of hydraulics.

Whilst the committee had been deliberating on the future of engineering in the University, Robert Beattie was evidently confident about the future of electrical engineering, at any rate with regard to student numbers, as he made an application to Senate for increased accommodation. At its meeting on 18th December 1919 Senate resolved:

That a committee consisting of the Vice-Chancellor, Professors Beattie, Bragg [Physics], Edwards [Metallurgy], and Mr Dewhurst [Engineering], be appointed to report on increased accommodation immediately necessary in Electro-Technics.

Several other departments were asking for increased accommodation at the same time, presumably because of the expected boom in student numbers in the wake of the war. At the same meeting at which the question of accommodation in Electro-technics was raised, Senate resolved:

That a Committee consisting of the Vice-Chancellor, Professors Dean, Dixon (Chairman and Convenor), Edwards, Lamb, Pear, Bombas Smith and Tout, with authority to consult any members of the teaching body, be appointed to report further on the more permanent accommodation required for the Departments of the University.

The Chairman, Professor Dixon, was Professor of Chemistry. Beattie succeeded in his plea for more accommodation. The tinsmith's and glassblower's workshops adjoining Bridge Street which are shown in the plans on pages 84-85 were not part of the department; they were the premises of two private companies, albeit ones whose main work was done for the university. Arising from Beattie's request the university did a deal with the two firms whereby they moved to nearby premises and the vacant space thus provided became part of Electro-technics. A second, small, lecture room was constructed to occupy part of the accommodation thus made available; it was on the upper floor overlooking Bridge Street and was always known thereafter as "the new lecture theatre". The Scientific Glass Blowing Company moved to 12-14 Wright Street (behind the Chemistry Department) and the Tinsmith, (Stelfox), moved to new premises further along Bridge Street (which was re-named Bridgeford Street in September 1953 [4]). These changes were not completed until 1926-27. Re-arrangements to the layout of the machines in the dynamo house were also made in the early 1920s. The accommodation available to Electro-technics then remained the same for the next eight years.

Appointment of Joseph Higham

The Electro-technics Department's staff of Robert Beattie (Professor) and Harold Gerrard (Senior Lecturer) increased to three in the autumn of 1920 with the appointment of Joseph Higham as Lecturer. He had graduated with first class honours in Physics under Rutherford in 1911 and had followed this, also at Manchester University, with a Teacher's Diploma (first class with distinction) and a Teacher's Certificate (also first class), both in 1912. In later years he used to tell his family that he was one of the few lecturers at the university who had teaching qualifications. He was born in Hyde in 1889 of a well-known local family; his father, Charles Joseph Higham, was a master letterpress printer and his father's father was Editor and Manager of *The North Cheshire Herald*. Before coming to the university as a physics undergraduate he had been educated in Hyde and then at Manchester Grammar School.

As an undergraduate, Higham was a member of the Officer Cadets Training Unit and a close friend of James Chadwick who was in the same year. After completing his teaching course he took up a post at the College of Mines in Johannesburg (now part of Witwatersrand University) and is thought by his family to have left for South Africa in 1914. In 1916 he volunteered for the Army and enlisted in the Kings African Rifles as a Second Lieutenant. Later he was commissioned in the Royal Engineers. His intention had been to return to South Africa after the war, but passenger shipping was virtually non-existent at that time and he took a post at Durham University instead. [5] But after a brief stay he accepted the Manchester lecturing post and in the following year he obtained his Manchester M.Sc. degree with a thesis entitled "The flickering of incandescent lamps on alternating current circuits". So began a tenure which would last 32 years. Following his appointment in 1920 the same three people (Beattie, Gerrard and Higham) comprised the Electro-technics staff until the summer of 1929, their status remaining unchanged. A sad event occurred at the beginning of the decade when Arthur White, the department's solitary technician who had trained in the Physics Department under William Kay, died of tuberculosis. Charles Richardson, then aged about 17, was appointed to replace him.

The undergraduate course in the 1920s

During this decade the undergraduate course in electrical engineering changed very little in basic essentials. The only significant development as the years went by was that more electronics was gradually introduced. The following course description is derived from the University Calendar of 1923-1924 and is representative of the decade:

The Honours School of Engineering.
I. Before the end of the second year of study students had to satisfactorily attend the following:

(a) Two courses of lectures in Engineering, each of three hours a week for a session. The subjects of the courses were: Surveying, Graphic Statics and Theory of Structures, Strength and Testing of Materials, Kinematics and Dynamics of Machines, Heat Engines, Hydraulics.

(b) Not less than three courses of lectures in Pure and Applied Mathematics, each of three hours a week for the session.

(c) Two courses in the Engineering Laboratory, each of one day a week for a session, including special courses of lectures on laboratory work.

(d) Two courses of Drawing and Design, each of one day a week for a session, including special courses of lectures on Design and Drawing.

(e) Two courses of not less than two hours a week to be chosen from the following four: Physics; Geology; Electrical Measurements and Theory of Electrical Machines; Electrotechnics, Metallurgy and Fuel.

II. During their Third year:

(a) Four courses, each of not less than three hours a week for a term, to be chosen from the following eight: Geodetic Surveying and Descriptive Civil Engineering; Higher Theory of Elasticity and Dynamics of Machines; Hydraulics with Applications in Machinery; Thermodynamics with application to Heat Engines and Refrigerating Machines; Civil Engineering and Higher Theory of Structures; Higher Theory and Design of Electrical Machines and Apparatus; Generation, Transmission and Distribution of Electrical Energy; Mechanical Equipment of Mines.

(b) Two courses, each of not less than three hours a week for the session, in Pure and Applied Mathematics.

(c) A course in the Engineering Laboratory of not less than one day a week during the session.

(d) *Either:* A course of not less than one and a half days a week for the session in Machine and Structural Design, including special courses on Design,

Or: A course of not less than one day a week for the session in the Electrical Laboratory, together with not less than half a day a week for the session in Mechanical and Electrical Design.

The Part I Examination normally taken at the end of the second year consisted of:
1. *Five* papers on the following subjects to be taken at the end of the second or third year:

1. (a) Strength and Elasticity of Materials. (b) Mechanics and Kinematics of Machines.
2. (a) Dynamics of Machines. (b) Theory of Heat Engines.
3. (a) Differential and Integral Calculus. (b) Analytical and Plane Geometry.
4, 5. *Two* of the following:
(i) (a) Surveying. (b) Theory of Structures.

(ii) (a) Structural and Mechanical Design. (b) Problems in Geometrical Drawing.
(iii) Continuous-Current Measurements, Apparatus and Machinery.
(iv) Alternating-Current Measurements, Apparatus and Machinery.

Part II. Five papers to be taken, normally at the end of the third year.
 (a) Three papers chosen from the following:
 6. Geodetic Surveying and Civil Engineering.
 7. Higher Elasticity and Dynamics of Machines.
 8. Hydraulics with Applications to Machinery.
 9. Thermodynamics with Applications to Heat Engines and Refrigerating Machines.
 10. Higher Theory and Design of Electrical Machines and Apparatus.
 11. Generation, Transmission and Distribution of Electrical Energy.

 (b) Two papers in Mathematics, covering differential and integral calculus, including simple types of differential equations; the elements of projective and co-ordinate plane geometry; outlines of analytical solid geometry; analytical statics; hydrostatics and dynamics.

The work done in the laboratories and drawing offices was also taken into consideration.

The syllabus at the end of the decade, as given in the 1929-30 calendar, was practically identical with that quoted above. Clearly the Electrical Engineering students were required to do a lot of civil and mechanical engineering, as indeed was still the case 25 or 30 years later, but this fact never held them back in any way. In some ways it was an advantage as they were able later to move very easily into other (non-electrical) branches of engineering if they so wished. The Department never failed to produce a stream of very capable and much-sought-after electrical graduates.

In Chapter 6 it was explained that after the founding of the Electro-technics Department, Honours Physics students were still able to take an Electro-technics option, (this has always been available, in some form, to the present day) so the students taught by the three electrical engineering staff were not limited to those taking the Engineering course. Civil and Mechanical engineers took an electrical course, and electro-technics was also available as an option to students in the Honours School of Metallurgy.

Some Electrical Engineering graduates of the period

A survey carried out by the Department in 1943 [6], in which questionnaires were sent out to all traceable students who had graduated since 1920, showed that most of the Department's graduates had done very well in their careers. Running briefly through the 1920s, here are a few examples, chosen to include at least one from each year. The year quoted is their year of graduation:

1921: F.S. Edwards: Joined the Metropolitan-Vickers Electrical Company and became a key figure in their world-famous high-voltage laboratory.
1922: H. de Boyne Knight: Joined British Thompson Houston, specialising in the design of gas-filled valves (thyratrons, etc.) and became a leading authority in this field.
J.W. Homer: Became Technical Director of an engineering company.
H.T. Aspinall: Lectured in a technical college.

1918-1930: Consolidation in the post-war years

T. Havekin: Took a Ph.D. at Birmingham University, lectured there, and in the 1940s and 1950s was a well-known part-time lecturer in Electrical Engineering at Manchester University.
1923: R.H. Evans: Turned to civil engineering and became Professor at Leeds University in 1946.
1924: E. Rushton: Went to the National Physical Laboratory where he rose to a high rank.
H. Shackleton: Went into the Electricity Supply Industry where he rose to a position in which he was responsible for a large area of the supply system in Lancashire and Yorkshire.
1925: T. Gill: Became an expert on technical problems associated with coal-mining.
W. Jackson: Became Professor of our own Department in 1938; was knighted, was President of the Institution of Electrical Engineers, and later created Lord Jackson of Burnley.
E. Swift: Went to the Lancashire Dynamo & Machine Co. and represented the Company in Canada.
F. Roberts: Stayed to do research and became a staff member in Electrical Engineering at Manchester.
1926: A.E. Starkey: Was an engineer with English Electric, representing the Company in Britain and India.
I. Stewart: Was a test engineer with Parsons.
1927: C.V. Vinten-Fenton: Became Principal of Newton-le-Willows Technical College.
1928: R.H. Dunn: A research and development engineer who became an expert on switching and signalling.
1929: A. Raven: Was an engineer with the Harrogate Corporation Electricity Department.
1930: H. Page: A senior research engineer with the B.B.C., specialising in the design of transmitting arrays and the study of propagation problems.
W.H.R.A. Coates: Was an engineer with Metropolitan-Vickers but was killed whilst serving as an air crew member during the second world war.
E.A. Jones: Was a radio engineer with Marconi after earlier work in radio with other companies.
C. Stead: Was an engineer with Marconi and the B.B.C.

None of the graduates of that period had any difficulty in getting jobs and the survey showed that they nearly all stayed in the technical (electrical engineering) area in which they were trained. A few moved into mechanical engineering but none of them entered completely non-technical areas, such as accounting, as present-day graduates occasionally do.

Staff of the Engineering and Physics departments in the 1920s

The 1920s, and to a lesser extent the 1930s, were periods of relatively small numbers of students and correspondingly few staff members. Electrical engineering students' time was divided between their own Electrotechnics department and the departments of Engineering, Physics and Mathematics. Because of the large 'engineering' content of their course the lecturers in engineering were as well known to them as the three staff members of their own department, so it is worthwhile to take a brief look at who the engineering staff members were. Taking the 1923-24 session as an example, being half-way through the period considered in this chapter, the ten people on the staff of the Engineering Department were:

A.H. Gibson (Professor); E. Sandeman (Associate Professor); C.B. Dewhurst (Lecturer and Assistant Director of the Laboratories); C.M. Mason (Senior Lecturer);

Eric Jones and H. Wright Baker (Lecturers); G.H.W. Clifford, J.S. Wrigley and G.F. Mucklow (Assistant Lecturers); and Julius Frith (Special Lecturer, i.e. came in on a part-time basis to give lectures on a particular subject).

Students in those days were a lively lot. Everyone knew everyone else and there was an easy social rapport between the staff and students. Herbert Shackleton, who was born in 1904 and graduated in Electrical Engineering in 1924, recalled in a letter some of his memories of the people whose lectures he attended as a student:[7]

"As you are aware I entered the university in October 1921 along with the final entries from men who had served in the First World War and were using their Gratuity money in a very sensible manner. A friend of mine, Eric Rushton of Haslingden (later National Physical Laboratory), was in the same group. A year later Willie (Willis) Jackson (Lord Jackson) and Tom Gill, both like myself from Burnley, became students in the department.

During my first year I took Applied Maths (Lecturer J.M. Child) and Physics (Lecturer Prof. W.L. Bragg). I also took Surveying; Lecturer C.B. Dewhurst who we understood was a champion ice skater. He seemed rather a dreamy sort of person and on one occasion during a lecture he said "I have now got the theodolite focussed on the church steeple" which was visible through a window of the lecture theatre, when all the time the cover was over the theodolite lens. I also had lectures during my first year on Strength of Materials - Lecturer H. Wright Baker* who also guided us in the laboratory. At one of the engineering department 'smokers' we parodied *Clementine* thus:

> *Oh my darling, Oh my darling,*
> *Oh my darling, Baker boy,*
> *In the lab you're just a nuisance,*
> *But at home you're mother's joy.*

I think he was present at the time.

During my second year the lectures on design and drawing were by Eric Jones, who I understand had something to do with the development of the depth charge used in World War I for attacking submarines. The lectures on Electrical Measurements and Theory of Electrical Machines were by Harold Gerrard - a very reserved sort of gentleman but very capable. Lectures on Metallurgy were by Professor Thompson.†

During my third year the lectures on Hydraulics were by Prof. A.H. Gibson. I thought he was the best lecturer of the whole bunch. He could put it over so simply and his own text book is a masterpiece. Lectures on Higher Theory and Design of Electrical Machines and Apparatus were by Prof. Robert Beattie who I understand was responsible

---

* H. Wright Baker (1893-1969), whose appointment had come about through the resignation of Professor Petavel (see p. 118), moved to the College of Technology (now UMIST) in 1939 on his appointment as Professor of Mechanical Engineering, a post he retained until his retirement in 1960. He achieved fame when his expertise was employed to 'open' the Bronze Dead Sea Scrolls. His wife, Kathleen M. Drew (d. 1958), lectured in the Botany Department for many years until the mid 1950s.

† F.C. Thompson (1890-1977) was Professor of Metallurgy in the university from 1920 until his retirement in 1959 and was then Honorary Curator of Coins at the Manchester Museum until 1976.

for the fluxmeter. He was very approachable. Lectures on Generation, Transmission and Distribution of Electrical Energy were by Joseph Higham. He was a first class lecturer and I would think he had good practical experience in his subject.

As regards a project during my third year, I had to design a generator from first principles and get the final design on the drawing board.

In answer to your question [on whether there was any electronics in the course] I can only remember having two or at the most three lectures on the thermionic valve and little else. Harold Gerrard would be the lecturer. Radio was in its infancy. On leaving University and starting working and living in digs I had a crystal set with a cat's whisker. In 1930 I managed to make an AC mains radio.

I am not quite sure whether Julius Frith lectured us during the second or third year. He dealt with Specifications, Estimates and Contracts. He was a Consulting Engineer, practising in the city. He was a very good lecturer. We used to pull his leg because before chalking on the board he always wrapped the chalk in paper so that he didn't get chalk on his fingers."

Mr Shackleton was one of the many young people of Burnley to benefit from an Edward Stocks Massey Scholarship, in his case £30 p.a. for the three years of his course. A large number of Manchester University students came from the local area in those days and most used to travel to the university daily from home. Mr Shackleton travelled in each day from Harle Syke on a £28/year rail contract, a special rate for under-18s, which was paid by Lancashire County Council. The tuition fees at the time were £45 p.a.

C.B. Dewhurst, who had been a staff member of the Engineering Department for many years, died during the 1925-26 session, and C.M. Mason replaced him as Assistant Director of the Laboratories (which meant he was responsible for much of the day-to-day running of the Department), a position he retained until his retirement in September 1955. Otherwise the engineering staff changed little during the 1920s, numbering eight at the end of the decade. The staff of the Physics Department in 1923-24 totalled 11, headed by Professor Bragg, excluding Arthur Schuster who was still named as Honorary Professor on the staff list but now lived in the south of England and had long since ceased to take any part in the department's activities. J.M. Nuttall, one of Rutherford's protégés, was now a Senior Lecturer and would remain so for the next thirty years. The Physics staff was boosted by the presence of D.C. Henry who was named not amongst the Physics staff but under a separate heading of 'Colloid Physics' (a cross between chemistry and physics), in which subject he was Lecturer at the time but later became Reader. He remained in post until 1960. His son, Tim, was a member of staff of the Mechanical Engineering Department from 1963 to 1993.

The Department of Mathematics comprised seven or eight staff members during the 1920s. Following Horace Lamb's retirement in 1920 the new appointee, Sydney Chapman, stayed only until 1924, to be replaced by L.J.

Mordell as Professor of Pure Mathematics, and the staff was strengthened in 1929 by the appointment as Professor of Applied Mathematics of D.R. Hartree, who in the 1930s would become well known for his work on the Differential Analyser, the story of which will figure more fully in the next chapter.

Research in Electrical Engineering

The period under consideration was not a very active one from the research point of view. Robert Beattie's paper on 'A permeameter for straight bars' published in 1916 when he was 43 seems to have been the last paper he ever published. A systematic trawl through the literature has failed to reveal anything later. It appears that once the war ended he confined himself to consolidating his new department by devoting his time to teaching and administration, duties he performed very ably. Resources at that time were minimal, which did not encourage any research effort, and by the early 1920s, when Beattie was 50 and with many administrative chores to cope with, he had probably lost his motivation and enthusiasm for research. He still lived in Buxton and travelled in to the university daily. In 1922 he and his family moved from "Whiteknowle" in Buxton to a detached Victorian house, "Fern Lodge", London Road, Buxton.

Surprisingly, Harold Gerrard appears to have shown little interest in research following his release from the war work in which he had distinguished himself. He published a paper in the *Philosophical Magazine* in 1920, 'On electrical disturbances due to tidal waves', jointly with F.B. Young and W. Jevons, but this paper, reporting on some of the interesting findings arising from the war work of himself and his co-authors, marked the end of Gerrard's publishing, though in 1920 he was only 32 and already a Senior Lecturer. The departmental annual reports of the early 1920s stated that he was engaged in research on current transformers, slip meters and harmonic analysis, and later in the decade it was reported that he was working on harmonic analysis and on the characteristics of thermionic valves.[8] However, nothing arising from the work was ever published except for one paper on harmonic analysis which Gerrard magnanimously allowed his research student, C.F.J. Morgan, to publish under his name alone, thereby giving the student all the credit. Gerrard used to say in later years that doing research in his subject in those days was impossible because of a complete lack of resources; even the nuts and bolts had to be counted out and costed. It might be argued that if there is sufficient motivation ways can be found of overcoming such problems but the true reason for his lack of research activity can only be left to conjecture. Nor did Joe Higham do much

1918-1930: Consolidation in the post-war years

research, though he was a fine teacher. It is very possible that Robert Beattie's refusal to allow anyone to take a Ph.D. degree in his department might have had an inhibiting effect on both Gerrard's and Higham's zest for research. Gerrard had obtained his M.Sc. in 1910 and Higham's was gained in 1921, so because of Beattie's edict concerning the Ph.D. it was not possible for them to obtain this degree themselves. All three staff members undoubtedly had a heavy teaching load. It is not surprising that they came to regard this as their main work and, without ever making a conscious decision not to do research, the likelihood of actually doing any significant work would gradually recede into the background as the years went by.

As none of the three staff members was engaged very actively in original investigations, there were no research groups in the modern sense at that time. The only occasions, therefore, when any major research work was done was when a graduate wanted to work for an M.Sc. The supervision was then shared between the three staff members. In the 1920s only three students stayed on to to work for an M.Sc. in Electro-technics. They were Willie Jackson (B.Sc. 1925; M.Sc. 1926); Frank Roberts (1925; 1927) and C.F.J. Morgan (1927; 1928). The research work of Roberts led to the publication of a paper read before the Institution of Electrical Engineers, as did that of Morgan, and both papers were awarded Premiums. In addition, R.H. Evans (1923; 1928) and C.V. Vinten-Fenton (1927; 1934) did their M.Sc. work in the Mechanical Engineering Department and P.E. Brockbank (1928; 1929) jointly with that department. The M.Sc. projects of Jackson and Roberts were on radio subjects ('efficiency of high-frequency transformers' and 'errors in valve voltmeters' respectively) and mark the beginning of what would later prove to be a distinguished record of electronics research and invention in the department. Morgan's work, referred to earlier, was an extension of that begun by Harold Gerrard, and was an investigation of an experimental method of harmonic analysis.

In addition to these three people who 'stayed on' and did an M.Sc. there were four others of this period who presented an M.Sc. as 'external students' several years after their original graduation, having worked in industry in the meantime. They were F.S. Edwards (B.Sc. 1921; M.Sc. 1939); H. de B. Knight (1922; 1947); Herbert Shackleton (1924; 1937); and H. Page (1930; 1940). It seems likely that, had the Ph.D. been acceptable to Professor Beattie, the number of students staying on to do research would have been larger. As it was, most of the best departed elsewhere once they had acquired their M.Sc., including Willie Jackson who, having obtained his degree in 1926, accepted a post as Lecturer in Electrical Engineering at Bradford Technical College.

The titles of the research projects of all the M.Sc. graduates named above are listed in Appendix 8 (A).

The end of the 1920s

An exception to the general exodus of the brightest students once their M.Sc. had been obtained was Frank Roberts. Following his graduation in 1925 he took a Teaching Diploma of the University in the following year (obtaining a First Class) and, after gaining his M.Sc. in 1927, stayed on as an Osborne Reynolds Fellow to do further research in electronics. His particular interest was in matters related to diode detection. By the late 1920s radio was an important aspect of electrical engineering, the design of radio sets having advanced considerably since the invention of the superheterodyne receiver in 1917-18. [9] Many householders already owned a radio set and most of those who did not had aspirations to acquire one as soon as possible. In the 1920s the limited amount of teaching in radio and related subjects in the Electro-technics Department had been in the hands of Harold Gerrard, who was an expert in machines, instruments, and similar long-established areas of electrical engineering. Radio was not his primary subject but he would have been quite capable of reading up enough of the topic to remain well ahead of his students. However, as the end of the decade approached it was clear that an electronics specialist was needed to enhance the department's expertise in this area and to promote research in electronics. To fulfil this need Frank Roberts was appointed Assistant Lecturer in 1930, thereby increasing the department's staff numbers from three to four.

The last session of the decade, 1929-30, was marked by some important arrivals. A 14-year old boy, Albert Cooper, was taken on as laboratory assistant to Charles Richardson, thereby starting an association with the department that would last for nearly half a century. Two young women, Beatrice Shilling and Sheila McGuffie, registered as Electrical Engineering students in October 1929; when they graduated three years later they would be not only the first women graduates in Electrical Engineering but also the first female graduates in any branch of Engineering at Manchester. Miss Shilling was a remarkable young woman of whom more will be said in Chapters 9 and 10. The Engineering intake for the session, which turned out to be a particularly gifted one, also included G.H. Kenyon who became Chairman of the University Council and J.A.L. Matheson who achieved distinction as a Professor of Engineering in Manchester and later as Vice-Chancellor of Monash University in Australia. But from the standpoint of our story the most noteworthy undergraduate arrival for the 1929-30 session was a youngster from Romiley, Frederic Calland Williams, whose future career would be as distinguished as it is possible to imagine and of whom countless people have said, with no prompting whatsoever, "He was the brightest person I have ever known."

References: Chapter 8

1. David Wilson. *Rutherford - Simple Genius*. Hodder & Stoughton, London, 1983; pp. 408-9.
2. Rutherford correspondence, Cambridge University Library.
3. David Wilson's biography of Rutherford (see ref. 1); p. 267.
4. The name change from Bridge Street to Bridgeford Street was approved by the Highways Committee of the Manchester City Council on 16th September 1953. It was changed in line with the policy to do away with duplicate street names. These details were provided in a letter dated 20th February 1997 from Mr R.C. Cordock, Director of Operations for Manchester City Council, in answer to a question on when and why the name was changed.
5. The family details relating to Joseph Higham have been kindly supplied by his son and daughter, Mr Joseph G. Higham and Mrs Jean Oliver. Their help is acknowledged. Durham University has been unable to confirm that Joseph Higham spent a brief period there at the end of World War I as their records do not cover the period adequately, but his family believe it to be the case.
6. Survey on the activities of graduates carried out by Joseph Higham in 1943.
7. Recollections communicated by Mr Herbert Shackleton to the author in February 1991 in telephone conversations and a letter. Mr Shackleton died in 1994, aged 89.
8. Annual Reports of the University Council to the Court of Governers, University of Manchester. (Contains the Departmental Annual Report). Year by year, 1918 to 1930.
9. Alfred T. Witts. *The Superheterodyne Receiver*. Pitman, London, 1939.

Note: The superheterodyne receiver did not come about as a result of a single 'Road to Damascus' revelation. Like most worthwhile inventions, it evolved as the final stage of a series of ideas, each of which built on earlier work. The outstanding names associated with the early development of the superheterodyne are:
(a) R.A. Fessenden, who invented the heterodyne receiver in the United States in September 1901.
(b) L. Levy, who used the heterodyne principle in a receiver designed for the elimination of atmospherics, in France, April 1917.
(c) W. Schottky, who was the first to describe a superheterodyne receiver intended to be a powerful and selective amplifier, in Germany, June 1918.
(d) E.H. Armstrong, who not only conceived the idea of the superheterodyne, but was the first to investigate the practical capabilities of this type of receiver late in 1918.

Chapter 9

1930-1938: Productive years, and the retirement of Beattie

The 1930s began with the country in the same state of deep economic depression that had prevailed during most of the previous decade. In the Electro-technics Department the old stalwarts of Beattie, Gerrard and Higham were content to limit themselves to teaching, which they did ably enough, and the country's economic straits ensured that virtually no money was available for research. This meant that little research would be done in the department unless someone with a large measure of vigour and enthusiasm appeared on the scene. A bright youngster was clearly needed. The call was answered initially by Frank Roberts, a 1925 graduate who had been appointed to the staff as an Assistant Lecturer in 1930, and later, with great distinction, by F.C. (Freddie) Williams.

The undergraduate course in the 1930s

With the appointment of Frank Roberts the scene was set for the much-needed revitalisation of the department's undergraduate courses. The basic structure of the course remained little changed from that described on pages 121-122, much of the course being concerned with civil and mechanical subjects, but the electrical content was modernised by the introduction of two new third year courses, 'Higher Frequency Currents' and 'Higher Pure Electricity and Magnetism', both of which were given by Roberts, with associated laboratory work. The other two third year courses available, 'Higher Theory and Design of Electrical Machines and Apparatus' (given by Harold Gerrard) and 'Generation, Transmission and Distribution of Electrical Energy' (given by Joseph Higham) were old-established courses which continued as before. Professor Beattie continued to give the second year electrical courses. He also did all the department's adminstration. Albert Cooper recalled that Beattie used to check every account and take the signed invoices across to the Bursar's office himself. [1]

The full syllabuses of the electrical courses of this period are given in Appendix 9. They did not alter much during the remainder of the decade. The updating brought about by the introduction of Roberts' courses, compared with the degree course of the late 1920s, was significant. Due recognition was now being given to electronics and the burgeoning importance of light-current subjects.

Alan Fairweather, who graduated in 1933, recalls some of the staff who lectured to him or with whom he was associated as an undergraduate: [2]

1930-38: Productive years, and the retirement of Beattie

"Robert Beattie: A kindly retiring man, concerned for the well-being of his students. Gave me a very generous testimonial. (I still have it.) Took the trouble to mention the fact when he had noticed sets of Heaviside's "Electromagnetic Theory" for sale at a much reduced price. Strict standards: did not think much of (Sebastian Z. de) Ferranti's invention of the power transformer as it had an open magnetic field. Did not recommend H. Cotton's *Electrical Technology* for supplementary reading, as he considered it to be based very largely on his own lecture notes - but, instead, a book by W.T. Maccall. Considered it impossible to teach electrical engineering in three years. Did not approve of the Ph.D. degree and did not accept candidates for it.

He had a large bushy moustache and a marked Scottish accent. An occasional student ploy was to ask him, with the air of an earnest seeker after truth, the rather odd question "What happens if you remove the excitation from a series motor?" The anticipated reply, to everyone's delight, and accompanied by much wiggling of the aforesaid moustache, was "It would rrrev up until it burrrst its arrrmature!"

Frank Roberts: A kindly, serious-minded man, by no means devoid of a sense of humour. In breaks between spells of work, he and his research students, R.F. Cleaver and I, would discuss a variety of subjects unconnected with science or engineering. His research interests were exemplified by the topics allocated to his students. Roberts introduced me to the study of non-linear phenomena. Lectures became notorious for the length of the equations involved, (as, for instance, in D.W. Dye's analysis of the transformer-coupled amplifier). They might extend right across the blackboard and part-way back again. On one occasion, before F.R. arrived, his class had arranged things so that a toilet roll was located at one end of each left-to-right row of desks, and unrolled initially as far as the other end of the row. As the lecture progressed, and as each set of symbols appeared, the class busily copied them down and then, on cue, pulled the paper along towards the free end. All this was taken in good part.

On another occasion, a lecture was given by (I think, an Alderman) Walker of the Electricity Generating Board. It was not a good lecture, and it fell to Roberts to move a vote of thanks. He remarked that "Mr Walker, like other walkers before him had got himself thoroughly lost . . . !"

Once, F.R. and his wife entertained R.F. Cleaver and me to tea (with strawberries and cream) at his home in Fallowfield and then took us to play billiards at a local saloon.

R.J. Cornish: Later became head of the Sanitary Engineering Department at the College of Technology. Taught us surveying and, rather ambiguously, the art of "chaining on a slope". And, at the Annual Departmental Party, invested a colleague with "The Order of the Plug and Chain" - a splendid brass chain with a china "Pull" suspended on the chest of the recipient.

R.W. James (Physics lecturer): Author of a monograph on "X-Ray Crystallography"; also a one-time Polar explorer with Shackleton - an experience on which he was occasionally invited to give a lantern lecture. Led a team working on the crystal structure of metals and alloys.

G.F. Mucklow (Mechanical Engineering): During a lecture on Heat Engines, he was challenged by a brash student, who just stood up, and announced "I don't agree!" G.F. Mucklow replied "I don't really mind", and continued - to the discomfort of the deflated student and the amusement of his fellows.

A.H. Gibson (Professor of Engineering): Occasional conflicts occurred between students from the Eng. Dept. and some from the Medical School on the opposite side of the street. At this time, A.H.G. had a son in the Dept. and he was harassed by some Medicals. Retribution was swift: the ringleader was duly abducted and dumped in the flume in the hydraulics laboratory.

G. Sutherland: Warden of Dalton Hall, taught Intermediate (first year) Physics. When discussing boiling under reduced pressure, expatiated on the difficulty of boiling an egg at the top of Mont Blanc. Solemnly asserted in almost a stage Oxbridge drawl that it was "because the temperature of ebullition of water at the reduced pressure was insufficient to coagulate the albumen of the egg. Cheers!

J.M. Jackson: Lectured in Applied Mathematics (later to move to Westfield College and thence to the University of St. Andrews). Analysing the propagation of sound in a near-exponential pipe, he included a discussion of the effect produced on the harmonic content of the wave by partial closure of the output end, and hence the action of the "mutes" used by jazz trumpeters. Provided a practical demonstration with the aid of his French horn: he first blew it in the normal way, and then with his arm stuffed into the bell, so producing the familiar rasping sound. General applause!

D.R. Hartree: Mathematics Department, famous for the Differential Analyser [see page 135]. A helpful and most unassuming man, completely devoid of any "side". Characteristic of him that when you asked him about a problem, he would start by saying "Of course, you know much more about this than I do" and then reach for a pencil and paper and after several pages you were much wiser and realised (which, of course, you had already done) that his opening remark was utterly untrue and intended to encourage you. He once, in all innocence, greatly alarmed a prospective Ph.D. student (Arthur Porter) by explaining to him that the difference between the M.Sc. and Ph.D. degrees was that the Ph.D. was expected to be "a real contribution to knowledge - rather like my Self Consistent Field Theory".

He had several children, and students working in the basement room housing the differential analyser were occasionally entertained by the youngsters waving to Daddy through the railings above the room at street level. Girl students had been known to take them walks and cynics said that they did it in the hope of getting better examination results! The Hartrees held occasional musical parties at their home on Sunday mornings, in which the staff of the [Mathematics] Department took part. For instance, J.R. Todd might play the piano, and J.M. Jackson the French horn. Mrs Hartree had, beforehand, taken the wise precaution of covering the carpet with newspaper to protect it from the drainage associated with wind instruments. It was said that, when recruiting new members of staff, preference was given to those who, in addition to their mathematical abilities, could play a musical instrument."

Notable 1932 graduates

It was noted at the end of Chapter 8 that the 1929 intake was a remarkable group. Out of 24 Honours engineering graduates that year, 12 were awarded Firsts. Two of the brightest, George Kenyon and J.A.L. Matheson were Mechanical Engineers, but F.C. Williams, who sailed though the course with flying colours, was an Electrical and was the top graduate of 1932. It was apparent to everyone that here was an exceptionally gifted young engineer.

1930-38: Productive years, and the retirement of Beattie

Equally notable, as an outstanding "personality" and a technically very able one at that, was Beatrice Shilling who graduated with a II(1) in the same year. Alan Fairweather, who graduated a year later, recalls her: [3]

"She was unique, a chain-smoker, and the possessor of a 490cc, o.h.c. Norton motorcycle equipped with large Brooklands-type fish-tail silencers. These machines are famous in the history of motorcycle racing. She rode it, dressed in full racing leathers, and usually set off with a running push-start as in racing. This was no mean feat for a woman of no more than average build. Needless to say, she was an immediate centre of attention wherever she stopped in Manchester. She reckoned to go to London in less time than that taken by the train - this before the motorways though with less traffic. She later became the first woman to lap Brooklands at 100 m.p.h. A very tough lady."

For her Brooklands exploit, which was achieved whilst she was a student, Beatrice (or Beaty as she was known to the other students) was awarded the Gold Star, which was a six-pointed star, rimmed in gold, with a blue enamel background, and gold "100" across it, with "MISS B. SHILLING" engraved on the back. It was quite small since it was worn beneath the standard British Motor Cycle Racing Club membership badge. Her motorcycle was a standard Norton (capable of 80 m.p.h. plus) which she modified and tuned herself until she lapped Brooklands at 106 m.p.h. Her future husband (whom she married in 1938) also achieved the Gold Star, on her bike. [4] She made most of the modifications to her machine in the Electro-technics workshop, where she was often to be found. Albert Cooper recalled that she had originally owned a Levis motor-cycle and sidecar, which was much safer than the Norton, but disposed of it in favour of the more lively machine. [5]

Beatrice Shilling, who graduated in electrical engineering in 1932. A remarkable young lady, she tuned her motor cycle herself and lapped Brooklands at more than 100 mph.

Although she graduated in Electrical Engineering, Beatrice's heart was in Mechanical Engineering and after her graduation she did a research project on diesel engines in the Mechanical Engineering Department, gaining her M.Sc. in 1933. Later she went to the Royal Aircraft Establishment where her technical expertise played a major part in helping the RAF to win the Battle of Britain.[6] More will be said of this and other aspects of her later work in Chapter 10. She was without doubt a pioneer of women's place in engineering and it is fitting that she, along with her contemporary Sheila McGuffie (also an 'electrical'), should be the first female engineering graduates at Manchester. The first engineering degrees of the newly-formed federal university had been awarded in 1882. It is indicative of women's non-association with engineering as a course of study that 50 years then elapsed before the first women graduated.

Research in the early 1930s

In the 1920s, it will be recalled, research was of a sporadic nature, occurring when an individual student wished to work for an M.Sc. With the appointment in 1930 of Frank Roberts, an electronics specialist, the pace of research accelerated. His paper on photo-electric cells published by the I.E.E. in 1932 was awarded a Premium and he gave external lectures on Michael Faraday, photo-electric cells, and television. More importantly, Roberts founded what might be regarded as the first research group in the department. He was fortunate, in 1932, to have the newly-graduated F.C. Williams staying on for an M.Sc. Roberts' interest at the time was straight line detection with diodes and this was the subject of Williams' M.Sc., which was duly completed in 1933.* A paper on the subject, with Roberts and Williams as co-authors, was published by the I.E.E. in 1934; it was Williams' first published paper.[7] Williams then began a 2-year College Apprenticeship course with the Metropolitan-Vickers Electrical Company at Trafford Park but abandoned it when he was awarded the Ferranti Scholarship of the Institution of Electrical Engineers in 1934 and joined E.B. Moullin at Oxford to do research on circuit and valve noise for a D.Phil. degree.

Fortunately for Roberts, other bright research students were available in the persons of R.F. Cleaver, Alan Fairweather and F.H. Moon, each of whom graduated with a First in 1933. All stayed on to do an M.Sc. under Roberts. Cleaver worked on high-frequency measurements, Fairweather on the stability and performance of a high-frequency amplifier, and Moon on frequency conversion in superheterodyne reception. All were awarded their

---

* A list of all the research projects carried out by Electro-technics graduates of the period covered by this chapter is given in Appendix 8 (B).

1930-38: Productive years, and the retirement of Beattie

M.Sc. degrees, in 1936, 1934 and 1935 respectively. Cleaver and Fairweather enjoyed successful careers, but the life of Fred Moon, who kept goal for the university's soccer team, came to an end at a tragically young age. In one of the university matches he sustained a broken leg and had to walk on crutches for a while. Soon afterwards he fell victim to tuberculosis, an ever present threat in those days, and died within two years of the award of his M.Sc. It was not known whether the accident triggered the disease.

The differential analyser

This machine, the brainchild of D.R. Hartree, Beyer Professor of Applied Mathematics (see page 132), was an important device in its day. The Electro-technics Department was involved in the machine to some extent in that it was of interest to the new crop of research students who would use it to solve particular problems in the same way that research students 20 years later would use the Williams/Kilburn machines. Following the award of his M.Sc. degree, Alan Fairweather worked for a year on the analyser. A review of the period would therefore be incomplete without a brief note concerning the machine, which was a sort of pre-cursor of certain later computers.

In modern terminology the machine was an analogue computer; in this case a wheel-and-disc integrator designed for solving differential equations and based on an earlier machine built by MIT.[8] Such machines have been described in a book by John Crank who was one of the Manchester research students at the time.[9] Alan Fairweather described Hartree's work and some of the subsequent machines that arose from it:[10]

"The Manchester machine came into being under the direction of Professor Hartree in 1935 and in two stages: First, a Meccano model built by Arthur Porter, then a postgraduate student in Physics, and secondly, a full-size replica of the MIT machine built by Metropolitan-Vickers. Work on this was done by a succession of postgraduate students under Hartree, and is described in M.Sc./Ph.D. theses by Arthur Porter, John Crank, Miss Nicholson, John Ingham and others. Ingham and I used the machine to solve some problems in subsidence transients in non-linear circuits. This was published in J.I.E.E. and was used for design purposes by Met.-Vick. for many years.

A description of the Manchester Analyser appeared in *The Manchester Guardian* in 1935 [11] and mentioned the development of an automatic curve-follower. My work on this is described in the literature;[12] also in an article in a popular technical journal, *Popular Mechanix*, about the same time, by Clifford C. Ashton, a former Manchester University student and then a journalist and press photographer.

Part of the original Meccano analyser is in the Science Museum but the curve-follower was eventually destroyed. Williams is thought to have produced an improved version. The large Analyser was later transferred to the Cambridge University Computing Laboratory (under M.V. Wilkes) where a similar one had been built. The basement room [in Manchester] which it had occupied was subsequently used for cosmic ray research by P.M.S. Blackett.

On leaving Manchester I joined the (then) Post Office Research Station at Dollis Hill, London, and - along with other work - developed a six-integrator Analyser, intermediate in size between the original Meccano model and the MIT machine but properly engineered. The Physics Department at Queen's University, Belfast (in the person of H.S.W. Massey, later Professor at University College, London and subsequently knighted) was also interested in the production of a small machine and the final form of my (wheel/disc) integrator was arrived at after discussion with him.

Sadly the small machine was never usefully employed and had an unfortunate history. Copies of it were made under the direction of James Grieg then at Northampton Polytechnic, London (and later Professor of Electrical Engineering at King's College, London); and also of Joergen Rybner at Copenhagen. This last machine was blown-up during the war. The Dollis Hill one was transferred to King's College and I do not know its ultimate fate."

A new A.C. machines laboratory for the Electro-technics Department

In the period 1934-35 the Electro-technics Department improved its facilities by the acquisition, largely as the result of urging from Joseph Higham, of a small a.c. machines laboratory containing a number of new machines (Schrage motor, etc.). It was situated in a basement room to the east of Higham's instrumentation laboratory and was approached down a flight of steps, the new basement space having been made available by excavation. It remained in operation for the remainder of the department's tenure of the building. According to Albert Cooper's account, Professor Beattie got himself into a little trouble with the university hierarchy over the manner of the application. [13] He made a case for the laboratory and submitted an application to cover the total cost of the machines and the work necessary to develop the site. This was duly granted, but when the work was nearing completion Beattie submitted an application for additional funds, the basis of the request being that the machines could not be used unless proper instrumentation was provided, for which the further money was needed. The committee considering the applications thought, according to Albert Cooper and others from whom the story has been heard, that Beattie's method of application had been 'tactical'. They were almost certainly correct. Beattie knew that if he had asked for the full amount in the first place the sum involved would have been too large for the grant to be approved; however, once the machines had been installed and Beattie had then pleaded that they were useless without the instrumentation, the committee had no sensible alternative but to approve the second application. It is said that Beattie had his knuckles rapped and was told not to do it again. At that time Beattie was about 61 and probably did not much mind. He was one of the university's elder statesmen, being the second longest serving professor[*]; he understood

---

[*] The longest serving was W.H. Lang, Professor of Cryptogamic Botany, appointed in 1909.

exactly how the university system worked and how best to use it to ensure that his Department got what it wanted. So Joe Higham acquired his new machines laboratory which proved to be a very useful and welcome addition to the department's resources.

The death of Frank Roberts

In 1935 the Electro-technics Department was stunned by the death of its youngest staff member, Frank Roberts, in a climbing accident. He was a keen and experienced member of the University Mountaineering Club and it was this interest which led to his death at the end of the summer term. The circumstances were reported in detail in the Manchester newspapers. The first intimation of trouble came on the back page of the 'Last Extra Edition' of the *Manchester Evening News* on Wednesday, 12th June 1935:

"**600ft Rock Hunt for Injured man. Fall at formidable Savage Gulley. All-day quest. From our Special Correspondent. Keswick, Wednesday.**

Rock climbers from Keswick and police from Egremont and Wasdale hurried to Pillar Rock to-day when it was reported that a man had been found seriously injured at the foot of Savage Gulley, one of the most notorious climbs on the rock.

It was about 7am that the discovery of the injured man was first reported. Late this afternoon definite news was still being anxiously awaited from the rescue party, which included Mr Stanley Watson, the guide.

Treatment on spot?

It was believed that they must have been giving the man treatment on the spot, and perhaps carrying him on a stretcher to Wasdale Head - a distance of several miles over a rough and stony mountain path. It is thought that the injured man must have been lying on the mountain side all night before he was discovered. While a number of parties of climbers would probably be on the Pillar Rock during the day, it is far removed from any human habitation.

600ft Mass of Rock

No calls for help during the night time would have a chance of being heard in the bleak Ennerdale Valley in which the huge 600ft mass of rock stands. Savage Gulley received its name because of the terrific obstacles it puts in the way of the climber. Few, even expert climbers, would attempt to tackle it direct, but there are climbs on the rock face near by from which it would be possible for a climber to fall into Savage Gulley."

The next morning's *Manchester Guardian* (13th June 1935, p.9) confirmed that the news was the worst possible:

"**CLIMBER DIES ON PILLAR ROCK. Manchester Science Lecturer. Thigh Broken.**

Mr. Frank Roberts, a lecturer at Manchester University, died yesterday as the result of a fall while climbing the Pillar Rock in the Lake District. The fall took place about 5pm on Tuesday, and Roberts died about 3am yesterday, still on the rock, from which his friends had not been able to reach him.

The accident was described to the Manchester Guardian by Mr. J.M. Marchington of Cheadle Hulme, who was a member of the party. He, Roberts, and about seven other members of the Manchester University Mountaineering Club were camping at Wasdale Head. On Tuesday, with a medical student, Mr. A. David, they went to the Pillar Rock to ascend the North Climb.

THE CLIMB DESCRIBED

The North Climb, in good conditions, is not of more than ordinary difficulty, except for the top pitch. Here it is necessary to work round and up a projecting rock known as the Nose.

When the party reached the Nose they were drenched with rain, and the rocks were very cold. Nevertheless, Marchington succeeded in leading over the Nose and reached a secure position beyond it. Roberts, with benumbed fingers, was unable to follow him, even with the help of the rope. There is an alternative to the Nose. A climber can be lowered, from the ledge by the Nose, down a crack (too difficult for all but the most expert to descend without help from the rope) into Savage Gulley, which comes up on the east side of the Nose. Both above and below the point of landing in the gulley the passage is interrupted by extremely difficult rock pitches, but by traversing eastwards across the gulley one can reach easy ground, and so work round to the top of the gulley, above the Nose, and above all difficulties on the climb.

LOWERED INTO GULLEY

It was decided that Roberts should take this course. He was lowered into Savage Gulley by the third man, David, and took the rope with him, in the belief that he could use it to safeguard his exit from the gulley. While still in the gulley he slipped, either from pure mischance or through mistaking the best way out. He was not in sight of his companions at the time.

He fell only about 15 feet, but broke his thigh and also knocked himself unconscious. Marchington could now see him from the top of the Nose. He moved once only and then lay still.

LEFT WITHOUT ROPE

It was about 5 o'clock. The position was a perplexing one. Even if Marchington could have reached Roberts he could not, single-handed, have given him any assistance. David was on the ledge by the Nose which, with the rocks so wet, he could not safely leave without the help of the rope - which was in the gulley with Roberts. He was, in fact, cut off on the ledge for five hours.

Marchington decided to go to Wasdale Head for help. Here he recruited five more members of the party and they returned to the rock with a stretcher, a splint and spare ropes.

It was nearly 10pm when they arrived and was nearly dark. A medical student called Charnley* was let down the gulley with the rope and found Roberts still breathing. Another rope was attached to him, and with the help of Charnley he was hoisted up the gulley to a good ledge. It was now dark, and it was impossible to move Roberts any farther. They made him as comfortable as they could, rescued David (who climbed up the Nose with the help of the rope) and waited for daylight.

Roberts never fully recovered consciousness; he moaned and mumbled from time to time, but was never coherent. When dawn came about half-past four, nothing more

---

* This must have been the John Charnley (later Sir John) who became a pioneer of hip-replacement. He graduated in Medicine from Manchester in 1935 and was the only person of that surname in the university at the time.

could be done for Roberts, and his friends, wet and worn by their vigil, left his body on the rock and returned to Wasdale.

BODY RECOVERED

A party of climbers, including Alderman George Basterfield, ex-Mayor of Barrow, set off before 8.30 in the morning to recover the body, which they reached about 11am. They were joined at the rock by Mr. Stanley Watson, the guide, with three others. They had to climb down about two hundred feet to the point where he lay, and climb back that distance, hauling the body with them. Watson, according to a statement by Mr. Basterfield, risked his life "at least a dozen times".

It took four hours and twenty minutes to get Roberts to the top of the rock. Meanwhile further relief parties had set out from Wasdale Head, and they got to the scene just as the body had been hauled to the top. Then began the hazardous journey over five miles of rough fell side to Wasdale Head. The journey took three and a half hours, and the party reached the hotel shortly after 6 o'clock.

Frank Roberts was 30 years of age and was born at Tyldesley. From an elementary school he won a County Scholarship to Leigh Grammar School. He took his B.Sc. Degree in 1924 [it was in fact 1925] and the M.Sc. two years later. He was awarded a Research Fellowship in 1927 and was appointed Assistant Lecturer in Electro-technics at the university about three years ago. He leaves a widow and one son."

The official cause of death was given as "From injuries received by an accidental fall while rock climbing at the Pillar Rock, Ennerdale on the 11th June 1935." [14] There was no post-mortem.

Naturally the tragic death of the youngest of its four staff members shocked everyone in the department and a fund was set up to help Mrs Roberts and her baby son. Robert Beattie decreed that dangerous activities such as rock-climbing by members of his department must stop. (Harold Gerrard, then 47, was also a keen outdoor man who enjoyed climbing; he once attached a rope to the girders in the roof of the dynamo house and showed the department's young technicians the correct way to climb it. [15]) At the Faculty of Science meeting held on 18th June the Dean referred to the death of Roberts and a letter of sympathy was sent to Mrs Roberts.

New appointments

After the death of Frank Roberts it was imperative that another staff member be appointed as soon as possible. The Department wanted F.C. Williams, who still had to do another year at Oxford to complete his D.Phil., and it was agreed with him that he would return to the department on completion of his degree. As an interim measure William Makinson, who had graduated in Physics with a First only that summer (1935), was engaged on a temporary (one-year) basis. Although he was a Physicist, Makinson was one of the small number who had specialised in Electro-technics so, having just taken Roberts' two lecture courses and done well in them, he was well equipped to give these courses in the 1935-36 session. He also found time to do some research on the input impedance of diode rectifiers fed from a tuned circuit. In the summer of 1936 F.C. ("Freddie") Williams joined the staff of the Department in Makinson's place as an Assistant Lecturer, as planned. He took over Roberts' courses and set up a variety of research projects, thus adding to the increase in liveliness initiated by Roberts.

When Albert Cooper joined the Department in 1929 as a 14 year old Laboratory Assistant he was junior to the then Laboratory Steward, Charles Richardson. In 1934 Richardson, then aged 31, left, and Albert, who had proved entirely satisfactory, became the department's Laboratory Steward at the age of 19. In 1935 he acquired an assistant, Joe McCormick, who had joined the Physics Department as a 14 year old in 1933, working under Dr J.M. Nuttall, but was transferred to Electro-technics when it was decided that an extra person was needed. Thus, from 1935 Albert Cooper, at 20, was the head of the department's technical staff, and had a 16 year old assistant. Joe stayed with the department until 1964 when he became Chief Technician (later re-named Laboratory Superintendent) of the newly-formed Computer Science Department. Albert enjoyed a cordial friendship with F.C. Williams from Williams' earliest days as an undergraduate student. In the early 1930s 'F.C.' (as he was often called) used to take Albert to his parents' home in Romiley from time to time, where, to Albert's wonderment, there was a maid "and they used to have a joint of meat to eat in the middle of the week!" Albert said that he taught 'F.C', who had bought a Morris 8, how to drive.[16]

Research in the period 1936 to 1938

With the return of F.C. Williams, now the possessor of an Oxford D.Phil., to the department in 1936, the newly appointed Assistant Lecturer launched himself into a continuation of his research work with a vigour that will be familiar to all who knew him. Shortly after his return to Manchester he devised an idea for a television system. It involved the use of a photocell, an amplifier, and two oscilloscopes running on the same time base. Using this equipment Williams was able to pick up an electrical image of a holiday photograph of his future wife, Gladys Ward (they were married in 1938) and display the picture on the screen of one of the oscilloscopes. At Williams' invitation John Logie Baird visited the department to see the system in action, but as neither Baird nor Williams had any money available for development nothing came of the idea and it was subsequently overtaken by the E.M.I. system.[17]

Williams soon recruited research students. One of the first was Robert Kyd Beattie, the son of the professor, who had joined his father's department in 1934 as an undergraduate student. Some might doubt the wisdom of his

1930-38: Productive years, and the retirement of Beattie

coming to the department where his father was professor but there was never any cause for concern about possible bias in young Robert's direction; there was no need for any such favours even had they been proffered as he was a very bright student, a fact that was obvious to everyone in the department. His sister recalls that when he returned home after attending one of his father's lectures for the first time, Robert announced in astonishment and incredulity that his father was a very good lecturer!

The atmosphere in the department at the time was relaxed. Williams, still in his mid-twenties, looked even younger than his years. Albert Cooper and Joe McCormick were the only technicians and according to Joe they used to lay a half-penny stake against Williams' experiments working first time. Joe says they did not win very often. Young Robert Beattie's partner in the laboratory before he graduated was a student named Alex Jackson. Both were big, strapping, six footers. On one occasion they set off a firework under one of Dr (as he then was) Williams' experiments. Joe remembers that the dressing down given to two tall young men by the diminutive Dr Williams was a masterpiece of invective.[18]

Young Robert Beattie graduated with a First in 1937 and was awarded the Fairbairn Engineering Prize, along with Joseph Bromilow (who was not an 'electrical'), and immediately enrolled for an M.Sc. under Williams' supervision. Another research student was A.E. Chester who graduated in

Engineering graduation photograph, 1937. The staff (front row, left to right) are: J.Higham, L.J.Kastner, F.C.Williams, G.F.Mucklow, C.M.Mason, A.H. Gibson, R. Beattie, H. Wright Baker, J.A.L. Matheson and Jack Allen.

the same year. Beattie's research was concerned with radio modulation and Chester's with an investigation of wide-band amplifiers.

In 1938 J. Highcock, a graduate of that year, joined Williams as a research student and was set to work on an investigation relating to the "shot" effect in diodes. Williams' topics of research at the time all involved the use of valves in interesting ways and even then he was acknowledged as an expert on the subject, having published several excellent papers. One of his lines of research was to produce and utilise several traces on the same cathode-ray tube. Some very good valves were obtained from Cossor, and Williams built what Albert Cooper described as "a super-amplifier". In addition to the work being done in collaboration with his own research students, Williams was doing other work at the same time with Alan Fairweather, who had left the department in 1935 after completing his work on the differential analyser. Fairweather explains: [19]

"My association and collaboration with Williams came about in this way. He was a year senior to me and we met whilst he was doing an M.Sc. with Roberts (on diode detection - later published). I also spent a year with Roberts and then, after another year attending lectures on theoretical physics and applied mathematics and working on the differential analyser I left to spend my professional life at the Post Office Engineering Research Station at Dollis Hill, London.

I regarded Williams as the outstanding man in his field, of his generation. His intuitive feeling for electrical systems, and gift for invention, were almost uncanny. He was probably the cleverest man with whom I ever had the pleasure of working. His early work on thermal agitation and shot noise is a classic example of his abilities and is discussed in a book by Moullin. [20] My collaboration with him usually took place between Manchester and London in this way. One of us would have an idea, and then one or both of us would do some theoretical work. The next step would be an experimental check and this might require the assembly or construction of special components (valves, inductors, etc.). This would be done by me at Dollis Hill. I would then load this gear into my rather elderly Ford "8" and drive up to Manchester. My parents were still living at our family home in Salford. Williams and I would then work like mad for two or three days (typically through a weekend). By then, the basic work would have been done and also a rough draft of the paper. I would then return to London and prepare the final version of the paper for publication (curve sheets, diagrams, oscillograms, typescript, etc.). When in Manchester I was usually entertained at Williams' home with his parents and when he came to London he stayed at my digs."

Williams was responsible for almost all the research in the department at that time, but in 1936-37 Joseph Higham made a rare foray into the world of research when he supervised the M.Sc. work of J.P. Wolfenden, who had graduated with a First in 1936 and did some investigations on mercury arc rectifiers. Wolfenden was awarded his M.Sc. in 1937 and the work was subsequently published in the Institution of Electrical Engineers' Journal.[21]

1930-38: Productive years, and the retirement of Beattie

The retirement of Robert Beattie and the appointment of Willis Jackson

At about the time that Robert Beattie's son entered the department as a student the family sold their home in London Road, Buxton and moved to "Bredon", 5, Hale Road, Altrincham. Professor Beattie and his son then travelled together on the train to and from the university. Whilst young Robert was at the university his younger sister, Janet, was enrolled at Altrincham High School for Girls, where she enjoyed a happy, and not undistinguished, secondary school education. [22] Robert Beattie's final years as Professor of Electro-technics, during which he was not very well, were quiet ones. He had developed Parkinson's disease and became rather stooped, and walked with shuffling steps. He continued to do his teaching and administration effectively enough but was not active in other ways. In short, his career was winding down. He and his family always spent the long vacations in Scotland, initially at the family's house in Friockheim (the former home of the Kyd family) and, after it was sold in the mid-thirties, at their flat in Montrose, to which destinations his examination papers and other pressing correspondence were sent.

Robert K. Beattie's B.Sc. graduation photograph, 1937. Professor Beattie is in the centre of the front row with F.C. Williams at his right hand and J. Higham at his left. R. K. Beattie is at J. Higham's left.

Robert Beattie retired at the end of September 1938, having served the university since 1896. Shortly afterwards he was elected Professor Emeritus. His daughter, Janet, has said that his salary never exceeded £1,000 p.a. throughout his 42 years at the university.[23] Beattie's retirement took place three months after his son's M.Sc. graduation ceremony. The two of them are seen together on the photographs shown on pages 141 and 143. Both were taken after the B.Sc. graduation ceremony in 1937 but the picture on page 143 includes only the 'electrical' students from the year's graduates. Professor Beattie's presence at his son's M.Sc. graduation in the summer of 1938 was one of his last official acts.

Beattie's tenure of the Edward Stocks Massey Chair of Electro-technics had not been spectacular but it had been efficient; he had established the department and presided over the years of economic depression when, in spite of the severe lack of money, he had produced a stream of highly able graduates over a long period. His years as head of electrical engineering had coincided with a period of tremendous change and technical advance. It is a remarkable fact that when Beattie was appointed to lecture in electro-technics in 1896, a motor vehicle still had to be preceded by a pedestrian carrying a red flag, and manned flight was not possible except by balloon. By the time he retired, it was a different world; a world of technology in which air transport was commonplace and technical innovation was proceeding at an ever-increasing pace. In particular, Beattie's own subject, electrical engineering, had changed out of all recognition, especially in regard to the rise in importance of electronics, as the impending war would soon prove.

The university had known of Robert Beattie's forthcoming retirement for a lengthy period so there had been plenty of time to appoint a successor. The new man, Willis Jackson, who was one of Beattie's own graduates (B.Sc. 1925, M.Sc. 1926, D.Sc. 1935), took over at the start of the 1938-39 session, immediately after the old professor's departure. The long Beattie era had ended and a very different one was beginning.

References: Chapter 9

1. Tape recorded conversation with Mr G.A. Cooper, 29th April 1987.
2. Correspondence with Dr A. Fairweather, 1991-92.
3. Further correspondence with Dr A. Fairweather, 1991-92.
4. Letter from Mr G.A. Naylor (Beatrice Shilling's widower), 15th September 1991.
5. Tape recorded conversation with Mr G.A. Cooper, 29th April 1987.
6. As reference 4, and also obituaries of Beatrice Shilling in the *Sunday Telegraph* (18th November 1990), the *Independent*, (24th November 1990) and the *Farnborough News* (16th November 1990).
7. F. Roberts and F.C. Williams. *Straight line detection with diodes.* J.I.E.E., **75**, p. 379, September 1934.
8. V. Bush. *The differential analyser: a new machine for solving differential equations.* Journal of the Franklin Institute, **212**, No. 4, October 1931. M.I.T. Research Bulletin No. 75.
9. J. Crank. *The differential analyser.* Longmans, London, 1947.
10. Dr A. Fairweather's account in correspondence, 1991-92.
11. Description of the differential analyser in the *Manchester Guardian*, 28th March 1935.
12. A. Fairweather. *A note on the development of an automatic curve follower for an integrating machine.* John Hopkinson Electro-technical Laboratories, Manchester, Feb. 5th - Apr. 15th, 1935.
13. Mr G. A. Cooper's account of how the department got its new A.C. laboratory. Tape recorded conversation, 29th April 1987.
14. Certified copy of the death certificate obtained from the Superintendent Registrar, Whitehaven, 20th May 1992.
15. Information communicated by Mr J. McCormick.
16. Mr G.A. Cooper's tape-recorded recollections, 29th April 1987.
17. Ibid.
18. Story communicated by Mr J. McCormick, 1997.
19. Information communicated by Dr Fairweather in correspondence, 1991-92.
20. E.B. Moullin. *Spontaneous fluctuations of voltage*, Oxford University Press, 1938.
21. J. Higham and J.P. Wolfenden. *Voltage regulation of the 6-phase fork-connected grid-controlled mercury arc rectifier.* J.I.E.E., **83**, pp. 171-175, August 1938.
22. Information from the Beattie family archive, reproduced by kind permission of Mrs Janet Paterson (Professor Beattie's daughter). Also Mr G.A. Cooper's tape-recorded recollections, as above.
23. Information from the Beattie family archive, reproduced by kind permission of Mrs Janet Paterson.

Chapter 10
1938-1946: Willis Jackson and the war years

Willis Jackson, who was born in Burnley on 29th October 1904, took up office as the new Edward Stocks Massey Professor of Electro-technics at the beginning of October, 1938, immediately after Professor Beattie's retirement. After graduating from the department with a first class honours degree in 1925 under his baptismal name of Willie Jackson, he obtained his M.Sc. the following year and then lectured in electrical engineering at Bradford Technical College until 1929. He then worked at Metropolitan-Vickers at Trafford Park as a College Apprentice for a year, leaving in 1930 to become a Lecturer at the College of Technology in Manchester where he stayed until 1933. In that year he moved to Oxford University where he lectured at Exeter and Queen's Colleges and worked with E.B. Moullin, gaining his D.Phil. degree, before returning to Metropolitan-Vickers in 1938 as a research engineer and personal assistant to A.P.M. Fleming who was the Director of Research. He was awarded his Manchester D.Sc. degree in 1935. Thus, at the time of his appointment to the Manchester chair at the age of 33, thirteen years after his original graduation, he had already enjoyed a varied and successful career. Jackson's main research interest was the investigation of the electrical properties of dielectric materials. This work necessitated a lot of high-frequency measurements, in which area he had consequently become an expert. In the same year that he was appointed to the chair at Manchester he married Mary Boswall, daughter of Dr R.O. Boswall who was a Lecturer in Mechanical Engineering at the Manchester College of Technology, and the couple set up home at 92, Knutsford Road, Wilmslow.

Jackson was an ambitious young man who disliked the name "Willie", which was his given name, not an abbreviation, feeling no doubt that it was not appropriate or helpful to the image of an aspiring man who wished to go far. Consequently he changed it, around the year 1933, and became Willis Jackson, which he remained for the rest of his life.[1] It is remarkable what a difference a single letter can make to one's image, but it goes without saying that he would always be known affectionately as "Willie" by his research students, when they were not in his presence.

Jackson's first year

One could have expected that Jackson, the young "new broom", would want to enliven the department, and so it proved. He told the Chief Technician, Albert Cooper, shortly after his arrival, "I'm going to put this department on the map!"[2] Albert told an amusing story concerning P.M.S. Blackett, the

1938-1946: Willis Jackson and the war years

Professor of Physics who had been appointed to his chair the year before Jackson. The Electro-technical laboratories were in effect an annexe of the Physics building, geographically, and shortly after Jackson arrived he noticed Blackett eyeing the dynamo house through the window at the "physics" end of the laboratory. Jackson knew exactly what was going through Blackett's mind, which was that the room would do very nicely as an addition to the Physics Department's cosmic ray laboratories. But one of Jackson's great virtues was that he was an excellent 'contact man' - he knew everyone who was anyone in the electrical engineering world, especially at Metropolitan-Vickers, whence he had recently come. Jackson promptly persuaded the company to donate to the department an old and huge transformer weighing many tons, which, with great difficulty, was manhandled to the far end of the dynamo house and deposited against the wall, thus obliterating Blackett's view of Jackson's laboratory. According to Albert the transformer was never used but it served the purpose for which Jackson acquired it, as a very effective means of blocking out Blackett's territorial aspirations.

The department which Jackson took over was still very small. The staff, apart from Jackson himself, comprised Harold Gerrard (Senior Lecturer), Joe Higham (Lecturer), both veterans of many years' standing, and F.C. Williams (Assistant Lecturer). Jackson knew Gerrard and Higham well enough from his own student days and was aware that they would take care of the teaching in the 'older' subjects of machines, instrumentation, basic theory, etc. but they would be unlikely to contribute to the research effort. But Williams, as well as giving the "light-current" lectures, was turning out original papers at a prolific rate, thus enabling the department to quote a number of publications each year in its Annual Report. In the years 1936 to 1939, both years inclusive, Williams published 20 papers. He was the sole author of eleven of them; the others were written in collaboration with R.K. Beattie (one paper), P.M.S. Blackett (one; on the differential analyser), A.E. Chester (2), Alan Fairweather (4) and J.P. Wolfenden (1). Clearly then, the future research would be in the hands of Williams and of Jackson himself, aided by any research students who came along. Two such students who had graduated (B.Sc.) the year before Jackson's arrival, R.K. Beattie and A.E. Chester, had already obtained their M.Sc.s. Two graduates of 1938 stayed on to work for an M.Sc; they were John Highcock, who worked under Williams on a problem associated with the "shot" effect in valves, and W.A. Cowin who worked with the new professor, Jackson, on the movement of ions and polar molecules in solid dielectric materials. They were awarded their M.Sc. degrees in 1939 and 1941 respectively. Another graduate from

1938, Percy Bowles, worked for his M.Sc. in the Engineering Department and obtained it in 1939. There were also one or two graduates of the Physics Department. The new regime was therefore getting off to a lively start.

In 1939 Williams was awarded his D.Sc. degree by Manchester University, only seven years after the date of his first graduation. But his association with the university was to be broken, temporarily, when he left the department at Easter 1939 and joined the Government Experimental Establishment at Bawdsey Manor, Suffolk. Perhaps he felt that the time had come for a change, and with the smell of war in the air, a war which would clearly depend heavily on Britain's technical supremacy if it was to be won, the work at Bawdsey offered great challenges. Also, it was no secret that he did not get on very well with Willis Jackson, not because of any personal antagonism or antipathy, but because they were very different people who approached their work in different ways. Willis Jackson was clearly an able researcher (he would not have gained his D.Sc. at the age of 30 otherwise), but his main strength was his gift of making use of his many contacts in the electrical world to profitable advantage. He was an excellent manager and committee man, very proficient at liaising and getting what he needed, not just for himself but also for his department. Williams, on the other hand, was not interested in committees and liked to be on as few of them as possible, though he was very quick to assess and sum up all the arguments when he did have to be on one. His great strength was his technical ability; he had brilliant inventive skills and a seemingly intuitive understanding of electrical and mechanical devices. His knack of quickly grasping the essence of a problem and promptly solving it was second to none. He liked nothing better than busily working in his laboratory.

With Williams gone, a replacement was required to give the electronics courses, and Raphael Feinberg was appointed as an Assistant Lecturer. An expert on ignitrons, he was a refugee from Hitler's Germany and possessed a Dr.Ing. degree from Karlsruhe. Four electrical engineering students were awarded "Firsts" in 1939, including Ron Cooper, later to have a distinguished career in the Electrical Engineering Department at Manchester. With the political situation in Europe tense to say the least, there was, by the summer of 1939, an air of great uncertainty about the future.

The electro-technics course at the end of the 1930s

The course had not changed in basic essentials from that described in the previous two chapters. It retained its former structure, wherein much of it comprised civil and mechanical engineering subjects. The electrical courses still had the same titles as those listed in Appendix 9. Any changes had been by evolution; i.e. by updating the syllabuses whilst keeping the same course titles. Jackson, when he arrived, had no wish to initiate a major restructuring of the course; he was content with the policy of gradual revision.

At that time there were no electrical engineering lectures in first year, but apart from the lectures on civil and mechanical subjects, there were courses in Pure and Applied Mathematics and Physics. The Physics course was still taken by Dr Sutherland, as had been the case for several years. Professor Gibson gave the 'Strength of Materials' course. He was a very good lecturer and was always polite with his students. He would enter the lecture theatre followed by a huge cloud of smoke from his pipe, which he would extinguish carefully and say "Good morning, gentlemen". He called the students "Mr" when using their names, as did some of the more traditional lecturers years later, and was in every way kind and helpful to students. Professor Beattie had given the second year electrical courses until his retirement, after which the duties were re-distributed, and the four Electrotechnics lecturers gave their specialist third year courses. Several 'civil' and 'mechanical' courses were available as third year options for the 'Electrical' students. Professor Gibson's 'Hydraulics' course was always popular, partly because he was a good lecturer but also because it was a shorter course than, say, Mathematics, which meant that there were less lecture notes to revise. Ron Cooper, who was a final-year student in the 1938-39 session, recalls that Williams gave his final-year 'Higher Frequency Currents' course at breakneck speed because he knew well in advance that he would be leaving at Easter, 1939.[3] Willis Jackson gave the third year 'Higher Magnetism and Electricity' course and, according to Cooper, was an excellent lecturer. Gerrard still gave the third year 'Electrical Machines' course and Higham the 'Generation, Transmission and Distribution' course. Higham also gave an electrical engineering course to non-engineers. There was provision in the time-table for Saturday morning lectures, but most lecturers avoided them by giving extra or longer than normal lectures from Monday to Friday.

The start of the war

Everyone's worst fears were realised when, on 3rd September 1939, Britain was once again plunged into war, 21 years after the end of the previous conflict. Readers will recall that in the First World War the Electrotechnics Department was left in the sole charge of Robert Beattie and almost ground to a halt as there were practically no students other than a few from overseas. This time things would be very different. The importance of technical supremacy was realised. Radar stations were operative when the war began and would play a crucial role in the Battle of Britain, and eventually it would be realised how important was the code-breaking work at Bletchley Park. The authorities knew that electrical engineers would be a vital part of this "technological war" and consequently the department was allowed to maintain its student intake. The Vice-chancellor reported in July 1940 on the general situation regarding call-up and deferment:

"The position of students has been much happier than was the case in 1914, when each individual had to make the difficult decision as to whether he should enlist or remain at his studies. On the whole, the students have been treated most fairly and sympathetically, and in their case there has been a real effort to use the man power of the country to the best advantage in the national interest. They have been given guidance as to the right course of action to take in order to support in the best way the needs of the country. Certain groups of students, who have completed a full year in the University, have been reserved and told they must complete their course, when, as trained chemists, engineers, and the like, they will be of greater service to the country in industry and in certain branches of the armed forces. The others are expected to pursue their studies until their age group is called up, and postponement has been allowed when a critical examination is to be taken within a reasonable time. Council and Senate have made special provisions for those who have to interrupt their studies and everything possible is being done to mitigate hardship." [4]

As events turned out, many talented people graduated in electro-technics during the war. Moreover, a lot of important research was done in the department during the war years, as will be described later in this chapter.

Tragic events in the Beattie family

Professor Beattie, now retired, was still living in Altrincham when the war started but his retirement in the autumn of 1938 was the prelude to a tragic period for the family. In the summer of that year his wife, Janet, had been diagnosed as having cancer. She was operated on in October 1938, but the operation was unsuccessful; her health never recovered and she died at the family home at 5, Hale Road, Altrincham, on 16th February 1939. Robert Beattie's own health began to deteriorate following the death of his wife and the symptoms of his Parkinson's disease became more marked. Meanwhile, his son, Robert, had enrolled for his Ph.D. in the department following the award of his M.Sc. in 1938. He was one of Williams' research students until the latter's departure at Easter 1939. Robert's sister, Janet, having sat her Higher School Certificate in the summer of 1938, had completed her first year as a History student at Manchester University in the 1938-39 session but following her mother's death she decided to transfer to Medicine and enrolled in the first year of the medical course at St. Andrews University in September 1939. The move to Scotland was because the medical course there was a year shorter than the one at Manchester, which she felt was an important factor, having 'wasted' a year (as she saw it) studying history.

1938-1946: Willis Jackson and the war years

During the summer vacation of 1939 young Robert Beattie, along with other young scientists, was sent to a Government Communications Establishment "somewhere on the East Coast", and, when the war started in September, his university research equipment was confiscated by the Post Office authorities as being a possible security risk. He was then sent home, only to be ordered to report to the Air Ministry Research Establishment in Dundee. He spent the first couple of months of the war at Scone near Perth, where there was a small airfield and scientific work was being done. Then, during that autumn, he was transferred to Barry Airport in South Wales, where a team was to work on the radio systems for the Fairey Battle bomber. He settled in with a local family who treated him very kindly. The three members of the Beattie family spent the Christmas of 1939 together in Altrincham; Janet then returned to her university hostel in Dundee and young Robert returned to Wales. Then suddenly, a few days later, Janet received a telegram from her father saying that Robert had been killed in South Wales. Several of the young scientists had gone out on a proving flight in an R.A.F. plane and the pilot had flown into the side of a mountain, killing everyone on board. The accident had happened on 7th January 1940 at Nant-y-moel Mountain, 20 miles north-west of Barry. Robert was 24.

Although the weather was bitterly cold and his health far from good, Professor Beattie insisted on going by himself to South Wales. The R.A.F. had made all necessary arrangements after the inquest and the coffin, escorted by an R.A.F. officer, was sent by train to Montrose. Professor Beattie travelled on the same train. Robert was buried in his mother's grave in Montrose. His father insisted on attending at the graveside in spite of the wintery weather. He remained in Scotland, in failing health, and, before the month was out, he died. His death took place on 25th January 1940, at the age of 66. He was buried in the family grave.

In these sad circumstances Janet had lost her whole family (mother, brother and father) in just over a year. She bravely put her energies into her medical course and duly qualified as a doctor. She has kindly given permission for the above details to be included in this book.[5]

Soon afterwards Janet received the following:

The Victoria University of Manchester

Copy of a resolution adopted at Meetings of the Council and of the Senate in February, 1940:

"That the Members of Council and Senate have heard with deep regret of the Death of Professor Robert Beattie, who, for some forty years, served the University, firstly as Lecturer, and from 1912 to 1938 as Professor of Electro-technics.

They remember with gratitude his services in developing his Department, and as a Teacher and Administrator during this long period. The success of his work is manifest in the careers of so many of his students, whose achievements and success in after life have been due so largely to his influence and training.

The members of Council and Senate offer to his Daughter their most sincere sympathy.

The letter accompanying the above was dated 2nd February 1940, and was signed by the Vice-Chancellor, Professor John S.B. Stopford, F.R.S.

Several people who knew F.C. Williams well have said that he always felt a sense of remorse about the death of his young protégé, as he had a hand in encouraging him to enter the wartime communications work in which he lost his life. But accidents happen in wars; indeed, they also happen in peacetime, and no-one could possibly do anything to prevent such tragedies occurring without the benefit of hindsight.

The early war years in the Electro-technics Department

The internment of Feinberg

Soon after the start of the war, Raphael Feinberg, the department's Assistant Lecturer, was interned as an alien, together with his wife, and incarcerated in the Isle of Man, along with countless others. The authorities had no reason to suspect him of not supporting the Allies' cause, but they did not have time at that stage of the war to decide which people from Germany were with us and which were against, so there was a blanket imprisonment of everyone classified as aliens. The Vice-chancellor and Professor Jackson made representations on Feinberg's behalf and in due course, when the authorities had satisfied themselves that he was indeed an innocent refugee victim of Nazi Germany, he was released and resumed his duties. He was back in the department by the latter part of 1941.

Temporary shortening of the B.Sc. course

At the end of 1941 it was decided to shorten the B.Sc. Engineering course, as a temporary measure, by abolishing the long vacation, whilst retaining the full content of the course together with the 9-term structure. The course would therefore last two years and one term. The change was brought about, with Britain then in its darkest hour, by the need to produce well-qualified engineering graduates as quickly as possible. The detailed proposals were set out in the Minutes of the meeting of the Board of the Faculty of Science held on 24th February 1942:

(a) "The Registrar and Professor Gibson, in anticipation of instructions from the Ministry of Labour, recommend that the following regulations for the Honours School of Engineering be approved to remain in force during the period of the war or until such

time as may be decided by Senate:
(i) That from September, 1942, the University session shall consist of four terms (the present three terms plus a term during the long vacation).
(ii) That the first-year course shall commence in September and conclude at the end of the third term (June). The Intermediate B.Sc. examination will be taken at this stage.

That the second-year course shall commence at the beginning of the fourth term (July) and continue until the end of the sixth term (March). The Part I examination will be taken at this stage.

That the third year course shall commence at the beginning of the seventh term (April) and continue to the end of the ninth term (December). The Part II examination will be taken at this stage.

That the present regulations of the Honours School of Engineering be modified, in so far as it is necessary, to comply with the foregoing scheme.

Approved and forwarded to Senate

It was noted that it was intended that students who do not satisfactorily complete the first three terms, should not be allowed to continue and that the same provision should apply to students during the second three terms.
(b) It was reported that it might become necessary to institute a similar arrangement in other departments in the Faculty.

Resolved: That a Committee consisting of Heads of Departments, together with one other member of staff from each department for consultation, consider possible arrangements and report to the Board of the Faculty.

Similar arrangements to the above were made in the Faculty of Technology but conditions differed in other Faculties. In Medicine a quota system was imposed by the Ministry of Health which allowed men and women to complete their full course provided certain conditions were observed. Men in the Faculty of Arts were normally only permitted to stay for three terms, but women, with a few exceptions, were allowed to stay for two years and those accepted by the Board of Education were permitted to stay for the Teacher's Diploma course in addition. This was because of the shortage of teachers, with many male teachers away in the armed forces.

Clearly, in Electro-technics, the new arrangement would put heavy extra teaching demands on the staff, especially as it was decided by the university that all students of Physics must take a course in radio. An extra full-time appointment was out of the question in wartime but Professor Jackson managed to secure a number of part-time appointees to augment the department's resident staff of four (Jackson, Gerrard, Higham and Feinberg), thus enabling it to to carry out its extra duties. By the start of the 1942-43 session the staff was boosted by the appointment of Dr Thomas Havekin (who had graduated from the department in 1922 and obtained his Ph.D. at Birmingham in 1935) as a Special Lecturer and by three Demonstrators, Peter Dénes, A.C. Normington and Waldemar Rosenberg, to help with the laboratory work and to give some lectures. Dénes (a refugee) and Rosenberg had graduated in the Department, in 1941 and 1942 respectively, and had stayed on to do research, and Normington was one of the research staff at Metropolitan-Vickers (later Head of Bolton College of Technology). By October 1944 Havekin had been joined as a Special Lecturer by J.M. Meek, a senior research engineer at Metropolitan-Vickers who was an expert on gas breakdown. (He was Professor of Electrical Engineering at Liverpool University from 1946 to 1978 and President of the I.E.E., 1968-69). Normington was still in post as a Demonstrator but Dénes and Rosenberg had been replaced by G.C. Collins, C.T. Baldwin and C.J. Braudo, all research engineers at Metropolitan-Vickers. John Durnford, another man from the same company (later a staff member of Liverpool University), also helped out. Professor Jackson's association with the local company had been very valuable, enabling as it did local talent to be brought in to maintain the department's teaching through a difficult period. Others who helped by giving lectures at various times were N.C. Stamford (a Communications Lecturer at the College of Technology) and Dr T.S. Littler, a physicist who was a Lecturer in the Department of the Education of the Deaf.

The 2¼ year course began in September 1942, as planned, and its operation was assessed in the Vice-chancellor's report in the summer of 1943.[6]

> "The most notable change to report is the introduction of a 4th term during the summer months of July, August and early September. Whilst it is fully realised that this extra term imposes a heavy strain upon everyone and is open to criticism on educational grounds, it became clear that in the present circumstances it was the best thing to do both in the interest of the students and in the national interest. The extra term has permitted the scientists to complete the full course of nine terms in the limited time now allowed and has enabled those who are to become teachers to attend the complete graduate and diploma course in three years instead of four. In practice the introduction of a fourth term has fully justified itself, and in spite of many difficulties which had to be overcome, has worked well and remarkably smoothly. One result of the speeding up of courses has been the need to hold additional degree ceremonies at unusual times. For the first time one had to be held early in May, at which a large number graduated and reduced the numbers at the ceremony in July.
>
> Another result has been the reduction of activities in the unions. Fortunately the curtailment has not been severe and most students have been able to enjoy reasonable social and recreational facilities."

Many talented students graduated in Electro-technics under the shortened-course regime. In spite of the Vice-chancellor's remarks about degree days being held at unusual times, some students (from various departments, not just Electro-technics) were not offered any degree ceremony at all. In 1940, for example, degree ceremonies were cancelled for

1938-1946: Willis Jackson and the war years

fear of the slaughter that would ensue if the Whitworth Hall were bombed whilst the graduands and their families were assembled there. One of the people who should have attended a ceremony wrote to the university 50 years later and asked if one could be held, and as a consequence the university arranged a special ceremony on 21st September 1990 at which wartime graduates could, at last, have their degrees conferred, if they wished. Eighty-two of them attended.[7] John C. West, who graduated in Electro-technics in the department during the shortened-course regime and subsequently enjoyed a distinguished career in four different universities, said that he never received anything to show that he had obtained his degree, other than the receipt he was given when he paid his 8-guinea graduation fee.[8] The university did not forget to collect its fees, even in wartime!

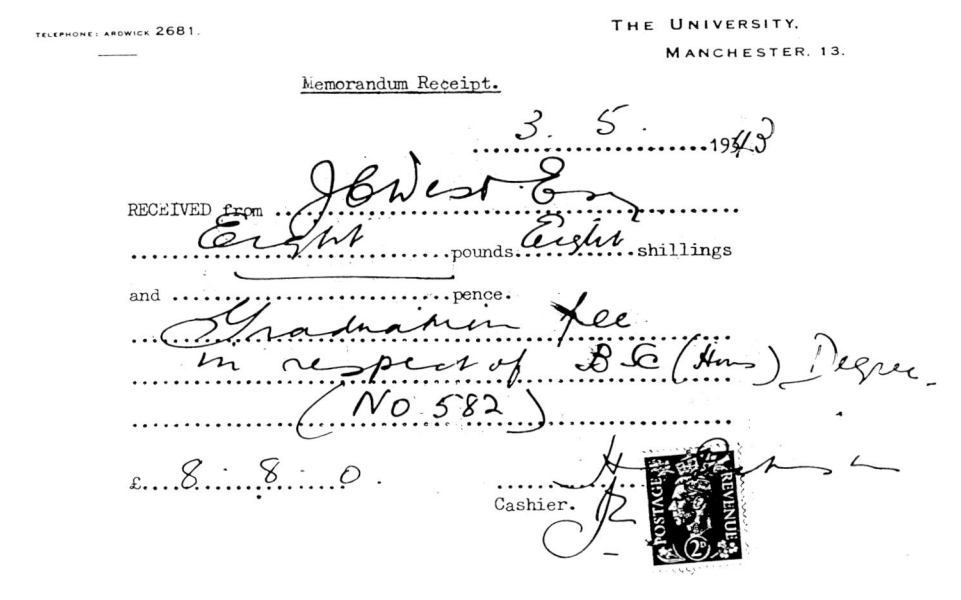

Professor J.C. West's receipt for his graduation fee. This was all he was ever given to prove that he had graduated.

The introduction of the new courses necessitated a lot of extra work for Joseph Higham who, over the years, had taken on a steadily increasing proportion of the department's day to day adminstration. As the department's quota of staff at that time was one Professor, one Senior Lecturer, one Lecturer and one Assistant Lecturer, it was not possible for Professor Jackson to procure a promotion from Lecturer to Senior Lecturer for him, but in 1942 Jackson arranged for Higham to be given an additional title. He became "Lecturer and Assistant Director of the Laboratories", a title he retained until his retirement.

Electro-technics research in the war years: "Jackson's Circus"

Between 1940 and 1945 an active research group of young engineers operated in the Electro-technics Department under the direction of Professor Willis Jackson and made a significant, but unsung, contribution to the war effort. One of the group was Ron Cooper, who later (in 1955) became a member of staff of the department. He was present during part of 1940 and continuously from the beginning of 1941 until 1946. The group was affectionately known in the service research establishments, with which there was close collaboration, as "Jackson's Circus". Much of the following account has been built up from information related by Ron Cooper, who is better qualified than anyone now alive to describe the wartime events. [9] Additional information was supplied some years ago by his namesake, the late Albert Cooper, who was the department's Chief Technician throughout the war. [10]

Conscription and deferment

When war was declared every man in the country within a certain age range became liable for military service, soon to be joined by young females for the ATS, WAAF, WRENS, etc. Groups according to age were called for medical examination. Those not medically fit were rejected but the rest were deemed available for military service and calling-up papers were sent unless the individual was in a so-called reserved occupation; for others the employer could appeal on the grounds that the person was doing "work of national importance" and deferment of call-up might be granted. Calling-up papers gave instructions about when and where the individual must report, and if he failed to do so the military police came and took him. The system on the whole was efficient and reasonable. Metropolitan-Vickers and similar companies carried a lot of weight in procuring deferment for obvious reasons, as did the coal-mining and agricultural industries. In the "technological war" just beginning it was clear that many young scientists and engineers would be more useful serving their country in the technological field than in armed combat.

During the war, the work of Cooper and many other young graduates like him was such that they were classed as "reserved". Cooper said that there was a permanent "tug of war" between the military authorities who wanted young scientists in the services to operate and maintain their technical equipment, and the various Ministries who wanted them to do important technical work in research establishments. Thus it was a question of the "best use" of a limited number of people with special knowledge, and, in the period 1942-43, controversy always raged in high places about the most efficient use of this special group. Their call-up for the armed forces would have cut down the training time of those operating the equipment and thus speeded up the deployment of radar sets then coming out of the factories, but it would have restricted further development. Cooper recalled that when calling-up papers were received the procedure was to go for the 'medical', which usually took place in the Dover Street High School, after which, on instructions from the employer, "do nothing". The authorities made the necessary arrangements for the granting and renewal of their deferment. Cooper said that he received calling-up papers several times during the war, each time ordering him to attend a medical. When this happened he had to attend it and then, on instructions he had been given, phone a certain number in Whitehall and tell them what had occurred. Nothing more would then happen for a few more months, after which the process would be repeated. Cooper said that as the war went on he and his colleagues "got used to it".

The build-up to the department's wartime research

At the start of the war no-one knew what was going to happen or what the department would be allowed to do during the war, but Willis Jackson, whose line of research was the properties of dielectric materials, was keen to get some nationally-useful research going with whatever support could be found. Ron Cooper, having graduated in 1939, had gone to Metropolitan-Vickers as a College Apprentice (their apprenticeship scheme at that time was world-famous), and was soon promoted to Research Trainee. When the war started he went to the university's recruiting office with a view to joining the Navy, but, completely by chance, bumped into Professor Jackson on the way in. On being told by Cooper why he was there, Willis Jackson told him off and said he would be of more use to the country by doing technical work. Jackson then phoned Tom Owen, who ran the M-V apprenticeship scheme, to arrange Cooper's starting date at M-V for the next day, if possible, and failing that, for not later than the following week, and instructed Cooper to get down to Trafford Park to see Mr Owen that very day.

Consequently Cooper remained at M-V for the time being, temporary deferment having been granted, but was soon back in the Department for a short spell. This came about because Willis Jackson, through his close ties with M-V, had organised a scheme whereby a small number of College Apprentices or Research Trainees could come to the Department, one at a time in succession, on a three-month stint to carry out some particular technical task as part of their research training. Each one took over at the point where the previous one had left off at the end of his three months' period. Cooper took over from Martin Oliver (later to become a Principal Scientific Officer at R.R.E., Great Malvern) and built a Hartshorne-Ward measurement set, and then, around June 1940, helped Willis Jackson with some research work on polythene. A.E. Chester, who was a graduate of the department (B.Sc. 1937; M.Sc. 1938) and a conscientious objector (he drove ambulances at night as his war effort [11]), was already in the department doing research on vacuum breakdown. At that time polythene was a brand-new material and was clearly of importance as a dielectric and Jackson arranged for a man name Tipping, who was a chemist, to come to the department as a Research Fellow to tell the research students about bakelite and other dielectrics and, if possible, to "brew up" rare dielectrics.

Dr John Cockroft was then Director of the Air Defence Experimental Establishment (ADEE) at Christchurch, Hampshire *, and in July 1940 he visited Jackson at Manchester. He spent some time talking to Cooper about the dielectric research, the Hartshorne-Ward set, and related matters. At the end of this he asked Cooper if he would like to join him at Christchurch on "a most interesting and nationally-important job." At that time the country was gripped in what Cooper called a profound secrecy phobia, and Cockroft would give no more information about the nature of the work, but Willis Jackson advised Cooper that it would be a good move if he went to Christchurch. Cooper arrived in Christchurch as a Scientific Officer in August 1940 "just in time for the Battle of Britain", as he remarked, when there were daily aerial dogfights over the Hampshire countryside.

Jackson assembles a small research team

Between August and December, 1940, Professor Jackson was very active in travelling around the country and having discussions with influential people with a view to setting up research in the department on a more organised basis. The result was that he won a grant from the Ministry of Supply to fund a small research team. Thus, from the beginning of 1941, the department's wartime effort was placed on an official footing. The arrangement was that Jackson could select two students from each set of graduates, year by year, to stay on and do research, but there were never to be more than about six in all at any given time.

In the autumn of 1940 Professor Jackson visited the research group at Christchurch and invited Cooper back to Manchester, with Cockroft's permission, to rejoin his research group. Cooper returned to Manchester on 23rd December, 1941 (the day of the great air raid) and on the 24th went to the department to see if it was still there. At that time the only holiday was Christmas Day and Boxing Day. His official transfer to the department,

---

* Re-named Air Defence Research and Development Establishment in 1941. It moved to Malvern in 1942.

on secondment from Christchurch, took effect on 1st January 1941 and he remained there for the rest of the war. But he was paid by Western Command (Radiolocation Experimental Establishment), which was the army's anti-aircraft radar organisation, throughout the war and was not part of the above "two graduates per year" arrangement. Thus, Jackson's team from the beginning of 1941 consisted of Cooper and four or five other research students, the latter personnel changing as the years went by.

Jackson's research team had no particular terms of reference. When Cooper came back to the department at the beginning of 1941, Jackson took him for toast and a cup of tea at Smallwood's café on Oxford Road and afterwards, in Jackson's office, they talked about the work going on at Christchurch, and possible developments. It was agreed that work should start on the development of hollow pipe technology for it could not be assumed that there would be appropriate developments in solid core coaxial cable technology to cope with the anticipated increase in radar power and frequency. Before 1900 Kelvin and others had shown that e.m. waves could be propagated down hollow metal pipes of practical dimensions provided the operating frequency was high enough. It was not until the late 1930s that suitable oscillators became available and at the end of 1940 almost no practical work had been done to develop the technology. So, in Cooper's words, "Albert, to his amazement, was asked to order enough brass pipes to start a euphonium factory", and research began in earnest. The outstanding scientific work which followed concerned the behaviour of junctions and the work was published at the end of the war. The authors were T.G. Cowling (Lecturer in Mathematics, who had acted as a 'mathematical consultant' to the research group), R. Cooper, and J.T. Allanson.

Cooper found, on his return to Manchester, that a contemporary 1939 graduate, Fred Horner, was already there, as was N.C. Stamford from the College of Technology. There was never a specific directive as to what anyone should do. Cooper explained that Jackson would sit on technical committees, meet technical people, find out what the problems were and what projects would be needed to solve them, and then decide which specific problems it would be possible for the department to tackle. It has to be remembered that resources were strictly limited. The department possessed only its pre-war conventional equipment, so if anything more was required it had to be begged, borrowed or made in the department or in the Engineering Department under the benign guidance of Jim Bamber, their Chief Technician (or Laboratory Steward, the title used at the time). Jackson, with his persuasive powers and wide range of contacts, was very proficient at inducing other organisations to donate or loan equipment, and managed to procure by one means or another much of what was needed, especially klystron oscillators, magnetrons, reflection oscillators, etc. The only technical back-up within the department was from the technicians, Albert Cooper and Joe McCormick. Ron Cooper recalled that, in connection with the cavity resonator work, Albert would make any pattern necessary for the phosphor bronze casting; it had to be turned very precisely and then sent to be silver-plated. Joe used to make wooden formers on which to wind coils, and between them the two technicians would do most of what was necessary.

The research team and the specific projects

Polythene had become available around 1938 and was a material of great importance as a dielectric for cables. Some of Cooper's early work at the time was on the electrical properties of polythene and polythene-insulated cables. Jackson's contacts with industry were once again invaluable, as he was able to obtain samples from ICI at Northwich. Besides the experiments on polythene alone, a lot of work was being done on the measurement of characteristic impedance, attenuation coefficient and phase shift coefficient of extruded polythene-cored coaxial cables of the type used in many radar applications. This work was of particular interest to cable manufacturing companies from which a number of people were sent to the department to learn about the techniques. Later, Cooper worked on the dielectric properties of sea water, which was relevant to the radar work. A lot of work with waveguides then followed - none of it published until the end of the war because of the 'secrecy' obligation.

Fred Horner left at the end of 1940 or early 1941 but came back later in the war in connection with the cavity resonator work. After the war he became Director of the Appleton Laboratory. He was replaced in the research group in 1940-41 by "Tommy" Taylor. Most of those who were part of Jackson's wartime research teams obtained an M.Sc. degree and a few gained their Ph.D. in circumstances that will be explained shortly. The following list summarises the main personnel involved in "Jackson's Circus". The dates in brackets give their period in the research team. Degrees are quoted when obtained in the department:

R. Cooper: (1940-46). B.Sc. 1939; M.Sc. 1941; Ph.D. 1946. Worked on various projects including the electrical properties of polythene and polythene-insulated cables; the electrical properties of sea water; a lot of work involving waveguides and waveguide junctions, etc., and the electrical strength of air at pulsed microwave frequency.

N.C. Stamford (1939-40): Aerial polar diagrams; attenuation and phase shift.

F. Horner (1939-41): B.Sc. 1939; M.Sc. 1941. Dielectric parameters using cavity resonators. Returned later in the war from NPL with J.A. Saxton to find out what the group had been doing since his earlier departure.

R.B. Quarmby (1940-43, approx.): B.Sc. 1940; M.Sc. 1941. Similar work as Stamford's,

and also worked on coaxial cables and helped Cooper with his sea water work.
T.A. Taylor (1940-42): B.Sc. 1941; M.Sc. 1944. Measurements with cavity resonators.
J.T. Allanson (1942-44): B.Sc. 1942; M.Sc. 1949. Waveguide measurements; permeability of iron at microwave frequencies. He was the N.U.S. President, 1942-44.
G. Saxon (1943): B.Sc. 1939; M.Sc. 1942. (Seconded from M-V research labs.) Dielectric properties of salt solutions; various microwave measurements.
J. Lamb (1943-46): B.Sc. 1943; M.Sc. 1945; Ph.D. 1946. Cavity resonators; attenuation of supersonic waves in liquids.
R. Dunsmuir (1943-46): B.Sc. 1943; M.Sc. 1945; Ph.D. 1946. Radiation from open-ended waveguides and electromagnetic horns.
W. Reddish (1944-46): B.Sc. 1942; M.Sc. 1945. Measurement of dielectric properties at high frequencies; high permittivity ceramics.
K.W. Plessner (1944-46): B.Sc. 1944; M.Sc. 1945; Ph.D. 1947. (Karl Plessner, like Feinberg, was a refugee from the Nazi regime). Measurement of permittivity, power factor, etc. in dielectric materials (especially ceramics) at microwave frequencies.
W.G. Oakes (1944-46): (Seconded from ICI). Electric strength of polythene.
J.G. Powles (1944-46): B.Sc. 1945 (but the course was completed in '44); M.Sc. 1946; Ph.D. 1948. Cavity resonators; high-permittivity ceramics at microwave frequency.
C.A. Whalley (1944/5-1946): (Seconded from TRE). Electric strength of gases (a continuation of Cooper's work).

In addition to the above, many people came from industry for short periods to learn techniques. A few important people were regular visitors. E.B. Moullin from Oxford University, for example, was seconded to Metropolitan-Vickers during the war and often visited the department to discuss the research work, and sometimes to give lectures. The work had its lighter moments. Ron Cooper claimed that it was he who invented the microwave oven. This came about because the department's errand boy used to be sent out to fetch hot "meat" pies (mainly potato and Oxo, according to Cooper!) from the nearby café for the research workers' lunches. By the time he got back the pies had generally gone cold, so Cooper and his colleagues used to use their microwave equipment to heat them up.

Organisation of the work

The organisation and co-ordination of the research effort was complex. Various bodies set up committees as appropriate to consider matters of urgent concern. The Ministry of Aircraft Production had its committees and two were concerned with cables; one considered scientific aspects of problems, the other was a commercial committee discussing matters such as how best to manufacture polythene cables. Both committees were attended by someone from Manchester, usually Willis Jackson, or Cooper, or both. There was a committee on waveguides and also one concerned with radar beam scattering by raindrops and hailstones. Cooper recalls that committees appeared to form more or less spontaneously as problems appeared and die accordingly as they were resolved. The Chairman of a committee would usually be an older, senior, scientist or engineer.

Some of the team register for the Ph.D. degree.

Whilst the Electro-technics Department was involved in its active research programme throughout the war, there was little corresponding work in the Physics Department. The Professor, P.M.S. Blackett, was away for the whole of the war and the teaching was maintained by stalwarts such as L. Jánossi, M. McCaig, J.M. Nuttall, G.D. Rochester and S. Tolanski. The differential analyser was still working under Professor Hartree's direction and a small Ministry of Supply team did calculations on electromagnetic phenomena inside a cavity magnetron, but other than that the department was fairly quiet. In the Electro-technics Department no thought had been given to the possibility of any of Jackson's research team being registered for Ph.D. degrees; the whole purpose of the research effort was to help win the war. However, Professor Jackson discovered, around the middle of the war, that one of two research students in the Physics Department who were conscientious objectors were engaged in cosmic ray research and were registered for Ph.D.s. Jackson, who always looked after his students very well as individuals and had a commendable sense of justice, did not think it right that his students should not be able to get Ph.D.s if the Physics research students were getting theirs, and promptly made arrangements for a small number of his current team to be registered for Ph.D.s.

The regulations for the degree at that time stated that a candidate "may be required to take certain special courses of instruction in departments other than the one mainly concerned with the course, as well as in the subject in which he proposes to pursue advanced study or research". Cooper was one of those who registered for the degree and took lectures in Physics and in Economics and Industrial Administration. Theses were not submitted until the war had ended. Ph.D. degrees were then duly awarded to R. Cooper, R. Dunsmuir and J. Lamb (all 1946), K.W. Plessner (1947) and J.G. Powles (1948). K.M. Entwistle, who graduated in Electro-technics in 1945 and was a contemporary of Powles, moved to the Metallurgy Department immediately after graduating and obtained his Ph.D. in that subject in 1948. Subsequently he became a staff member in the Metallurgy Department and later Professor of Metallurgy at UMIST. Previous to this group of Ph.D. graduates, the only Ph.D. awarded in the Department had been that of A.E. Chester, a B.Sc. graduate of 1937, who obtained his Ph.D. in 1942 on a project (electrical breakdown in vacuum) started before the war. Members of Jackson's wartime research team who had already left before the opportunity to register for a Ph.D. was introduced were, perhaps, unfortunate.

1938-1946: Willis Jackson and the war years

The end of the war

In August 1945, after the dropping of the atomic bombs on Hiroshima and Nagasaki, the war came to an abrupt end and with it the department's wartime research contribution. During the war the department's pioneering work, mostly done in support of the country's radar effort, had been of the highest standard and it attracted many "brain picking" visitors from industry and from the U.S.A. It was an outstanding group of people. Four of the team became Professors in British universities (five if a visiting professorship is counted), one became a Reader, and the rest achieved high positions in industry. Ron Cooper maintains that much of the work done in the department during the war was outstanding and says he is proud to have been involved. He feels that although Willis Jackson received some credit for his war work, the group as a whole, "Jackson's Circus", never got the recognition it deserved.

Fire watching

A discussion of the war years would be incomplete without a mention of the fire-watching duties in the university. Manchester suffered many severe air raids and it was clearly necessary to take all possible steps to protect university premises, particularly against the effects of the numerous incendiary bombs. The university operated a small fire brigade which was run by the staff of the Engineering Department, it being thought they were best fitted to maintain and operate the two rather rudimentary Coventry Climax pumps. The smaller of these had to be manhandled, but an ancient Lanchester car was available to tow the other. Practices were held periodically, the main effects of which were to douse everyone in water. Fire squads were set up in which all the the university's lecturing staff, technicians and research students were called upon to take part. A hierarchical system operated in which the head of the squad was a professor, a chief technician would be his deputy, and the research students were at the bottom of the pecking order. Once every six nights each individual would stay all night. The participants would sign on and don overalls. The Government provided a subsistence allowance and this was just enough to pay for a supper and a breakfast provided in the Refectory. The breakfasts were said to be excellent despite food rationing.

As people from several departments were on duty at any given time there were many interesting discussions throughout the night. Cooper recalled long chats with Professors Thompson (Metallurgy), Mordell (Mathematics), Wood Jones (Anatomy) and many others. Fire-watching was thus a social and mind-broadening exercise. Cooper recalled that there were never any serious fires when he was on duty. The only incident of note occurred when the fire brigade telephoned one day to report that they had put out a fire in the Engineering Department and instructed his fire-watching squad to go and clean up the resulting mess. The fire was not a result of enemy action but was caused by hot slag from a welding operation falling onto rubbish.

The university was very fortunate in suffering relatively little damage during the wartime raids, except that many of the glass roofs in the older buildings, e.g. Alfred Waterhouse's building, were frequently shattered by shrapnel. The surrounding area was not so fortunate. A land-mine fell in Hulme, behind the university, killing many people and making a lot more homeless. Albert Cooper recalled that on the night after the raid the university opened the Medical School building to the homeless and there were hundreds of people sleeping on the floors. All Saints Church, built in 1820, was damaged beyond repair in the blitz of December 1940 and was demolished at the end of the war. It was never rebuilt.

Evacuation of children

Another wartime feature worthy of brief mention is that in the early years of the war a scheme was set up whereby the children of university staff could be evacuated to Canada for the duration of the war if the parents so wished. Many staff took up the offer. Tim Henry, for example, a former member of staff in Mechanical Engineering and son of D.C. Henry, Reader in Colloid Physics, was evacuated to the Toronto district at the age of four with his elder brother and remained there until the war was nearly over.

A new high-vacuum laboratory in the Electro-technics Department

In 1944 it was decided to clear the dynamo house of its machinery and to install modern high-vacuum equipment there. This came about largely because of the shortage of people with expertise in valve technology and the consequent need for a centre where training courses could be held. The laboratory, which was partly financed by industry, was thus set up not only for the needs of the Electro-technics Department but also those of the Physics Department and of people from industry who were sent to attend courses to learn the latest technology. So, after a history of about 44 years, the dynamo house (see pictures on page 61) was put to a new use. Most of the machines taken out were by then very old. Albert Cooper recalled that one of them had a "Royce" starter, made by the famous Manchester engineer. The high-vacuum laboratory was used until the building was vacated in 1954 and was then set up in the new building in Dover Street, where it remained until the mid 1980s, though its usefulness tailed off after the late 1970s.

Mid-war plans for the post-war era

Even in the darkest days of the war, thought was being given to ways in which the university could be improved after the war. In his Annual Statement of 1942 the Vice-chancellor, John Stopford, reported:

" . . . From what I have stated [a review of university work during the war] it will be clear that both staff and students are very fully occupied but this has not prevented us giving thought to post-war reconstruction and development. We wish to be prepared and avoid the mistakes which were made during the period of reaction and excitement succeeding the last war, and we believe it is good for our souls to think about future policy at a time when the opportunities for development and expansion are so restricted. The Senate has received a first report and after general discussion remitted the whole subject to a Committee of Deans to be fully thrashed out. This Committee has had a number of most interesting meetings and many fundamental points of great significance and importance are under consideration . . . So far these discussions have been confined to the teachers in the University and in wartime we cannot expect to get much help from those outside our community with special knowledge and experience in industry, commerce and the professions. We very much hope that at a later stage we may have the benefit of advice from our many friends in these spheres. It is realised that such advice is of supreme value, particularly in certain fields. Engineering is a good example where we have two large schools, one in the Faculty of Science and the other in the Faculty of Technology, situated in the midst of perhaps the largest engineering centre in the world. With these resources and a much closer association between the two departments and industry it ought to be possible for us in Manchester to create the finest engineering school in the country. What has been said with regard to engineering is applicable to several other subjects and we are determined to take full advantage of our exceptional resources and opportunities." [12]

It appears from the above remarks that consideration was being given to either amalgamation or close collaboration between the Engineering Department and the corresponding department at the College of Technology, along the lines of the discussions that took place on the same subject in 1920, as reported on pages 118-119. However, nothing had come of the earlier plan and it appears that the same applied in 1945, as the Engineering Departments in the Faculties of Science and Technology continued to operate much as before, except that both steadily expanded. After 1946, it would be another 42 years before the matter was raised again.

Events in the Physics Department after the war

When the war ended, Professor Blackett returned to the Physics Department and a great flurry of activity began there. One of the early arrivals was A.C.B. Lovell, accompanied by a large 'horse box' containing an ex-service radar set, with which he intended to detect radio emissions from heavenly bodies. Lovell found there was so much interference in the university environs, some of it emanating from Cooper's 30A, 30kV, ½MW modulator, that eventually a plot of land at Jodrell Bank belonging to the Botany Department was acquired, thus founding the now famous research station. Blackett intensified his cosmic ray research and was awarded the Nobel Prize for Physics in 1948.

Electro-technics after the war and the departure of Willis Jackson

In the Electro-technics Department research work was still continuing with great vigour when the war ended. No-one knew what the future held, but as the work the team had been doing for five years was all war-related it was obvious that it could not continue much longer and members of the group began to wonder about their next job. Consequently, the completion of experiments and the writing of theses and papers for publication took priority and, as a result, the vigour departed from the practical work. This continued until the summer of 1946. Professor Jackson then announced that he was leaving in order to become Professor of Electrical Engineering at Imperial College, London, and invited all his team, and the two senior technicians, to come with him. It is not difficult to understand why Willis Jackson wanted to move to London. He was a great committee man who knew everyone in his field, and much of his work centred on London where most of the meetings were held. Moreover, many of his activities were associated with the Institution of Electrical Engineers (he became President in the session 1959-60) which necessitated frequent trips to London; this meant a journey of four or five hours in those days.

So, in the autumn of 1946, Jackson's wartime team dispersed. Bob Dunsmuir had accepted a job with BTH Research, Rugby, and Wilson Reddish went to ICI Research Laboratories at Welwyn. Several of the research team decided to accept Jackson's offer to join him at Imperial College, but the technicians stayed in Manchester. In the autumn of 1946 a van departed for London with a large amount of the team's research equipment. Ron Cooper reported that Albert Cooper, the Department's Chief Technician, was "disgusted" because it seemed to him that the whole of the department was disappearing into the van. Ron Cooper and John Lamb left for London by train on 18th September and stayed for the permitted three nights at the Youth Hostel in Euston Road before finding accommodation near Imperial College. Cooper recalled that when they set eyes on the Imperial College premises in Exhibition Road they felt like turning round and going back to Manchester. There had been a lot of bomb damage, most of the windows were blown out and boarded up, the place was dirty and run down and seemed to be completely "dead". During the war it had been kept going through a very difficult period by Professor C.L. Fortescue, who had retired when the war ended.

Bill Oakes, Karl Plessner and Jack Powles made their own transport arrangements and it was not until the end of September that what was left of the Manchester group was reunited at Imperial College. Later they were joined by Stuart Dryden, an Australian, who had been sent to acquire the technologies which had been developed at Manchester and also a Ph.D. if possible, and W.A. (Bill) Prowse, a lecturer on a one year's sabbatical leave from Durham. Cooper was awarded a generous grant by the Electrical Research Association to continue his electrical breakdown studies and also received a fee as a Special Lecturer. Financial arrangements were also made for the others. However, a year later Cooper and Plessner left to take up new appointments at AEI Research Laboratories, Aldermaston, and the remainder of the group eventually dispersed and went their separate ways. After two years at Aldermaston, Cooper went to Liverpool University and thence to Manchester; Lamb and Dunsmuir eventually became Professor and Reader respectively at Glasgow. Powles, who joined Jackson at Imperial College later than the others, became Professor of Physics at Kent, and Reddish and Oakes enjoyed illustrious careers in industry.

Willis Jackson, who was always an excellent manager, had been a very good and effective professor at Manchester from the year before the war started until the year after it ended. He stayed at Imperial College until 1953 before taking up an appointment as Director of Research and Education at Associated Electrical Industries (which incorporated Metropolitan-Vickers). An arrangement was made whereby he could return to Imperial College when he wished, as a professor, which he did in 1961. He was elected a Fellow of the Royal Society in 1953, was knighted in 1958 and created Lord Jackson of Burnley in the New Year Honours of 1967. He suffered a heart attack at his desk at Imperial College on 16th February 1970 (he was then Pro-Rector of the College) and died the next day at the age of 65.

Professor John Lamb's memories of the years 1939 to 1946

In 1991 Professor Lamb of Glasgow University was invited to recall his period in the Electro-technics Department during the war years, and to include his memories of Professor Jackson and of any particular anecdotes relating to that period. In response, he generously wrote an account, though he was very ill at the time it was written, and subsequently died. [13] His narrative gives such an interesting first-hand portrait of the department, and of Professor Jackson and some of the other leading personalities of the years 1939 to 1946, that most of it is reproduced here as an overall "summing up" of the Electro-technics Department in the war years. It is complementary to the earlier account based largely on Ron Cooper's recollections.

"My first contact with Willis Jackson was in May 1939 when I visited Manchester University along with a group of fellow pupils from Accrington Grammar School. We were taken by coach and were given the opportunity to visit the University Departments of interest to us. Along with two or three other pupils I went to Engineering and ended up in the Electrical Engineering Department, which was my personal interest. If I remember correctly the department was designated "Electrotechnology". The professor saw each of us individually for an interview lasting 10 minutes or so and our small group was then taken round the electrotechnical laboratories on a conducted tour.

My first impression of Willis Jackson was that he stimulated interest in young men such as myself and although I had already decided that my future career was to be in electrical engineering, meeting him confirmed my decision. He had a glint in his eye with the pupil of one eye considerably smaller than the other and with his small moustache he appeared to be dapper. I learnt from him later that he grew his moustache to counteract his youthful appearance.

In the following year I took the Manchester University Open Scholarship examinations and collected a scholarship from Hulme Hall for an annual contribution towards residence in Hall. I did have offers from other universities but the £60 per year Beckworth Scholarship from Manchester which supplemented other scholarships plus my favourable impressions of the Department took me to Manchester University as an undergraduate in October 1940 - a step which I have never regretted.

The engineering curriculum in the first year was a general course for all engineers - about 80 in total in my year of whom about 15 were destined for electrical engineering. We were lectured to in Physics by 'Jock' Sutherland and in Mathematics by W.W. Sawyer who was a fine teacher and provided me with an interest in mathematics which I have never lost. He wrote the popular Penguin edition "Mathematics for the Million". I mention these extraneous factors simply to emphasise that the teaching to engineers and scientists at Manchester was distinguished by the excellence of some of the lecturers and professors.

Contact with the electrical engineering department increased by the second year, and by the third year students were deeply immersed in their chosen brand of engineering, with mathematics and one subject chosen from those offered by other branches.

We had electrical machines from Mr Gerrard, transmission of electrical power from Mr Joe Higham, radio and electronics from Dr Feinberg and Peter Dénes (a refugee from Czechoslovakia), and high frequency transmission lines from Willis Jackson. Willis Jackson was a born lecturer who prepared his material thoroughly and delivered it with a flair for attracting and keeping the attention of his class. His writing was round and immaculate though a complete page tended to dazzle the eyesight. We never saw Willis Jackson in the laboratory and quite frankly he was not an experimentalist but somehow had the knack of deciding what experiments were worthwhile for others to perform; this comment applies to research. Strangely he was not intimate with his students - quite contrary to first impressions which I gained as a school pupil.

During my first summer vacation, 1941, he arranged for me to work at Metropolitan-Vickers, Trafford Park - an enjoyable and unforgettable experience but not directly relevant to this account. This proved to be the only 'long vacation' of the war years and during the following summer (1942) we attended university and finally graduated at Easter 1943. Prior to that, all engineering students in their final year were interviewed by representatives of the Ministry of War to determine where they should go upon graduation, i.e. Army, Navy or Air Force. However, Willis Jackson had established a small research group under the auspices of the Ministry of Supply and unknown to me he had requested that I should join his team along with Bob Dunsmuir, a fellow student. Clearly he had anticipated the results of the degree examination.

After a 2-week holiday, I joined the "Ministry of Supply Extra Mural Research" group under Willis Jackson. Other members of the group were Ronnie Cooper, Jack Allanson, Wilson Reddish and for a short time Fred Horner on short-term secondment from the radio division of NPL . . . The following year we were joined by Jack Powles.

Our research was directed to the study of microwave propagation and devices and particularly to the measurement of dielectric materials involved in microwave propagation - polymers, ceramics, magnetic films, water (rain storms) and ice. Willis Jackson arranged the contracts but there were regular meetings in London of co-ordinating groups to which he sent the members of his team. In this way we became involved in projects and had considerable freedom in the selection of projects and in their execution. We rarely had a group meeting with Willis Jackson to discuss what we were doing and leadership was given by Ronnie Cooper who is a born experimentalist and from whom, as a young research worker, I learnt a great deal. We went to Willis Jackson when we had any worthwhile results to report but at the stage when these had to be written as a report or published paper he was meticulous and would accept nothing short of what he termed perfect. So it was from him that I learned how to write scientific reports.

Willis Jackson was a stickler for prompt attendance of his work force. We were fairly conscientious about this but since we spent our nights working our attitude was more flexible than his. Hence, periodically, he would stand at the front door surveying his gold watch and stamping his feet in frustration. On such occasions, the word soon spread and we entered the Department by the back door so that when he came to check that we had not appeared for work at 9.15 a.m, there we were as busy as ever in the lab. He never discovered that we had a key to a back door which was kept permanently locked. The above episode was therefore repeated every three or four months.* At one stage I recall that we got tired of the war-time refectory food so that Ronnie, Jack Allanson and I, who shared a large laboratory, decided to fend for ourselves. Mostly we had sandwiches or pie and chips, the chips being brought in by one of the lab. boys. But occasionally we had a 'fry-up', as a result of which the corridors were permeated by a delightful aroma of frying bacon and egg. Willis disapproved of these activities but did not actually say that we should not indulge ourselves. However, it all came to a head on Pancake Tuesday when the corridor was full of a cheering crowd watching Cooper, Lamb and Allanson tossing their concoctions. Thereafter we resumed our patronage of the University refectory and formed a friendship with Mr Kay, steward of the Physics Department and one-time laboratory assistant to Rutherford.

Although we saw little of him, Willis Jackson looked after our career interests and arranged with the university for me to submit for higher degrees with an external examiner approved by the Ministry of Supply and appointed by the university, with the

---

* Professor Cooper tells a similar story, in his recollections, of Willis Jackson standing at the front door of the Department, watch in hand.

thesis retained by the university but not available in the library nor as published titles until such time as war-time secrecy restrictions were lifted.

My overall impression of W.J. was that he was a first rate teacher and a master of the written text. Sadly he was unable to mix easily with his younger research workers and even when we had ourselves reached academic distinction, at no time were we invited to address him by his assumed first name which he adopted early in his life - for reasons which one can understand.

At the end of the war I approached him with a view to working for British Thomson Houston Co. at Rugby, whereupon he told me in confidence that he was to take the Chair of Electrical Engineering at Imperial College and invited me to join him as an Assistant Lecturer, which I accepted. This was in October 1946. Ronnie came with me as a Research Fellow, as later did Jack Powles. He knew what had to be done at Imperial College and he knew how to do it."

Beatrice Shilling's "Battle of Britain" war work

Though it is not relevant to events in the Electro-technics Department, the war work of Beatrice Shilling, the young woman who graduated from the Department in 1932 and was famous for her motor cycle exploits, played such a significant part in the winning of the "Battle of Britain" by the R.A.F. that it deserves a brief mention in this chapter. She joined the Royal Aircraft Establishment in 1936 and was renowned for rolling up her sleeves and getting her hands dirty. Workshop technicians respected the fact that she could braze a butt joint between two pieces of copper with the skill of a fitter. During the war she was in the Engine Department and it was there that she did her most publicised work - modifying S.U. carburettors to stop the Rolls Royce 'Merlin' engines in fighters cutting out under 'negative g', as was prone to occur when, for instance, the aircraft was put into a steep dive. Because of this fault pilots were obliged to turn their planes onto their backs to dive in combat, otherwise the engines would splutter or cut out altogether. The defect was critical, since the Daimler-Benz engine powering enemy Me 109s permitted their pilots to perform the manoeuvre unhindered.

The senior staff said the problem was a 'weak cut', i.e. the engine cut out due to insufficiency of petrol, but their modifications did not work. Beatrice said it was a 'rich cut' due to an excess of fuel and proved it by mathematical analysis. She soon designed a device to cure the problem. It was a metal disc about the size of a present-day 5p coin, with a small hole in the middle. Nicknamed irreverently "Miss Shilling's Orifice", it was brazed into the fighter's fuel pipe, and when the pilot accelerated in a dive the disc restricted the flow of fuel and prevented flooding of the carburettor, which, as she had shown, had been the cause of the problem. By 1941 the device had been installed throughout Fighter Command, sufficing until replaced by an improved carburrettor.[14] This was only one of her many achievements but it

1938-1946: Willis Jackson and the war years

was the most famous. She was awarded the O.B.E. in 1948 for her important contributions to the war effort. Later she excelled in research at Farnborough on frictional effects on runways. Her investigations proved invaluable in explaining why an 'Elizabethan' aircraft crashed on take-off in slush at Munich in 1958, killing most of the Manchester United football team.

New appointments in Electro-technics

One of Willis Jackson's last acts before leaving had been to appoint two young Assistant Lecturers, C.N.W. Litting and J.C. West. Both had graduated with first class honours during the war in the period of shortened courses, John West in Electrical Engineering in 1943 and Colm Litting in Physics in 1944. Subsequently West had served in the Royal Navy and Litting had been a radio valve engineer with the Cosmos Manufacturing Company.

Following Jackson's departure the priority was to appoint a successor as Professor of Electro-technics. Joe Higham confided to his friends that he knew "just the right man", F.C. Williams. The Appointments Committee, which included the influential P.M.S. Blackett, was of the same opinion. Williams was duly selected, the appointment taking effect, strangely, on Christmas Day, 1946, which meant, in reality, that he would start his duties at the beginning of 1947. So began the Williams era which would last for 31 years.

Two photographs of Willis Jackson:

Left: 20 years old, on graduation day, 1925

Right: Professor Willis Jackson (left of picture) viewing equipment at the B.T.H. Research Laboratories at Rugby in 1948, shortly after leaving Manchester

171

References: Chapter 10

1. The name was quoted as "Jackson, Willie" in the M.Sc. graduates' list in the university calendars up to and including the 1932-33 session, and it was given as "Willie Jackson" in the College of Technology staff lists whilst he was a lecturer there (1930-33). In the M.Sc. list in the university calendar of 1933-34 the name was given as "Jackson, Willie (now Willis Jackson)".
2. Mr G.A.Cooper's tape recorded recollections, 29th April 1987.
3. Information communicated by Professor Cooper verbally, in correspondence and telephone calls, and in a tape-recorded interview.
4. Vice-chancellor's Annual Statement, 24th July 1940. University Calendar, 1940-41, p. 325.
5. Information from the Beattie family history, kindly offered by Professor Beattie's daughter, Mrs J.E. Paterson, and reproduced with her permission.
6. Vice-chancellor's Report published in the Annual Statement; University Calendar, 1943-44, p. 357.
7. Report in *The Daily Telegraph*, 22nd September 1990.
8. A photocopy of his Graduation Fee Certificate was kindly donated by Professor West and is reproduced with his permission.
9. Information supplied by Professor Cooper, including that contained in a tape-recorded interview (see ref. 3).
10. Information supplied by Mr G. A. Cooper, including that contained in a tape-recorded interview (see ref. 2).
11. Information supplied by Mr J. McCormick.
12. Vice-chancellor's Annual Statement, 22nd July 1942. University Calendar, 1942-43, pp. 341-342.
13. Professor John Lamb's recollections of Willis Jackson and the Electro-technics Department in the war years, written in 1991.
14. Letter from Mr G.A. Naylor (Beatrice Shilling's widower), 15th September 1991; also, obituary of Beatrice Shilling in the *Sunday Telegraph* (18th November 1990).

Chapter 11
1946-1950: Williams, Kilburn, and the world's first stored-program computer

F.C. Williams took up his appointment as Professor of Electro-technics at Manchester at the beginning of 1947, at a time when he was already recognised as an exceptionally talented electronics expert. His pre-war work has been summarised in Chapter 9. There is insufficient space here to describe his war work in great detail, nor would it be appropriate, since he had no formal connection with Manchester University during those years. However, it is relevant to state that his work on circuitry at TRE during the war was outstanding and earned him the Order of the British Empire which was conferred in 1946. He produced the device now known as the operational amplifier and developed the 'velodyne' in conjunction with A.M. Uttley. He designed a large number of extraordinarily clever circuits. The following extract from an account by a co-worker, J.R. Whitehead, gives a flavour of Williams' work and methods during the war years:

"He was most prolific, enthusiastic and unselfish in his creativity. His sole concern was to see the desired electronic function performed elegantly, efficiently and reliably. His ideas were transmitted during informal, usually intense, sessions. He was notorious for his tangled breadboard circuits which often drooped over the edge of the bench towards the floor - a unique mixture of conceptual elegance and material chaos. It was thus that we saw the birth of the whole range of feedback timing circuits which brought precision into radar circuit design by use of inherently linear instead of exponential timing waveforms. These circuits carried typical Williams names, created on the spur of the moment in the laboratory. They included the Phantastron (fantastic!), the Sanatron ("sanitary" was his favourite description of a well-behaved circuit) and the hybrid Sanaphant. I remember improvements and extensions of these circuits such as the cathode-coupled phantastron and the sanatron stepping divider.

Most of these were made in the Pavilion of Malvern College, which, with F.C., assisted by N.F. Moody, S.W. Noble and F.J.U. Ritson, was a mecca for electronic circuit types everywhere. Advice and help were given with F.C.'s unfailing, even unwitting, generosity. Thus, in a very few years, his influence became apparent wherever circuits were designed. His contributions to the "Waveforms" volume of the Radiation Laboratory "five-foot shelf" [1] were as crucially important as his earlier direct solutions to immediate circuit problems for IFF, ground and airborne radar, G, H, and many other systems." [2]

'Designability' was the motivating force of Williams' circuit work. In his important paper 'Introduction to Circuit Techniques for Radiolocation', published in 1946, Williams wrote:

"In approaching the problems of precision, reliability and producibility, the author and his colleagues have, rightly or wrongly, assumed that basically these requirements reduce to a single requirement, "designability"; i.e. to the development of circuits whose operation can be predicted accurately before they are built." [3]

More detailed information about Williams' circuit work during the war is given in the obituary written by Kilburn and Piggott. [4]

Williams brought with him to Manchester Tom Kilburn, a 25 year old mathematician from his TRE team. Kilburn, born in Dewsbury and educated at Wheelwright Grammar School, Dewsbury, and Sidney Sussex College, Cambridge, had been actively involved in mathematics since the age of eight and had graduated with a First in that subject from Cambridge in 1942. He was then drafted to TRE and spent the next four years there with Williams, working on radar and related projects. In the course of this work he was able to combine his mathematics with a thorough knowledge of electronics. Williams' and Kilburn's partnership would, within 18 months, again prove extremely fruitful.

'Electro-technics' becomes 'Electrical Engineering'

One of Williams' earlier initiatives on his return to Manchester was to set in train the process necessary to change the name of his Department from 'Electro-technics' to 'Electrical Engineering'. The change was not effected immediately but was completed during the 1947-48 session. [5] However, Professor Williams retained the title of "Edward Stocks Massey Professor of Electro-technics" for the remainder of his career.

The world's first successful stored-program computer
The Williams Storage System

Professor Williams remarked, in 1974, that his interest in computers began at TRE and was directly caused by the atom bomb:

"Substantially overnight this event converted a mass of radar experts with endless problems for which they were seeking solutions, into a mass of experts with endless solutions and no problems . . ." [6]

So, when the war ended, he had to turn his talents to something other than radar. As part of his duties as Editor of the Radiation Laboratory's volumes, Williams had paid two visits to the USA whilst still on the staff of TRE. During his first visit, in November 1945, he heard rumours about storage on cathode ray tubes, and on the second visit in June of the following year he heard of further developments. The storage was concerned with permanent echo cancellation and was short term, being dependent on the charge on the inside face of the cathode ray tube, charge which leaked away in about 0.2 s. This did not permit the long term storage required by computers. On his return to TRE, Williams began work on cathode ray tube storage. In an off-the-cuff unprepared chat Tom Kilburn recalled:

"At TRE Williams had taken the idea which Bell Labs were using of a cathode ray tube with a mesh on top. They were using it in an analogue manner to eliminate ground echos on radar. They were doing this by passing signals between two cathode-ray tubes and subtracting the ground echos. When he came back from the States he set this up (using two CRTs) in August 1946. He had the idea that the gap in the trace produced a pulse (called the "anticipation pulse"), which "anticipated" this gap and you could make a digital store (which is what he was trying to do) but on one tube. He worked on this idea and by December 1946 one digit had been stored. But not much was known about why." [7]

The ability to store information on one tube, the "Williams Tube" as it became known, was revolutionary and a patent application for the 'Williams Storage System' was filed on 11th December 1946 by the Ministry of Supply. The protection afforded by the patent was all-important, because large organisations such as IBM were soon clamouring to use the "Williams Store" and it was essential to derive some income thereby. Through his inventions Williams' name is immortalised in the *Oxford English Dictionary*, as was that of Osborne Reynolds several generations earlier. The 'Williams Tube' and inventions such as the Sanatron have earned a lengthy entry in the dictionary. [8] An advantage of the Williams store over other stores of that time was that Williams Tubes allowed random access to word locations, in contrast to the sequential access mechanism inherent in delay-line stores. [9]

The development of the store

That was the stage the work had reached when Williams and Kilburn arrived in Manchester at the beginning of January 1947, Williams as Edward Stocks Massey Professor of Electro-technics and Kilburn as a research worker, not a member of the university staff at that time but seconded from TRE. The reason for Kilburn remaining on TRE's staff, temporarily, was purely strategic. When Willis Jackson moved to Imperial College with members of his research group in 1946, most of their research apparatus had gone with them. Much of it was, in any case, unsuited to what Williams wished to do, and in consequence the department was almost totally devoid of electronic equipment. Tom Kilburn recalls that there were hardly any valves there at the time. The ploy of allowing Kilburn to remain on TRE's staff whilst working in Manchester meant that he could draw on their resources, with their permission, for what components were required. This arrangement continued until November 1948.

On their arrival in Manchester, Williams and Kilburn immediately set about developing the store, helped by Arthur Marsh who was seconded from TRE, but Marsh left after about three months because he could not see any future in computers. He was replaced in July 1947 by G.C. ('Geoff') Tootill, also on secondment from TRE. Williams, of course, had departmental responsibilities as the newly-appointed Head, but Kilburn and Tootill were able to devote themselves full-time to the work. Professor M.H.A. Newman, who had been appointed Professor of Pure Mathematics at Manchester in 1945, was a contemporary. He was a leading figure at Bletchley Park during the war, having at one stage led the team which produced the first version of the COLOSSUS code-breaking computer, and was also a member of the university committee which appointed Williams. Newman had been awarded a capital grant from the Treasury, through the Royal Society, of £20,000 over a period of five years, and an annual grant of £3,000 for the same period, to enable the university to set up a "Calculating Machine Laboratory."[10] In July 1945 Newman appointed I.J. Good as Lecturer in Mathematics and D. Rees as Assistant Lecturer (but promoted to Lecturer the following year). Both came from Bletchley Park. Two years later (on 29th September 1948), A.M. (Alan) Turing was appointed Deputy Director of the Computing Machine Laboratory (the word "calculating" had been replaced by "computing") with the status of Reader in Mathematics. He came to Manchester from NPL, but was best known for his wartime code-breaking work at Bletchley Park.

Newman's group were members of the Mathematics Department and had their offices there. They had no direct influence on the actual design and building of a computer store, nor would they have had the knowledge of electronics necessary to do it; their interest was purely in the theoretical aspects of the subject. A "Newman Laboratory" never existed in a physical sense and the 20-ft square room where Williams and Kilburn built their machine, whose brown tiled walls belonged to the architectural style described as 'late lavatorial', became *de facto* "The Laboratory", a rather grand title for the dingy room overlooking Bridge Street (now Bridgeford Street) where the pioneer work was carried out.* Though Newman and his assistants took no active role in designing Manchester computers, they later helped in programming matters and were supportive about what Williams and Kilburn were doing. I.J. Good left Manchester in 1948 and D. Rees in 1949, both having discovered that programming computers was more difficult than they had initially thought.

To develop their store Williams and Kilburn used the commercially available type CV 1131 12-inch diameter cathode ray tubes. The basis of this early work has been described by Simon Lavington:

"The principle of a two-state electrostatic store can be visualised from the following simple experiment. Start with a focussed CRT beam and turn the beam current on (thus producing a charged 'dot') and off again repeatedly. Negative voltage pulses will be

---

* There is now a blue plaque on the outside wall of the building in Bridgeford Street commemorating the historic event that took place inside.

induced by capacitive coupling in a pick-up plate placed close to the outer surface of the CRT screen. Now move the beam whilst it is on so as to write a 'dash' on the screen, then move the beam back whilst the current is off, and then switch on the current again. This time a positive voltage pulse is induced. With dots and dashes representing logical 0's and 1's, readable as negative and positive voltage signals, a binary storage system is available. Other representations such as a 'focus/defocus' system were also used. Although the electrostatic charge leaks away in about 0.2 s, automatic refreshing (re-writing) of the information in less than 0.2 s is a simple matter electronically. Since the refresh rate is rapid, long term drifts in electrode supply voltage, etc., are not critical and a robust store can be made from standard components. In contrast, the mercury acoustic delay-line stores chosen by other workers had to be constructed to close physical tolerances. The biggest advantage of the CRT store was that it allowed random access whereas other contemporary systems were sequential." [11]

Progress in developing the store from the "anticipation pulse" method to the methods described above was rapid. Kilburn had suggested to Williams that the "anticipation pulse" method was not the best system, because of problems with astigmatism, fringe effects and variable focus in a cathode ray tube. The best way, he thought, was to look at the other pulse which was formed when the beam was switched on after the anticipation pulse. This scheme was adopted and jointly patented, and the "anticipation pulse" was never used in a computer. By the autumn of 1947 the team had successfully stored 2048 digits for several hours. [12] The way was thus clear for the construction of a prototype computer "to subject the system to the most searching tests possible". [13] But there were many problems to solve. Kilburn remarked:

"We were developing a store for not only serial machines but for parallel machines, and in a parallel machine you jump to a position rather than scan across, so there was the 'philosophical' problem of how do you jump to a gap? It was one thing to store 2048 digits on a tube, another to store digits whilst you are changing some of them. It was the difference between a static store and a dynamic store, which is what is needed in a computer. So we started to think of building a test gear for this. To devise a test gear which tests all the things a computer can throw at a store is just as hard, or harder, as building a computer. That is why we built a computer. We were still building a store but we built the computer to make sure that the store worked. So the aim was not initially to build a computer. But it was a nice bi-product to have a computer." [14]

In mid-1947 the National Physical Laboratory, which was trying to build a computer of its own, floated the idea that the Manchester store could go into their computer. This suggestion did not appeal to the Manchester team, when they heard of it, and nothing came of it. Instead, they pressed on apace on their own, and by June 1948 they had built a working computer to test their store. Kilburn was responsible for much of the logical design, which is described by Lavington:

"The baby machine", as it was called, had a specification which may be expressed in modern terminology as follows:

> 32-bit word length
> serial binary arithmetic using two's complement integers
> single-address format order code
> main store: 32 words, extendable to 8192 words, random access
> computing speed: 1.2 ms per instruction.

The instruction format had three bits assigned to the function field, 13 bits to the address field and the remaining 16 bits were unused. The main store consisted of a single CV1131 Williams Tube, with each 32-digit line occupying about 10cm on the screen and being scanned in 272 μs. A complete 'beat' of 306 μs consisted of 32 X 8.5 microsecond digit periods plus a four digit fly-back time. The rhythm of the whole processor was synchronised to this store beat. There was notional provision for extending up to 256 Williams Tubes to yield a total storage capacity of 8192 words. The arithmetic unit was based on a serial subtractor and the logic employed EF 50 pentode tubes, used widely for wartime applications. Using this technology, flip-flops (bi-stable circuits) were extremely costly and temporary storage throughout the central machine was implemented with Williams Tubes wherever possible. Thus the accumulator and control register could be viewed on a monitor CRT during or after a computation - so providing a simple output mechanism. Input for the prototype was via a 32-position keyboard and operator's control switches." [15]

The first program is run

The machine ran its first program, one written by Tom Kilburn, on Monday, 21st June 1948, an event reported in a letter to *Nature* dated 3rd August 1948. [16] The first successful run was described in 1974 by Williams:

"A program was laboriously inserted and the start switch pressed. Immediately the spots on the display tube entered a mad dance. In early trials it was a dance of death leading to no useful result, and what was even worse, without yielding any clue as to what was wrong. But one day it stopped, and there, shining brightly in the expected place, was the expected answer. It was a moment to remember . . . nothing was ever the same again. We knew that only time and effort were needed to make a machine of meaningful size. We doubled our effort immediately by taking on a second technician.

This machine solved a variety of problems including checking that 31 is the highest factor of $2^{30} - 1$ by the clumsiest routine we could devise. This took 52 minutes during which 3.5 million operations were successfully performed . . . It required no human intervention during the solution of the problem . . . For us, this was the breakthrough and sparks flew in all directions." [17]

Though the machine was small, it contained all the elements of a stored-program computer, with the manual input from a keyboard and the output to a monitoring screen.

1946-1950: Williams, Kilburn, and the world's first stored-program computer

Tom Kilburn (left) and F.C. Williams shown at the control panel of the Manchester University Mark I computer

Within a couple of months of this triumphal and historic moment - the first time a stored-program computer had worked anywhere in the world - the prototype was under intensive engineering development in order to produce a more realistic computing facility. [18] Two aspects of the revised design were particularly noteworthy. One was the Manchester invention of index registers, a feature now available on all computers. The other was the early combination of a small, but fast, random-access store backed by a slower, but larger capacity, sequential store. [19] In September 1947 the design team of Williams, Kilburn and Tootill had been joined by two research students, A.A. Robinson and E. Boardman. Robinson said that the atmosphere at that time was "like that of a 19th-century inventor's workshop with regard to the surroundings and the enthusiasm." [20] In September 1948 two more young research students who had graduated that summer from the adjacent Physics Department, D.B.G. ('Dai') Edwards and G.E. ('Tommy') Thomas, were added to the team. Both were to have distinguished careers in computing. They began work on improving the prototype, a machine which was known to everyone as the 'Mark I'. Two other research students, J.C. Plowman and R.E. Harvison, joined at the same time.

Commercial development

The new computer immediately aroused interest outside the department. It was seen as early as July 1948 by Sir Henry Tizard, the Director of NPL, who "considered it of [such] national importance that the development should go on as speedily as possible, so as to maintain the lead which this country has thus acquired in the field of big computing machines, in spite of the large amount of effort and material that have been put into similar projects in America." [21] Professor Blackett, (Physics), a very influential man who worked a few yards away, saw it working and contacted Sir Ben Lockspeiser*, the Chief Scientist to the Ministry of Supply, who wrote:

"I was alerted by Blackett to Freddie's computer when I was struggling with the problems of control and stability of guided missiles in the early days. We were firing experimental rockets and telemetering the results to the ground, but the processing of the data took so long that I jumped at the chance of drastically shortening the time involved." [22]

In October 1948 Sir Ben made an unannounced call to the Electrical Engineering Department whilst visiting Manchester University to see Professor Blackett. Geoff Tootill was working on the machine at the time and had encountered a slight technical problem. He had no idea who Lockspeiser was, but told him he would demonstrate the machine "as soon as I have fixed it". This he duly did, and Tootill's demonstration of the prototype computer so impressed Lockspeiser that he immediately initiated a government contract with the local firm of Ferranti Ltd. to manufacture a production version of the machine, thus beginning a long association between the Department and Ferranti. The letter from Lockspeiser to Eric Grundy, manager of the Ferranti Instrument Department, is quoted below:

Shell Mex House
London, W.C.2.
26 October, 1948

Dear Mr Grundy,

I saw Mr Barton yesterday morning and told him of the arrangements I made with you at Manchester University. I have instructed him to get in touch with your firm and draft and issue a suitable contract to cover these arrangements. You may take this letter as authority to proceed on the lines we discussed, namely, to construct an electronic calculating machine to the instructions of Professor F.C. Williams.

I am glad we were able to meet with Professor Williams as I believe that the making of electronic calculating machines will become a matter of great value and importance.

Please let me know if you meet with any difficulties.

E. Grundy, Esq.,　　　　　　　　　　　　　　　Yours sincerely,
Ferranti Ltd, Hollinwood, Lancs.　　　　　　　B. Lockspeiser

---

* Sir Ben Lockspeiser, born in 1891, died in 1990, four and a half months short of his 100th birthday.

1946-1950: Williams, Kilburn, and the world's first stored-program computer

Williams credited Sir Ben with "two world records for this event:
(i) Speed of response by a Civil Servant
(ii) Brevity of specification.
Small as it was, this machine was big enough to make a Government Department move so fast as to commit £100,000 within two months to the construction of a machine the total specification being contained in the words 'to the instructions of Professor F.C. Williams'." [23]

The cost of the contract has been estimated at £113,783. This figure is a reconstruction by the Ferranti Archivist, Cliff Winpenny; the original documents have been lost.

It has often been said that F.C. Williams never wrote a program in his life and Tom Kilburn wrote just one - the world's first. When asked in an unrehearsed and unprepared interview whether this was true, Kilburn said that it was, substantially, but explained:

"Computer programs are written to do something. There was nothing I wanted to do - except build computers. I wrote the first program. Any computer program is made up of inner loops. The only reason a computer is viable is that every instruction that is written is obeyed more than once. If every instruction were obeyed only once there would be no point in computers. It's obeyed more than once because there's a transfer of control that sends you round in a loop. When you look at any computer program you find there are loops in it which determine everything about the program. At least 99.999% of the design of a computer is concerned with what is happening in the inner loops, because that is where the computer spends most of its time, obeying instructions. I wrote hundreds of inner loops but not in order to do anything except to find out how to design computers. I had a typical engineer's view of this, which is that when an engineer makes anything - a rotating machine - a bridge - whatever; if you build a bridge, for example, you don't spend your time walking up and down the bridge; you watch other people doing that. You think about how to build the next one." [24]

Alan Turing has become a well-known figure since his untimely death in 1954. Although he held the nominal title of 'Deputy Director of the Computing Machine Laboratory', with the status of Reader, he was located physically in the Department of Mathematics and his contribution to the Williams/Kilburn machines was essentially of a theoretical nature. He was a mathematician, not an engineer and his only direct influence on the computer from a physical point of view was some work he did in collaboration with D.B.G. Edwards and G.E. Thomas on the input and output tape systems in 1949.

Outside users benefit from the machine

The computer soon became a useful facility, used by numerous people from outside the department and indeed outside the university. Outside users were initially charged £20/hour by the university. Professor Newman suggested some problems that could be solved by its use, and the first

Electrical Engineering at Manchester University

The Manchester University Mark I, built out of war-surplus components

Some of the design team: (left to right) D.B.G. Edwards, F.C. Williams, T. Kilburn, A.A. Robinson and G.E. Thomas

'realistic' problem to be solved, an investigation into Mersenne prime numbers, was run in early April 1949. By that time several improvements had been made to the original machine. There was constant development, resulting in the fact that three versions of the machine appeared between 1948 and 1950, a period in which 48 computer patents emanated from Manchester. The improvements in the machine led not only to an enhanced capability but also to an improvement in reliability. On 16th/17th June 1949 an overnight error-free computing run of nine hours was recorded.[25]

The Manchester Mark I computer excited interest internationally and in June 1949 Williams was invited to visit IBM's headquarters in the USA, "all expenses paid". There was much concern at that time about British personnel and ideas being lost to the United States; Williams was already on board the *Queen Mary* ready to sail when he received an urgent telephone call from Lord Halsbury of the National Research and Development Corporation informing him that the NRDC would pay all his expenses and he was not to accept a penny from IBM.

The computer built by Williams and Kilburn captured the imagination of the general public, to whom it was an "electronic brain". Consequently it received a lot of publicity in the national press, including an article in the *Illustrated London News*.[26] The computer was popularly christened MADM (Manchester Automatic Digital Machine) by a journalist though to everyone in the university it remained the 'Mark I'.

Few outside users of the computer had any knowledge of computer programming because, in the absence of computers until that time, the subject had not existed. In the summer of 1949 Cicely M. Popplewell, a mathematician who had been a contemporary of Tom Kilburn at Cambridge, was appointed to the staff of the Computing Machine Laboratory to help users with their programming. This marked the beginning of what would become the Computer User Service in which Miss Popplewell provided a very helpful role.

Development of the production version

Following the signing of the contract with Ferranti, the university and the company collaborated closely on the development of the production version of Mark I. G.C. Tootill spent four months with Ferranti from August 1949 and by the end of November the logic design was virtually finished. A.A. Robinson, who had been in the department since 1947 as a research student, gaining his Ph.D. in the process, joined Ferranti in April 1950 and the computer was commissioned on the factory floor by the end of that year. At the same time a special computer building was being planned at the university, the existing 'Royal Society Computing Machine Laboratory' being totally inadequate for housing the proposed equipment. The new building was to be financed from the Royal Society 5-year grant of 1946 and an architect was approached in December 1948. The necessary building permits had been obtained by July 1949 (not as straightforward as it might seem as many post-war restrictions were still in force) and despite various delays all was ready by the end of 1950. The university Mark I was dismantled in August 1950 to make room for further computer developments. One of the last users was Dr D.G. Prinz of Ferranti Ltd. who used it to compute Laguerre functions in connection with the control of guided weapons.[27]

The question is sometimes asked whether Williams and Kilburn realised the importance of the machine they had created, at the time they produced it. An answer, in part at least, was given by Tom Kilburn in a recent short television interview:

"We knew that something exciting was going to happen but we thought of it in terms of scientific computation rather than the general explosion that has occurred in computers."[28]

In retrospect, the invention of the first stored-program working computer was clearly an event of the utmost importance, in view of the way that computers have changed our lives. It was achieved in the Electrical Engineering Department by two men and an assistant, on limited resources and without fuss. Many of the Manchester inventions were used under licence in commercial computers. It is tempting, in this book, to review computing in the Department during those historic years in much more detail than has been the case here. But the book is concerned with the whole of electrical engineering over a lengthy period and it would be inappropriate to allocate a disproportionate amount of space to the subject. However, readers who wish to follow the history of Manchester computers in more detail will find what they need in the references on pages 190-191. The books and papers of special interest are indicated in the 'General' references section.

Staff changes

After the appointment of F.C. Williams as Professor of Electro-technics in 1946 the staff of his department comprised:

Professor: F.C. Williams   *Senior Lecturer*: Harold Gerrard
Lecturer and Assistant Director of the Laboratories: Joseph Higham
Special Lecturer: Thomas Havekin
Assistant Lecturers: Raphael Feinberg, C.N. W. Litting, J.C. West.
Demonstrators: A.C. Normington, G.C. Collins, S.T. Baldwin, J.F. Smee

In addition to the above, Tom Kilburn was present in the department on secondment from TRE, as explained above. The four demonstrators, all

1946-1950: Williams, Kilburn, and the world's first stored-program computer

from Metropolitan-Vickers, came in to the department on one or two afternoons each week to help with laboratory supervision. For the start of the 1947-48 session Dorothy Garfitt, also from Metro-vicks and better known later as the wife of Professor J.D. Craggs of Liverpool University, replaced Smee as a Demonstrator, and in December 1947 L.S. Piggott joined the staff as a Lecturer. Feinberg resigned in 1948 and went to Ferranti but continued to give lectures in the department for two more years as a Special Lecturer. Also in 1948, Tom Kilburn was appointed to the university staff (Computing Machine Laboratory) with the status of Lecturer and D.B.G. Edwards and G.E. Thomas became members of the same Laboratory as Assistant Lecturers. All four Demonstrators resigned during 1948 and were replaced in the 1948-49 session by two new ones, Eric Rawlinson and F.F. Heymann, both from Metro-vicks. The staff remained the same in 1949-50 except that Litting and West had been promoted to the Lecturer grade.

1950 was a significant year as it saw the appointment of two young men who would establish active research groups in the department; D.R. Hardy was appointed Lecturer with the intention of his instituting some high-voltage research, and E.R. Laithwaite was appointed Assistant Lecturer. Although Laithwaite, who graduated from the department in 1949, had been engaged as a research student in a problem connected with computing, his heart was in machines research and he would soon make a name in that area. In the next few years Hardy and Laithwaite were to be very successful in their aim to originate research in their respective subjects. Also in 1950, Frank Harlan (from Metro-vicks) replaced Heymann as Demonstrator and Feinberg resigned as Special Lecturer. So, in December 1950 (the end of the period covered in this chapter), the Electrical Engineering staff comprised:

Professor: F.C. Williams   *Senior Lecturer*: Harold Gerrard
Lecturer and Assistant Director of the Laboratories: Joseph Higham
Lecturers: Tom Kilburn (Computing), L.S. Piggott, C.N.W. Litting, J.C. West, D.R. Hardy.
Special Lecturer: Thomas Havekin *Assistant Lecturers*: E.R. Laithwaite; D.B.G. Edwards and G.E. Thomas (Computing).   *Demonstrators*: Eric Rawlinson, Frank Harlan.

The lecturing staff assisting Williams therefore comprised three relatively elderly gentlemen (Gerrard, Higham and Havekin) and eight relatively young ones (Kilburn, Piggott, Hardy, Litting, West, Laithwaite, Edwards and Thomas). Havekin was an unusual and distinctive man. Having graduated from the department in 1922 aged 32 he had obtained his Ph.D. at Birmingham, worked at Ferranti, and taken Holy Orders in the Roman Catholic Church. He always wore a black clerical suit complete with 'dog-collar' and was known to all students as "The Bishop". He lectured to first-year electrical engineering students from the first day of the course and it was usual, when students entered the lecture room for the first time and saw a clerical gentleman standing there, for them to believe they had somehow got into the Theology Department by mistake. They were soon relieved of this misconception as voluminous amounts of accumulated wisdom were dispensed, not only on basic magnetism, electricity and circuits, but on the mysteries of the more obscure measuring devices. The outcome was a larger pile of lecture notes from Dr Havekin's course than from any other. His manner was rather austere but underneath he had a dry sense of humour. When the attendance sheet was passed round for signature during the lectures in my year, someone always added the spurious name "Charlie Briggs". Havekin never said anything about it, but at the end of the very last lecture he wished us well in the forthcoming examinations and added, "If attendance is anything to go by, Mr Briggs should be well to the fore." In similar vein, Joe Higham deleted the name "Einstein" from his attendance list for weeks until he discovered there really was a student of that name.

Between 1946 and 1950 the technical staff, which, during the war, had comprised only Albert Cooper and Joe McCormick (assisted from time to time by a 'lab boy'), increased to seven by the addition of A. Gledson, B. Ward (a wartime bomber pilot), F. Sykes, R. Hewitt and A. Vaughan.

Meanwhile the Engineering Department, which was still giving a sizeable proportion of the electrical engineering students' course, had by 1950 taken on several young members of staff, namely J.K. Alderman, J. Allen, F.J. Edwards, R. Mathieson, J.K. O'Sullivan, J.R. Spooner, and B. Standing. Professor A.H. Gibson, who had held the Engineering chair since 1920, retired in September 1949 and was succeeded at Easter 1950 by J.A.L. Matheson who had graduated with F.C. Williams in 1932. In the 'interregnum' between the departure of Gibson and the arrival of Matheson, F.C. Williams was the only Engineering professor.

The undergraduate course in Electrical Engineering

When F.C. Williams arrived at the end of 1946, the course was not very different from that outlined in Chapter 9 (page 130, with syllabuses in Appendix 9, pages 334-338). During the next four years the basic structure remained the same but new electronic and other material was introduced into the existing courses. One of the most notable introductions was a course on circuit techniques given by Williams, which opened the eyes of the students to his remarkable ingenuity in this area. He also gave a new course on servo-mechanisms, then a novel subject created in part by himself. In later years it was given by his young Assistant Lecturer, J.C. West. C.N.W. (Colm) Litting introduced a course on 'High-vacuum Techniques and Engineering Acoustics' into the second year. The 'Acoustics' part of the course was given by Dr Littler, a physicist from the Education of the Deaf Department who had helped out with Electronics lectures during the war. He was the man who designed the first NHS hearing aid.[29]

At the end of 1950 the course consisted of:

First year: Elec. 1A (Dr Havekin): General Magnetism, Circuit Theory, Measurements.
Elec. 1B (Mr Gerrard): D.C. Machines. There were also compulsory courses on Descriptive Mechanical Engineering (Mr Alderman), Strength of Materials (Mr Standing/Professor Matheson), Metallurgy (Dr Entwistle) and on Specifications and Estimates (Mr Buckingham - a Consulting Engineer). All students had to attend a Drawing Office course which took place on Saturday mornings and was supervised by Dr Allen. In addition, all students took courses in Pure Mathematics and Applied Mathematics given by various lecturers, and Physics (Dr Lovell). Students were required to pass a test in German.
Second year: Further lectures on Circuit Theory, Electrical Machines, Electronics, and High-vacuum Techniques. There were compulsory courses given by the Engineering Department's staff on Heat Engines, Thermodynamics, Power Plant Testing, Theory of Structures, and Hydraulics. Students also took courses in Pure Mathematics and Applied Mathematics.
Third year: Courses were available in: Higher Theory and Design of Electrical Machines, Generation, Transmission and Distribution of Electrical Energy, Higher Frequency Currents, Higher Magnetism and Electricity, Modern Physics, Servomechanisms, Theory of Structures, Hydraulics, Thermodynamics and Mathematics. Students took five papers at the end of the year and in some cases two courses were combined in one paper for examination purposes.

All the above was supplemented by a large number of laboratory experiments, not only in the Electrical Engineering Department but also in the Engineering Department. There were so many experiments that they would not fit into the academic year; consequently, at the end of the first-year summer term, after examinations, students were required to stay on for what was described by Joe Higham as a "solid week" of electrical laboratory work in which ten experiments were completed. The laboratory experiments taken in the course (c. 1950) are listed in Appendix 10.

Student intake

At the end of the second world war priority in admissions was given to ex-servicemen and this system continued, to a steadily decreasing extent, for several more years, during which a quota system determined the ratio of 'ex-service' and 'non ex-service' entrants. Between 1946 and 1950 the annual number of entrants in Engineering did not change much. The numbers, year by year, were as follows ('electrical' students in brackets):

1946: 52 (17).   1947: 47 (14).   1948: 52 (26).   1949: 53 (27).   1950: 54 (27).

The uncomplicated nature of the admissions procedure in those faraway days is illustrated by my personal experience. In the Spring of 1949 I applied for entry the following September to the Electrical Engineering course, having already achieved appropriate Higher School Certificate grades (A-levels had not yet been invented). A letter was then received from

Mr Higham inviting me to come for interview on 26th April. The "interview" took the form of a friendly chat. Mr Higham struck me as a very nice man, rather like a kindly, benevolent uncle. He said, "If I offer you a place, will you accept it?" (He knew from what I had told him that I had applied to two other universities as well as Manchester, one of which had already offered a place, but I had placed Manchester first). I said that I would. He said that in that case he would put a letter offering me a place in the post that day, and when I received it I should write a formal acceptance. The letter came the next day as promised, I accepted the place by return, and that was that. UCCA did not appear on the scene for many more years.

Mr Higham told me at the interview that Professor Williams normally liked to be present at the interviews but he was away that day. When a friend from my year, Harold Wroe, had his interview, Williams was indeed present, with Joe Higham, and was at pains to make clear how difficult the course was. He remarked: "I want you to know that the Electrical Engineering course is one of the hardest in the University - that's right, isn't it Joe?" Joe agreed that it was, indeed, one of the hardest. A glance through the lecture syllabus and associated laboratory work rather bears out that view. Williams was always keen to impress on his students the fact that they would have to work very hard to get a good degree and for years used to tell new students this in his welcoming talk. There was no room for slacking!

Research (other than in computing)

Raphael Feinberg was interested in electronic circuits, and between 1946 and 1948 he published a number of papers on multivibrators, magnetic amplifiers, transductors and related subjects. C.N.W. Litting's main research interest then was on cathode ray tube storage and he contributed to the work being done by Williams and Kilburn on the computer. He was also interested in secondary electron emission from insulators, carrying out research work on the subject, and was in charge of the High-vacuum Laboratory. The other young staff member, J.C. West, contributed to the design of the computer's drum store and initiated an active research programme on servo-mechanisms which was to prove profitable over many years. Each of these three staff members was assisted by research students.

The appointment of Eric Laithwaite and D.R. Hardy in 1950 heralded the hoped-for future work in machines and high-voltage respectively, but it would be a couple of years before their efforts bore fruit.

Extensions to the building

Because of the increase in the numbers of electrical engineering entrants in the post-war years compared with pre-war (about 25 per year as opposed to about 15 pre-war) there was an urgent need for more space. The situation was exacerbated by the welcome amount of new research being done. During the years 1946 to 1949 the department increased the space available partly by digging underground to make room for new laboratories and partly by transfer of rooms previously used by other departments. Preliminary changes were made in 1946-47, then in the 1947-48 session work began on an extension to the machines laboratory and the provision of a glass-blower's workshop. Much of this work was completed during 1948-49, and by 1950 the department had acquired premises previously occupied by the Metallurgy Department, which, in turn, had moved into what used to be the old Dental Hospital (the white building at the corner of Oxford Road and Bridgeford Street). The added premises now available to the Electrical Engineering Department allowed the construction of a large and spacious workshop on the ground floor of the building. This also acted as a useful social centre, because all the staff, from the Professor down, and the research students and technicians, had their coffee and tea there. A lot of bright technical ideas originated in the workshop during refreshment breaks.

Plans for a new building for the Electrical Engineering Department

Although the extensions to the department had alleviated the problem of space, it was obvious that even more space would soon be needed, which could only come from the provision of a new building. The reasons were twofold:

1. The department was becoming increasingly lively and successful in research, a trend which looked set to continue.
2. Electrical Engineering was seen as an important subject area in the technological climate then prevailing. Thus, a steadily increasing demand for student places was anticipated, an expectation borne out by the increasing number of applicants.

The University Senate and Council did not need much persuasion that a new Electrical Engineering building was needed and by 1948 a provisional decision had been made to construct it on a site in Dover Street owned by the university. However, the final approval to proceed had still to be given.[30] Joseph Higham was given the task of setting out the specification for the building and by 1949 considerable progress had been made in this respect. Authorisation to proceed soon came, and a firm of architects, J.W. Beaumont (who had designed the New Physical Laboratory opened in 1900) was instructed to draw up a design for a building appropriate to an undergraduate intake of 50 students per year with a large number of rooms for research, and a laboratory capable of housing a 1 MV generator in which the voltage would be derived from cascaded transformers feeding metal rectifiers with the output smoothed by capacitors. There was an interruption in the planning work for a few months when the same architects were asked to design and oversee the new building in Coupland Street where the commercial version of the Mark I computer was to be housed. This diversion was only temporary and the computer building was finished early in 1950. In the early part of the same year the broad design of the new Electrical Engineering building had been established and detailed specifications were proceeding. It had been decided that the building should be H-shaped to let in as much light as possible to the various rooms. It would consist of a basement, ground floor and three other main storeys, and should be of Georgian appearance. By the end of 1950 work was under way.

References: Chapter 11

General:

Readers wishing to learn of the history of Manchester computers in more detail are referred, in particular, to the two books by Simon Lavington (References 9 and 11 below), to Professor Williams' paper *Early Computers at Manchester University* (Reference 6) and to a paper by Simon Lavington entitled *The Early Days of British Computers* (IEE *Electronics and Power:* Part I: November/December 1978; Part II: January 1979). Professor Lavington's two books each include extensive bibliographies.

I am indebted to Professor Lavington for kindly giving permission for material from his books to be quoted or included in this chapter.

Specific to the text:

1. The "five-foot shelf" was a major work undertaken at the Radiation Laboratory under the general editorship of Louis N. Ridenour. It was published by McGraw-Hill, New York, in 24 volumes over the period 1947-49. Williams acted as editor to Volume 19 on *Waveforms* and Volume 20 on *Electrical Time Measurements*, and contributed chapters to these volumes.
2. T. Kilburn and L. S. Piggott. *Frederic Calland Williams: 1911-1977*. Obituary published by the Royal Society. Biographical Memoirs of Fellows of the Royal Society, **24**, November 1978, p. 590. Note that one of the authors of this obituary, Tom Kilburn, modestly omits any reference to his own significant contribution to the development of computers at Manchester University.
3. F.C. Williams. *Introduction to Circuit Techniques for Radiolocation.*
   J. Instn. Elect. Engrs. **93**, Part IIIA, 1946, p. 289.
4. As reference 2, but pp. 584-590.
5. Electrical Engineering Department Annual Report for the session 1947-1948.

6. F.C. Williams. *Early Computers at Manchester University*. Paper presented at the Colloquium on the 25th Anniversary of the Stored Program Computer, held at the Royal Society, London, on 12th November 1974 and published in *The Radio and Electronic Engineer*, **45**, No. 7, July 1975, p. 327.
7. Tape-recorded informal interview with Tom Kilburn, 1st July 1992, in which he recalled the early years of Manchester computers. The various quotes attributed to Tom Kilburn in this chapter are from this source except where stated otherwise.
8. *Oxford English Dictionary*, current (full) edition. See under "Williams".
9. Simon Lavington. *Early British Computers*. Manchester University Press, 1980, p. 36.
10. University of Manchester Calendar, 1946-47. Vice-chancellor's Annual Statement: 'Grants' section, p. 417.
11. S.H. Lavington. *A History of Manchester Computers*. National Computing Centre Publications, Manchester, 1975, p. 7.
12. T. Kilburn. *A Storage System for Use with Binary Digital Computing Machines*. Ph.D. thesis, University of Manchester, 1948.
13. F. C. Williams, T. Kilburn and G. C. Tootill. *Universal High-speed Digital Computers: A Small-scale Experimental Machine*. Proc. I.E.E. **98**, Part 2, No. 61, Feb. 1951, pp. 13-28.
14. As reference 7.
15. As reference 11, but pp. 7 to 8.
16. F. C. Williams and T. Kilburn. *Electronic Digital Computers*. Nature, **162**, 25th September 1948, p. 487.
17. As reference 6, but p. 330.
18. T. Kilburn, G. C. Tootill, D. B. G. Edwards and B. W. Pollard. *Digital Computers at Manchester University*. Proc. I. E. E., **100**, Part 2, 1953, pp. 487-500.
19. As reference 9, but p. 37.
20. As reference 11, but p. 6.
21. M. H. A. Newman. *A Status Report on the Royal Society Computing Machine Laboratory*. Prepared for an internal committee of Manchester University, 15th October 1948.
22. As reference 2, but p. 593.
23. As reference 6, but p. 331.
24. As reference 7.
25. As reference 11, but p. 12.
26. *The Illustrated London News*. Saturday, 25th June, 1949.
27. As reference 9, but p. 38.
28. Tom Kilburn, interviewed on B. B. C.'s *North West Tonight*, 11th February 1997.
29. The author is indebted to Dr Litting for this information.
30. Electrical Engineering Department Annual Report, 1948-49.

Chapter 12
1950-1964: Golden Years

It is always tempting to look back on one's youth and see it as a Golden Age, but that is not why the title 'Golden Years' has been selected for this chapter. It was chosen because the period 1950-64 was indeed Golden as far as the Electrical Engineering Department was concerned. There are several reasons. The years of rampant university expansion, student unrest, increasing time-consuming bureaucracy, assessment exercises, accreditation, and the need for numerous committees and the like had yet to begin. The department was led by F.C. Williams, still relatively young and respected universally within the department, the university, and outside. His technical brilliance and liveliness, coupled with an unfailing cheerfulness, was an inspiration to everyone. He appointed young staff whose technical abilities and personal qualities he trusted, and let them get on freely with their job. The consequence was that a lot of high-quality research was done over these years in several subject areas in a relaxed atmosphere of high morale. Most of the era was spent in a new building which contributed to the general feeling of well-being. There was a stream of good undergraduate and postgraduate students, and many of the departmental staff and graduates of that time later became leaders in other universities and industry. Consequently it was a particularly happy and successful time. It has been said that the department in those days was like "a big happy family" and the description is not an exaggeration. The period ended with the departure of the computer staff to form the new Computer Science Department, which, for the Electrical Engineering Department, was surely a major landmark.

Staff changes

At the start of the 1950s an expansion in student numbers in electrical engineering was planned because of the generally-accepted importance of the subject. The intention was to increase the intake from about 25 per year to 50 - hence the need for the new building, which by 1950 was in course of construction, and additional staff to cope with the larger number of students and the hoped-for increase in research activity. This increase in student numbers duly took place. Several new young staff members were therefore appointed to teach them and, in the natural course of events, there was a considerable turnover of younger staff between 1950 and 1964. No fewer than 34 staff members in electrical engineering were appointed during these years. It would be pointless to name them all here as there are so many, but they are listed in Appendix 1 with their dates of appointment and

1950-1964: Golden years

departure. Of these 34, eight resigned before the end of 1964 (Hardy, Laithwaite, Chaplin, Grimsdale, Thomas, Somerville, Barraclough and Lanigan), 10 moved over to the staff of the Computer Science Department when it was set up in 1964 (see page 212 and the list in Appendix 1), and the remaining 16 were still staff members of the department at the end of 1964.

Back row, left to right:
  J.K. Alderman, J.C. West*, C.N.W. Litting*, E.R. Laithwaite*, D.B.G. Edwards*, F.J. Edwards
Middle row:
  R.K. Livesley, J. Allen, B. Standing, J.K. O'Sullivan, J.R. Spooner, L.S. Piggott*
Front row:
  G.E. Thomas*, H. Gerrard*, C.M. Mason, Prof. F.C. Williams*, J. Higham*, T. Kilburn*, R. Mathieson.

Staff of the Engineering Department, 1950
('Electrical' staff are indicated by an asterisk)

During the years 1950 to 1964 eight people were appointed who were to be Electrical Engineering staff members for at least 25 years each. They were (year of appointment in brackets) G.R. Hoffman (1951), R. Cooper (1955), D. Tipping and T.E. Broadbent (1957), G.F. Nix (1960), G.W. McLean and D.E. Hesmondhalgh (1962) and P.L. Jones (1963).

The retirement took place in the early 1950s of two long-serving members of staff - Joseph Higham and Harold Gerrard. Joe Higham suffered a severe stroke in 1950 whilst at work; Professor Williams took him home in his car and he was absent from the department for a lengthy period.

193

Eventually he returned (in 1951) but his health was much diminished and he was no longer well enough to give any lectures; a sad state of affairs as he had been such a good teacher. He was compelled to retire at the end of the 1951-52 session because of his continuing poor health, and went to live in a cottage in North Wales. Unfortunately he suffered further strokes and died in 1956 aged 66. A 3rd-year student prize was later instituted in his memory.

Harold Gerrard, who had been a staff member since 1912, retired from the full-time staff in 1953. He continued to give lectures as a Special Lecturer until 1959 when, at a dinner marking his complete retirement, he spoke of the opening of the Department (see page 83). He retired to Conway but occasionally visited the department. In the early 1970s he had a stroke. After making a partial recovery he died in November 1975 aged 87.

Thomas Havekin had never been a full-time staff member but his lectures had imprinted themselves on the minds of students from 1942 when he began giving lectures as a Special Lecturer until 1955 when he retired. After his retirement he lectured part-time in Physics at a local Roman Catholic school. In the early 1960s a student entered the Electrical Engineering Department who had been taught physics at school by Havekin. He told us that Havekin never had much time for heat, light, sound, etc., but plenty for magnetism and electricity, so that by the end of the course the pupils were very well versed in the wonders of the Wagner Earth, the Kelvin Double Bridge, and similar esoteric devices. Havekin died in 1965 aged 75.

In the Engineering Department, Professor Matheson resigned in 1960 to become Vice-chancellor of Monash University in Melbourne, Australia, and was succeeded as head professor in the department by Jack Diamond, who had been Professor of Mechanical Engineering since 1953. By 1961 there were three Professors of Engineering; Jack Diamond, W.B. Hall (Nuclear Engineering) and M.R. Horne (Civil Engineering). C.M. Mason retired from the Engineering staff in 1955 having been associated with the university for 50 years. He was usually known to the students as Colonel Mason, this rank having been achieved through his work as Officer Commanding the University of Manchester contingent of the Officers' Training Corps. He died in 1967.

Alan Turing, though based in the Mathematics Department, was Reader in the Theory of Computing in the early 1950s and was often to be found in the Electrical Engineering Department through his work on the Mark I computer. He was interested in logic and puzzles of all kinds, and used to share this interest with Eric Laithwaite, then a young Electrical Engineering Lecturer. Laithwaite said that Turing passed on to him what must be the longest palindrome in the English language. It related to an imaginary

American who consulted his doctor (whom he called 'doc') on the subject of dieting and obesity. The fictional conversation went thus:

"Doc note, I dissent, a fast never prevents a fatness. I diet on cod."

Laithwaite believed Turing had made it up himself.[1] Sadly, the interesting conversations between Turing and his friends were destined not to last long, for on 7th June 1954 a visitor to his home ("Hollymeade", 43, Adlington Road, Wilmslow) found him dead in bed. The coroner's inquest concluded that he had committed suicide by poisoning. A half-eaten apple was found nearby and cyanide, a chemical used in one of his home experiments, was found in the house.

Revision of the undergraduate Electrical Engineering course

In the early 1950s, after the retirement of Joseph Higham and the partial retirement of Harold Gerrard, a revision of the course was undertaken - the first for very many years. It is doubtful whether the total subject matter passed on to the students changed much, but the material was redistributed within courses, a number of new, shorter, courses were introduced, particularly in second year, and the courses were given new titles. Consequently some courses, such as 'Higher Theory and Design of Electrical Machinery' and 'Generation, Transmission and Distribution of Electrical Energy', which had been given by Gerrard and Higham respectively and had been mainstays of the course for just over and just under 40 years respectively, disappeared. In second year the main effect of the revision was to do away with the large blocks of syllabus simply described as 'Continuous Current Measurements', and 'Alternating Current Measurements' and replace them by more numerous but shorter, clearly defined, courses. After the restructuring the outline of the course was as follows:[2]

First year: Strength of Materials, Heat Engines, Engineering Drawing, Descriptive Engineering, Elementary Electricity and Magnetism, Principles of Electrical Machinery, and Electrical Measurements and Electronics. Students also attended courses in Physics and in Pure and Applied Mathematics. At the end of the year students took the Intermediate B.Sc. Examination in Engineering, Mathematics (Pure and Applied) and Physics.

Second year: All students attended courses in A.C. Machinery, D.C. Machinery, Electric Circuit Theory, Electronics and Radio Engineering, Materials of Engineering (Metallurgy and Ferromagnetic Materials), and Mathematics. They also attended at least three courses chosen from Electricity Supply (a shorter version of the old third year course, 'Generation, Transmission and Distribution of Electrical Energy'), Electronic Circuit Techniques, High-vacuum Technique, Strength of Materials, Theory of Machines, Mechanics of Fluids, and Heat Engines. The Engineering and Electrical Engineering courses were mostly of about 22 lectures' duration. At the end of the session students took the Part I Examination consisting of six papers in which students were examined in the courses taken during the year.

Third Year: Courses were available in Electrical Machinery, Electro-magnetic Waves, Transmission of Electrical Energy, Servomechanisms, Electronic Circuits, and Electron Physics.

Certain courses offered by the Engineering Department were also available as options, including Applied Thermodynamics and Heat, Hydraulics and Hydraulic Machinery, Applied Mechanics, and Fundamentals of Nuclear Engineering. Most of the courses were of about 33 lectures' duration. Five papers were taken in the Final (Part II Honours) Examination. Students were also given a short course on Research Topics in the second half-session.

The above courses were supported by appropriate laboratory work in each year including, as always, a project, which began about Christmas in the final year. Weekly first-year tutorials were introduced into the course in 1962. Until then any such small-group staff-student contact had been on an informal basis.

This Honours Electrical Engineering course changed a little between its introduction in 1952 and the end of the period covered in this chapter in that sometimes optional courses were dropped in favour of other optional courses, but the general pattern remained much the same. The content of all the courses changed slightly from year to year by evolution.

New Physics and Electronic Engineering degree course

This new course, directed by Dr Litting, was instituted in 1963 to fill a need for a coherent and balanced course bridging the gap between pure physics and electronic engineering. It was the first of its type in the country but was later copied by similar courses in other universities. It was given jointly by the Electrical Engineering and Physics Departments and was the first new course available in the Electrical Engineering Department since its foundation 51 years earlier. A typical syllabus throughout the period was as follows:

First Year: Physics: Quantum Physics, Dynamics, Relativity, Vibrations and Waves, Electricity and Magnetism, Properties of Matter, and Statistics. Electronic Engineering: Electronics, Mathematics, Computer Programming, and laboratory work.

Second Year: Physics: Quantum Physics, Thermal and Solid State Physics. Electronic Engineering: Electronics, Feedback and its Applications, Network Theory, Transmission, Mathematics and laboratory work.

Third Year: Students took four courses. Two were compulsory (Solid State and Semiconductor Physics; Electronics). At least one of the other two was chosen from: Control Engineering, Computer Systems Design, and Communications. The other could be chosen from these or from a further list including Computer Science, Mathematics, Engineering Management, Biophysics and other branches of Physics. The third year laboratory work consisted of a project.

The new Electrical Engineering building

As mentioned in the previous chapter, plans for a new building to house the Electrical Engineering Department were approved in the late 1940s and by 1950 contractors were digging a hole for the foundations; a very large hole, as the building was to have a basement and a sub-basement and foundations sufficient to support several storeys and heavy internal loads. According to the Chief Technician, Albert Cooper, who was heavily involved in the day-to-day planning, once the principle of having a new building was approved, money was not too important a consideration initially. The architect, J.W. Beaumont, said that for a building of this size he would normally plan according to a figure of 4s - 6d / square foot of floor area, but in this case he was going to submit a figure of 9s - 0d. This was approved without demur and an initial estimate of £335,000 for the cost of the building was accepted. This enabled Beaumont to produce what Albert called "a gold-plated job", with such luxuries as marble-covered walls. [3]

So far as the internal planning of the building was concerned, i.e. the utilisation of the various rooms and areas, F.C. Williams did not wish to do it himself, and delegated the task to Joe Higham. He and Albert spent hundreds of hours on this work, which included many meetings with contractors. The overall plan included a 1MV high-voltage laboratory housed in a 40 foot cube at the south-east corner of the building, because, according to Albert, Higham had "always wanted a 1MV high-voltage lab." [4]

On May 31st 1951, during the construction of the new building, Queen Elizabeth (now the Queen Mother) visited the university as part of the celebrations marking the centenary of the founding of Owens College. In the preceding days oases of plants, tubs of trees and shrubs sprang up to brighten the main quadrangle, surrounded by black, smoke-grimed buildings (they were not cleaned until the early 1970s), and were as quickly removed once the event was over. Students were given a day's holiday.

By the middle of 1952 constructional work on the new Electrical Engineering building was well advanced. Adding a personal note, I sat most of my Finals papers in 1952 in the old Economics building on Dover Street, opposite the new Electrical Engineering building, and there was a constant din from the building site throughout the examinations.

Higham's illness

After Joseph Higham's stroke in 1950 several months elapsed before he could resume his planning work, but he was able to do so during the 1951-52 session. However, after his enforced retirement in 1952 owing to continuing poor health, Professor Williams gave the responsibility of the internal planning to J.C. West, then a 30 year old lecturer. Higham's university career, which had extended from 1920 to 1952, had been concerned with machines, measurements and transmission systems. He had never been actively involved in electronics, nor in research. It is perhaps not surprising therefore that when West took over Higham's set of plans he concluded that the usage of rooms which Higham had envisaged was rather old-fashioned, with a lot of space being allocated to work involving such devices as ballistic galvanometers at the expense of more modern technology. Consequently, during the later stages of the construction of the building, West spent much of his time on the site, instructing the workmen to "knock down that wall you've just built and build one here instead", so as to make the shapes of the rooms appropriate to their revised intended usage. [5]

Completion of the building

By the end of 1953 the building was almost complete, and in the spring of 1954 the first research students started to move in. I was then half-way through my three years of Ph.D. work and believe I can claim to be the first person to actually do any useful work in the building, for the high-voltage group's research students who occupied the north-east corner of the ground floor were the first ones to have their equipment moved over from Coupland Street, and mine was the first to be set up and working. Work on the building was at that stage by no means complete and we had to enter the building through the north-east fire-escape door by walking across a plank over a "moat". In the following weeks the building took on a much more "finished" appearance and other research staff gradually moved across from Coupland Street, but the undergraduates remained in the old building for all their lectures and laboratory work until the end of the 1953-54 session. From the start of the 1954-55 session all their 'electrical' lectures and laboratories were held in the new building but they still had to spend a lot of time commuting between Dover Street and Coupland Street in order to attend the engineering and physics lectures in the respective departments, and to the main building for lectures in mathematics. The trek across Oxford Road for these lectures continued until the Departments of Engineering, Physics and Mathematics moved into new premises. The Mathematics

The New Electrical Engineering building, completed in 1954

1950-1964: Golden years

Department moved into the Williamson building when it opened in 1959 and thence to the Mathematics Tower on its completion in 1969; Engineering moved to the new Simon Laboratories in 1962-63 and Physics to the new Schuster building in 1967. The Roscoe building opened in 1964 and Chemistry in 1965. The roof of the Electrical Engineering building afforded a grandstand view of the constructional work on most of these other new buildings. Of particular interest during early work on the Roscoe building was the exposure of a sizeable underground river, the Corn Brook, which had been culverted a century earlier.

200kV research laboratory, the first working room in the new building

The fact that Electrical Engineering was the first department to acquire a new building was a measure of the importance attached to the subject at the time, but in one way the department paid the penalty for being the first building on the new science area. In the late 1950s and early 1960s expansion of the whole of the university system was very much to the political forefront, especially after the publication of the Robbins Report in 1963, and consequently more money was allocated to building programmes. The result was that the departments whose new buildings came later obtained much larger structures than that of the Electrical Engineering Department. One has only to look at the Electrical Engineering building dwarfed between those of Engineering and Chemistry to realise this fact.

The new Electrical Engineering building was fully operational by the start of the 1954-55 session, except for the main high-voltage laboratory where further work on the 1MV generator was still required. No official opening ceremony ever took place. As the first new building of the "science area" to the east of Oxford Road, it stood in solitary state amongst old terraced houses until the start of work on the buildings mentioned above. Consequently the flat roof afforded excellent views to north, east, west, south east and south west. The clear view of the Victoria Tower of the main building in the photograph below was no longer possible once constructional work on the Simon building reached the third floor in 1962.

The surrounding area at that time was interesting and varied. Rumford Street ran NNW from Ackers Street to Rosamond Street (near All Saints) and bordered the east side of the Electrical Engineering Department. It has now succumbed to university building, though a vestige remains - the entry road to the Dover Street car park, SSE to the Medical School. Opposite the Department, on the other side of Rumford Street, a convent stood which was occupied by an order named "Little Sisters of the Poor". There were shops of many different kinds on Oxford Road, and on Brunswick Street and Upper Brook Street where the Physics Department now stands.

Back row, left to right: D.R. Hardy (Lecturer), D. Drew (Technician), T.E. Broadbent (Research student)
Front: I.F. Chrishop, H. Latham (Research students)

Members of the high-voltage research group on the roof, July 1954, before the Simon Engineering building obscured the view of the Victoria Tower

1950-1964: Golden years

Description of the Electrical Engineering building

The building is steel-framed, brick-clad, with a total floor area of over 54,000 sq ft. The final cost was just over £600,000, nearly twice the original estimate. It was planned to accommodate about 200 students, with 50 entering annually, and places, initially, for 24 research students. Lecture rooms 'A' and 'B' were intended to seat all the students in a year (even more in the case of 'A') whilst 'C' and 'D' were for third-year classes. Floor by floor, the description of the building and the original usage is as follows:

Sub-basement: This has an area of about 1,500 sq ft and contained a large calorifier which received steam from the heating station across the road and delivered hot water for warming the building. All heating was by radiation from panels concealed in the ceilings. The rest of the floor space was used for storage.

Main basement: The floor area is nearly 16,000 sq ft and contained two large machine laboratories (a.c. and d.c.), each of about 4,000 sq ft, and a large lecture room ('A') to seat about 150 students. The d.c. laboratory contained a servomechanisms research bay at the north end. There was also a substation which, it was intended, would eventually serve the whole of the projected science area. There was also a large workshop.

Ground floor: This has an area of more than 10,000 sq ft and was divided between measurements and a number of high-voltage laboratories. The main high-voltage laboratory was a 40 ft cube with equipment intended to generate a direct or impulse voltage of 1MV. There was also a transmission laboratory, two lecture theatres ('B' seating 88 and 'C' seating 50) and a measurements laboratory supplied by a 15kW sine wave generator set installed in the basement.

First floor: This included high-vacuum and acoustics laboratories, research rooms, a students' common room, dark room, library and a small lecture theatre ('D'), identical with room 'C'.

Second floor: This was devoted entirely to electronics and electronic research with large second-year and third-year laboratories.

Third floor: This was reserved for electronic digital computation, with rooms for the Mark I computer, the successor to Mark I, and experimental computers including (at that time) a germanium-diode transistorised machine. The Mark I computer, when transferred from the old building, continued to work 24 hours/day, 7 days/week, except during routine mainentance and fault rectification.

Fourth floor: This floor, of smaller area than the third, was not for public access. It contained plant rooms, the battery room, two charging rooms, lift machinery, ventilation plant, a gas compressor and a vacuum pump.

The electrical installation in the new building started with the substation in the basement which was fed by the North Western Electricity Board at 33kV and could be extended to 2 MVA. Ultimately it was intended to run a ring main from the substation feeding all the other new buildings in the science area. In the Electrical Engineering building supplies were taken from the substation to each floor by a rising service duct. All horizontal wiring was in steel floor trunking covered by teak blocks to match the rest of the floor. Only in the machines laboratories was it necessary to install trunking on the wall in addition to that in the floor. There was a switchroom on each floor and the batteries on the fourth floor supplied a voltage of up to 120v to plug boards on each floor. The two main batteries were rated at 800 Ah while the two d.c. rectifiers in the substation were rated at 80kW, 200-0-200 V. The main lighting to all laboratories was fluorescent and in each room the load was distributed equally among the three phases. The majority of the laboratories and research rooms were electrically screened either to keep radiation in or to prevent electro-magnetic radiation from entering. This was very necessary because of the close proximity of, for example, the 1MV high-voltage laboratory to the digital computers. The screening consisted of 1 inch mesh copper gauze embedded in the plaster and the doors, and removable screen cages for the windows.

At the time of its completion the new Electrical Engineering building was regarded as a very fine structure. It was featured in an article in the *Electrical Review* in 1955.[6]

The 1MV high-voltage laboratory

In the early 1950s there were many difficulties concerning the design of the new 1MV laboratory. Joseph Higham, who had instigated it, had taken advice from outside authorities. Metro-vick had told him, "Ask for a 40ft cube and £20,000".[7] This was duly done and a 40ft cube was allocated by the architect. However, when everything was approved and Metro-vick were asked to quote for manufacturing the equipment, they declined, probably because installing high-voltage laboratories in universities was not very profitable. Williams then invited Ferranti to quote and eventually they agreed to construct the generator provided they were paid the money at the beginning. This was agreed. The design involved two 50Hz transformers connected in cascade feeding a voltage doubler using metal rectifiers in porcelain housings, the direct voltage output being smoothed by a stack of four 0.05μF, 250kV, output capacitors in series. However, many problems occurred. Originally a glass-fronted viewing gallery and an overhead crane were planned; work started on these features but they had to be removed when it was decided that the space available was inadequate. Matters were set back further by the deaths of two key Ferranti employees. Eventually it was decided that the only way the equipment would fit into a 40 ft cube, whilst still leaving the necessary spark clearances all round and space to put the equipment being investigated, was to stack the four capacitors upwards on the floor in the usual way but suspend the transformers and rectifiers by insulators from the roof. Moreover, part of the loading bay had to be 'robbed' to accommodate the control room. The laboratory was eventually completed in the mid 1950s, some time after the other parts of the building were in use, and was able to generate a direct voltage of just under 1MV with provision for the generation of impulse voltages up to 1.25MV.

Developments in computing

As soon as a new computer enters service, and usually before, its designers are thinking about the design of the next one. So it was with the Manchester computers. Between 1950 and 1964 (the end of the period covered in this chapter) three major computers designed in the Electrical

Engineering Department were produced commercially by Ferranti and brought into service. Their names and the periods of commercial service of these particular computers designed and installed in the Department were:

Mark I: 1951 - 1958
Mark II ("Meg") (Production version "Mercury"): 1957 - 1963
"Muse"(Production version "Atlas"): 1962 - 1971

Similar machines, or variants, supplied to other organisations often had longer lifespans; for example, at least four Mercurys were said to be still working in 1970. [8] The computers named above were the big production machines, but smaller computers were also designed in the department for experimental purposes, in particular, the transistor computer of the early 1950s, the aim of which was to gain experience of the newly-introduced transistor and to find how it would operate in computers. The history of the various machines is described briefly in the following sections.

Mark I

The Ferranti Mark I, based on the Mark I which had been running in the Electrical Engineering Department since 1948, was delivered in February 1951 and installed in its new building a few yards to the south-west of the front door of the Electrical Engineering Department in Coupland Street. After some teething troubles it was performing reasonably reliably by the date of the Inaugural Conference in the Department on 9th -12th July 1951. About 150 representatives of firms and departments interested in computing machines attended. Ferranti sold a similar machine to Toronto University and seven modified machines, known as Mark I Stars, were delivered between 1953 and 1957, two of them being exported to Holland and Italy. The Ferranti Mark I appears to have been the world's first commercially available computer. The machine was technically similar to the 1949 university version of the Mark I except for the following:

   extended function repertoire, including B-line arithmetic
   eight B-lines (20 bits each)
   main store: 256 x 40 word bits on eight Williams tubes
   drum backing store: 3.75K words, extendable to 15K
   add-time: 1.2 ms., multiply time: 2.16 ms

A fast multiplier, designed by A.A. Robinson, used a parallel technique which took up nearly one quarter of the computer's 4,050 thermionic valves. The good performance of the Mark I, which must have been one of the most powerful computers in the world at the time, was attributed to its random access main memory and its fast multiplier. [9]

The machine was used by numerous people from inside and outside the university. Cicely Popplewell helped users with programming problems and in 1950 Alan Turing wrote the first programming manual. However, by then his interests were drifting to other matters - in particular the subject of morphogenesis (the growth and form of living things) and although he used Mark I to solve problems in connection with this, he took little part in subsequent programming developments. In 1951 R.A. Brooker joined the computing team and developed the Mark I Autocode system which was available by March 1954. Its purpose was to provide a relatively simple and easy-to-learn programming language for computer users. It was generally felt to be an improvement on the machine-code system used previously. [10]

Mark I proved to be a very successful computer. When the department moved to the new building in Dover Street it was transferred to a large laboratory at the south west corner of the the third floor. From its inauguration early in 1951 it ran for nearly eight years, maintenance finally being discontinued in December 1958. The Electrical Engineering Department's annual reports quote performance figures year by year; in most years the machine ran for about 5,000 hours (100 hours/week) at a serviceability level of better than 80%. An exception was the session 1954-55 when the machine still managed to operate a total of 1,700 hours in spite of the time taken to close it down, transfer it to Dover Street, and set it up again. The machine was finally shut down on 24th December 1958. In its life it gave the following performance figures: [11]

Total useful operating time:	30,720 hours
At fault for:	5,838 hours
Scheduled engineering:	5,956 hours
Average efficiency over the years:	84%.

After its de-commissioning the machine was removed within weeks and donated to Ferranti. Plans for it to be placed in a proposed museum did not materialise and Ferranti eventually disposed of the remains.

Professor Williams' interests move towards 'machines'

The period 1950-51 was significant for it was then that Professor Williams started to turn his attention to the subject of electrical machines, and from that time on he was not actively involved in the design of new computers. This came about through Eric Laithwaite, then a young Assistant Lecturer in the department. Although Laithwaite had obtained his M.Sc. in 1950 on a subject related to computers, [12] his real interest was in machines and he had an idea to develop a linear induction motor which he had made as the drive for the shuttle of a loom. Williams turned his considerable talents to this and other projects in the machines field, leaving future departmental developments in computers in the capable hands of Tom Kilburn. However, although he had ceased to be involved actively in the design of computers as such, Williams was always available for chats over coffee with his friend, Tom Kilburn, and was responsible for a number of ideas which helped the computing team. He also contributed directly to the design of the delay lines that were being developed for the 'Meg' computer.

The Mark II computer ("Meg")

In 1951 Kilburn's group began work on the hardware of a new computer, the Mark II, often known as the megacycle machine ("Meg"). Its first program was run in May 1954. The design aim was to produce a faster, more compact computer than Mark I, using miniature thermionic valves. For user convenience it would include floating point arithmetic. The way in which these aims were achieved in Meg is described by Lavington as follows:

"The digit period of the serial arithmetic unit was shortened to one microsecond and parallel access was made to the store in 10-bit short words at a time. Floating point hardware was provided using a 30-bit mantissa, a 10-bit exponent and a base of two. The floating point add time was 180μs and floating point multiplication took 360μs using a parallel paired-multiplication technique. Eight 10-bit registers were provided, B-arithmetic orders being executed in 60μs. Instructions were 20 bits long with the same single-address format as the Mark I. Since magnetic core stores had not become generally available in 1952 the Meg prototype used a CRT main store, with provision for switching to cores for the production version. The use of semiconductor ('crystal') diodes helped to reduce the number of thermionic cathodes by 57% as compared to Mark I. This and the use of the smaller type 6CH6 pentode tubes also produced a reduction in power consumed (12kW instead of Mark I's 25kW) and in the physical size of the processor. Distributed electromagnetic delay lines produced a cheap, flexible implementation for many of Meg's internal registers, etc. A single drum backing store was provided for the prototype." [13]

The original papers describing the machine are listed in the references at the end of the chapter. [14]

The commercial version of Mark II ("Mercury")

Continuing co-operation with Ferranti led to the production version of Meg, known as the Ferranti Mercury. [15] This differed in design from Meg only in the details of the order-code repertoire and in the main store technology. The first Mercury was delivered to the Norwegian Defence Research Establishment in August 1957. Ferranti had promised to have a production 'Meg' ready for the Electrical Engineering Department by 1956 and in anticipation the prototype version was scrapped in that year, leaving a room empty, but there were delays and the department's production Meg did not arrive until late in 1957. It had passed its acceptance tests by 5th February 1958. It was paid for partly from Mark I earnings augmented by design information supplied to Ferranti in lieu of money, and partly by a £48,000 award from the University Grants Committee. The department's

'Meg', followed by 'Mercury', occupied the room facing the top of the stairs on the third floor of the building. When Mercury was taken out of service in January 1963, on the commencement of the Atlas service, it was returned to the UGC free of charge and was installed by the UGC at Sheffield University. Ferranti sold 19 Mercurys between 1957 and 1961, six of them going abroad. In 1958 Mercury was regarded as one of the most powerful computers in Britain. The only market rival was the IBM 704 which was a faster machine but cost five times more than Mercury.[16]

The Mercury computer in the Electrical Engineering Department provided a service to outside users as Mark I had done, but towards the end of its life it was used mainly by internal departments of the university. Industrial users were charged £50/hour. The software developments during the 1950s were in the hands of R.A. Brooker and it was he, assisted by a small team, who devised the high-level Mercury Autocode language during the years 1958-62.[17] From 1960 D. Morris assisted in the software work.

The transistor computer

Work on the design of Meg, starting in 1951, was accompanied by work on a parallel product, a relatively small and economic computer. It was decided at an early stage that it should be a transistor computer, intended to discover how the newly available transistor could be used in computer circuits in place of valves. Only germanium point-contact devices were available at that time. The following description is by Lavington:[18]

"Two versions of the prototype computer were commissioned in 1953 and 1955 respectively.[19] Both versions had a pseudo two-address (or 1 + 1) instruction format consistent with the fact that the only store was a single drum, with consequent latency problems. As with some other delay-line storage computers, the address of the next instruction was contained within each instruction, thus facilitating optimum programming. One track of the drum was used for the main accumulator and one track for the instruction register, in order to simplify the processor circuitry. The word length was 44 bits, divided into four 'syllables' for an instruction. The first computer (1953) had a simple seven-function order code and one track of 64 words for main storage. In the second computer (April 1955) the order code and storage were extended and a hardware multiplier included. One drum track formed an 8-word B-store which was also available as a set of 'fast' general registers. Arithmetic was serial, with a pulse rate of 125,000/s. The instruction times were related to the 30ms drum revolution time, thus being slow compared to the Mark I performance."

The 1955 machine, which utilised 200 point-contact transistors and 1300 point-contact diodes had a power consumption of only 150 watts. Early batches of transistors were unreliable, causing frequent breakdowns of the machine, but when junction transistors became available they were used in place of the point-contact devices with a resulting improvement in reliability. Thermionic tube amplifiers had to be used in the drum since the frequency response of junction transistors available at that time was inadequate for the required purpose. The operation of this computer provided much valuable experience to its designers in the use of transistors, and the machine commissioned in November 1953 is thought to have been the first transistor computer to come into operation anywhere. [20]

The design of the Manchester transistor computer was subsequently taken up by Metropolitan Vickers (which later became AEI and then GEC) who converted all the circuits to the improved junction transistors which by then had become available. They built a "production" version, the MV 950, of which six were made. They were not sold to outside users but operated successfully and reliably in various departments within the company for about five years. [21]

The design of "Muse"

Because of the delay in the arrival of the Ferranti Mercury following the scrapping of the department's Meg in 1956, the design team were left somewhat at a loose end and with an empty room. Tom Kilburn, in consultation with his team, had to decide whether to utilise their time by building a new computer, one which would be a very significant advance on the previous computers. It is perhaps worth noting that the need for any more new large computers was by no means universally acknowledged. Many leading scientists believed that the Manchester Mark I and the Cambridge EDSAC 2 [22] which was about to be commissioned (this occurred in 1957) were more than sufficient for all the country's needs, then and in the forseeable future. However, other voices took the contrary view, that there was indeed a need for a better computer, and if such a machine were to be produced it would need to be a large, fast computer to rival the latest American machines which IBM, with their vast resources, were starting to produce. The Manchester Mark I and the Cambridge EDSAC 2 were much cheaper than the American machines but were also slower.

Tom Kilburn and his team decided to design and build a large new computer, an all-transistor machine. Transistors were purchased from money which had been accumulated in a fund from Mark I user earnings, and planning began. Later, more money came in from earnings from Mercury. Kilburn said that the main problem in the early design work was not so much the lack of money but the need for ideas. [23] This sentiment will be echoed by all good researchers.

Kilburn's team, by the autumn of 1956, had initiated the outline design of such a machine, to be known, during the design process as the "Muse" (microsecond) computer. In the early stages of the design of Muse, Tom Kilburn met many potential users (Atomic Energy Authority, large commercial organisations, etc.) in order to establish what the ideal specification of the computer should be from the user's viewpoint. The result of these discussions indicated a required design specification of an instruction speed approaching one order per microsecond and the need to attach a large number of peripherals of various types, with an immediate access storage capacity far in excess of anything then available. It was clear that the proposed computer would be large and expensive, and the Electrical Engineering Department would not have the resources, on its own, to complete such a project. Kilburn spent 18 months attempting to get financial support for Muse whilst, in parallel, design work went on. Ferranti initially felt unable to support the project because of the financial uncertainties involved, but the Manchester design team, on its own, produced a prototype limited version of Muse with D.B.G. Edwards responsible for hardware and R.A. Brooker for software, using departmental resources and money from computer user earnings. However, in January 1959, Ferranti confirmed their willingness to become involved in the project and in May 1959 NRDC awarded Ferranti £300,000, repayable through a levy on eventual sales. By the start of 1959 the computer had been re-named "Atlas" by Ferranti, and from then on was developed as a joint University/Ferranti project under Tom Kilburn. About 13 university people were involved in the hardware design and six from Ferranti; the software designers numbered about seven from the university (mostly on compilers) and two (on operating systems) from Ferranti.

Michael Lanigan (left) and Tom Kilburn in 1959
with part of 'Muse', the forerunner of 'Atlas'

1950-1964: Golden years

Atlas, situated in a laboratory on the top floor of the Electrical Engineering Department (south west corner) in the room previously occupied by Mark I, was officially inaugurated on 7th December 1962 by Sir John Cockroft of the Atomic Energy Authority. It was considered to be the most powerful computer in the world at the time; Ferranti salesmen equated it to four IBM 7094s. It is described in 25 published papers. [24] A brief technical description has been given by Simon Lavington, as follows: [25]

> Atlas had a 48-bit word, single address instruction format, allowing for double B-modification and a user virtual address space of one million words. There were 127 half-word B-registers, mostly held in a fast 0.7µs cycle time core store. System routines and some frequently-used library routines were held in an 8K read-only store having an access time of 0.3µs. The core-plus-drum one level store used fixed-size 512-word pages, 32 page address registers forming an associative store for virtual-to-real address translation. A 'drum learning program' held in read-only memory attempted to optimise page swapping by accumulating statistics on page utilisation. The Manchester Atlas had the following storage hierarchy:
>
> main store: 16K*, 2µs cycle time four-way interleaved. Backed by 4* drums, each 24K, 12ms revolution time, capable of transferring one block of 512 words every 2ms.
> system working store: 1K*, 2µs cycle time.
> system read-only store: 8K, 0.3µs access
> magnetic tapes: 8* decks (on 8 channels), each 90kHz giving one word every 88µs.
> file disc (added in 1967): 16M word
> (* increased on subsequent production Atlases)
>
> The interrupt structure allowed for up to 512 peripherals. The Manchester Atlas had about 17 conventional input/output devices. Two high-speed data links, an on-line X-ray crystallographic unit and an experimental speech input/output device were added later. An idea of the instruction speed may be obtained from the following orders, measured on 7th December 1962:
>
> fixed-point B addition   1.59µs
> floating-point add, no modification   1.61µs
> floating-point add, double modification   2.61µs
> floating-point multiply, double modification   4.97µs
> floating-point division   10.66µs min. to 29.80µs max.

A user service using a Supervisor without multiprogramming capability but with Atlas Autocode and Mercury autocode as the only high-level languages was available in 1963. The full supervisor was available from January 1964 and languages such as Algol and Fortran followed later. Once the machine was fully operational with appropriate programming facilities it ran for up to 20 hours per day. Ferranti charged up to £500/hour to use the machine but it was part of the university's agreement with Ferranti that a share of the income from outside users should be paid to the University Computer Earnings Fund. Atlas was also available to five remote users via high-speed data links; they were Jodrell Bank radio telescope laboratory, Edinburgh University, Nottingham University, Ferranti's London Bureau and the Nuclear Power Group.

Ferranti sold two full versions of Atlas, one to a joint London University/ British Petroleum consortium and the other to the National Institute for Research into Nuclear Science, at Harwell. Ferranti also provided Cambridge University with certain units of Atlas hardware on special terms in return for aid in developing a simplified version of Atlas. This led to the construction of the Cambridge 'Titan' machine which became operational in the summer of 1963. [26]

Establishment of the Computer Science Department

In 1960 Tom Kilburn, who had been in charge of computer developments in the department for almost a decade, was promoted to Professor of Computer Engineering. He felt that he and his staff had, over the years, been in a relatively fortunate position in that they had been able to devote most of their time and energies to the design of computers. Although this was of great value to the department and the university, it had meant they had been sheltered to some extent from the hurly-burly of student courses. The staff had, of course, all given lectures, but less of them than most staff. Kilburn felt that this could not go on indefinitely; computers were now of such acknowledged, and increasing, importance that there was a need for a full course in computing, rather than the individual specialised courses in certain aspects of computing which had previously comprised part of the electrical engineering course. Much thought was given to the best way to proceed. One possibility was to introduce a composite course, along the lines of 'Electrical Engineering with Computer Science'. The other main possibility was to take a major initiative and request that the 'computing' section should be allowed to leave the Electrical Engineering Department in order to form a new department offering a course dedicated entirely to computing. [27]

The Electrical Engineering Department in the early sixties was steadily growing and computer staff formed about a third of the total. The computing group was in effect a department within a department. So, on the one hand, there was logic in forming a new department dedicated to computing with the freedom to proceed as it wished. On the other, there was the danger of denuding the Electrical Engineering Department by the departure of its now-famous computer group which comprised a third of the department's staff. Clearly a major decision had to be made. Kilburn felt that the balance of the argument was in favour of setting up a separate department and put the matter to his friend, Professor Williams. After much heartsearching Williams accepted Kilburn's arguments and called a meeting of senior science professors to discuss it. They were persuaded by the arguments. The proposal to set up a new department, to be named the Department of Computer Science, was accepted and the steps necessary to form it were initiated.

As the foundation of the new department drew near the Electrical Engineering staff were given the option of staying in their department or moving to Computer Science. The Electrical Engineering staff in the 1963-64 session, immediately prior to the founding of Computer Science, comprised:

Edward Stocks Massey Professor of Electro-technics: F.C. Williams
Professor of Computer Engineering: Tom Kilburn
Reader in Electrical Engineering: R. Cooper
Senior Lecturer and Tutor: L.S. Piggott
Senior Lecturers: C.N.W. Litting, E.R. Laithwaite, D.B.G. Edwards, G.R. Hoffman.
Senior Lecturer in the Theory of Computing: R.A. Brooker
Lecturers in Electrical Engineering: T.E. Broadbent, Dennis Tipping, J.F. Eastham, David Aspinall, D.C. Jeffries, J.A. Turner, G.F. Nix, E.M. Dunstan, K.F. Bowden, D.E. Hesmondhalgh, P.L. Jones, Derrick Morris, G.F. Turnbull.
Lecturers in Computing: Cicely M. Popplewell, R.B. Payne, F.H. Sumner.
Assistant Lecturers in Electrical Engineering: G.W. McLean, J.M. Townsend, C.T. Elliott, D.J. Kinniment.

Some of these staff were primarily concerned with computers and would obviously move to the new department. Others were involved in very different fields (machines, high-voltage, measurements, servomechanisms, etc.) and would stay. But there were others working in electronics who were peripherally but not directly involved in computers and it was they who had to make a choice. In this latter category two senior staff members, G.R. Hoffman and C.N.W. Litting, opted to stay in Electrical Engineering, thereby ensuring that the department would not be deprived of all its senior electronics staff.

The Department of Computer Science opened officially at the start of the 1964-65 session, before a suitable home was found. Tom Kilburn said that he would need an area of about 100,000 square feet; initially he was offered a brick building, already in existence, on the north side of Dover Street near its junction with Oxford Road but this was regarded as unsuitable and the department remained, geographically, in the Electrical Engineering building for a further year. The new department then moved to part of the old Whitworth Laboratory in Coupland Street which the Engineering Department had occupied from 1909 to 1962-63 when the Simon Laboratories were opened. The Atlas computer remained in the Electrical Engineering Department, physically, after the transfer of personnel to the new Computer Science Department, and remained there until the end of its life in 1971, continuing to earn money for the university in the process.

The staff of the Computer Science Department at its opening comprised:
Professor of Computer Science: Tom Kilburn
Readers: D.B.G. Edwards, R.A. Brooker.
Lecturers: D. Aspinall, K.F. Bowden, E.M. Dunstan, Derrick Morris, Cicely M. Popplewell, F.H. Sumner, Geoffrey Riding.   *Assistant Lecturer*: D.J. Kinniment

It will be seen from these lists that ten of the 11 staff were direct transfers from Electrical Engineering. David Kinniment, who had been appointed Assistant Lecturer in March 1964, was the youngest, having graduated in Electrical Engineering in 1962. 'Geoff' Riding was a new appointment. He took over the running of the Computer User Service when Cicely Popplewell resigned the following year. It will be noticed that the transfer was accompanied by a change of title for Tom Kilburn from 'Professor of Computer Engineering' to 'Professor of Computer Science', and promotion from Senior Lecturer to Reader for Edwards and Brooker.

Many people had mixed feelings at the time regarding the merits of the removal from the Electrical Engineering Department of the computing section - one its most successful components, and one which, over many years, had been very popular with the university authorities since it had brought large amounts of money in through royalties from patents and computer user earnings. From the point of view of the new Computer Science Department it was almost certainly a good move, but inevitably there was a weakening of the parent department, for a time at least. The move was probably inevitable in view of the subsequent burgeoning interest in and importance of computers and the large demand for places in the undergraduate course. At the time of writing (1997), the Computer Science Department has 47 full-time staff members whilst Electrical Engineering has 21. The offspring has long since outgrown its parent.

Research other than in computing

Geographically, Machines and Servomechanisms research were located (from 1954) in the basement of the new Dover Street department; high-voltage, dielectric breakdown and measurements were on the ground floor; high-vacuum on the first floor, and electronics on the second floor. The work of these groups will be briefly reviewed here in that order.

Machines:

After Eric Laithwaite completed his M.Sc. project under Tom Kilburn in 1950 he turned his attention to research in electrical machines, something he had wanted to do for years. He brought to the department (then still housed in the old building on Coupland Street) a linear induction motor stator which he had made, with the intention of using it to drive a shuttle for a loom. F.C. Williams was interested in Laithwaite's idea and joined in his research, which was soon widenened to encompass variable-speed a.c. machines of different types. The first idea involved a stator, excited from a polyphase supply, facing a conducting disc pivoted at its centre. The part of the disc under the stator would move at approximately the same speed as the exciting field. The disc would therefore rotate faster or slower according to whether the stator was moved nearer to or further from the centre of the disc. An elaboration of this idea followed leading to a practical machine, the spherical motor, in which concave spherical stator elements faced the surface of a spherical laminated iron rotor having embedded conductors lying in slots running not only longitudinally but latitudinally as well. The stator elements could be turned so that the air gap remained unaltered and the excitation wave travelled at a controllable angle to the equatorial plane. The ultimate version of the machine had a single movable stator in which the slots had 'pre-skew', so that turning the stator one way from the central position increased the speed and turning it the other way decreased the speed. It was a machine of great originality and was said to have sprung from an idea Williams had whilst watching yachts sailing at an angle to the wind.

Professor Williams explains a variable-speed a.c. motor to an interviewer, with Eric Laithwaite in the background

Research on electrical machines was aided by the appointment to the staff of D. Tipping (1957), J.F. Eastham (1958), G.F. Nix (1960), G.W. McLean (1962) and D.E. Hesmondhalgh (1962). Several highly original machines were invented and a lot of pioneering work was done on linear induction motors. Laithwaite's idea of driving a shuttle using such a motor was developed and he also turned his attention to other possible uses for the linear motor, in particular, transport systems. A feature of the early 1960s was a 'railway' in the basement a.c. laboratory of the Electrical Engineering Department in which a 'test vehicle' powered by a linear induction motor hurtled from one end of the laboratory to the other, transporting Laithwaite or other researchers. Important work was also done on magnetic levitation. The 'machines' work that Laithwaite and Williams introduced and developed was a significant addition to the department's research and was enhanced by the enrolment of several capable research students. Some of the main research projects undertaken by the group are summarised in chronological order as follows (the dates are those of completion):

1954: A brushless variable speed induction motor.
1955: A self-oscillatory induction motor for shuttle propulsion.
1956: Other brushless variable speed induction motors.
1957: New developments in electro-magnetic shuttle propulsion. Variable-speed electric drives.
1958: Unusual induction machines. Continuously variable-speed brushless induction motors.
1959: Development and design of spherical induction motors. The logmotor - a cylindrical brushless variable-speed induction motor.
1960: A brushless variable-speed induction motor using phase-shift control. The application of a linear motor to conveyors.
1961: Further developments of the self-oscillating induction motor. Electromagnetic forces in slotted structures.
1962: An oscillating synchronous linear machine. A theory of the operation of cylindrical induction motors with squirrel-cage rotors.
1963: Pole-changing motors using phase-mixing techniques.
1964: Analysis and design of pole-changing motors using phase-mixing techniques. Analysis of the properties of 2-phase servo motors and a.c. tachometers. Rotor winding for induction motors with arc-shaped stators. Linear induction motors for high-speed railways. An experimental impact extrusion machine driven by a linear induction motor.

This list is by no means complete; the above titles have been selected merely to give a representative idea of the type of work being done. Several Premiums were awarded by the Institution of Electrical Engineers for papers published by the group between 1950 and 1964. Eric Laithwaite resigned in 1964 on his appointment to a Chair of Electrical Engineering at Imperial College, London, but machines research continued under the supervision of Professor Williams and the newer staff members.

Servomechanisms

J.C. West joined the staff as an Assistant Lecturer in 1946. He was interested in servomechanisms, having gained experience as a lieutenant in the Royal Navy during the war of such matters as the position control of gun turrets. Some of his early work at Manchester was on the drum servo system for the Mark I computer. Soon he was publishing papers on servomechanisms and taking research students. One of his first investigations was on the measurement of the speed/torque characteristics of 2-phase induction motors and Magslips by a speed control servo system. In the 1950s West,

1950-1964: Golden years

with a number of different collaborators, completed research investigations on a range of problems connected with servomechanisms of which the following are a selection:

1951: A system utilising coarse and fine position elements simultaneously in remote-position control servomechanisms.

1952: Connection between closed-loop transient response and open-loop frequency response.

1953: Step-function response of an R.P.C. servomechanism possessing torque limitation. Effects of the addition of some non-linear elements on the transient performance of a simple R.P.C. system possessing amplifier limitations. The frequency response of a certain class of non-linear feedback systems.

1954: Subharmonic generation in non-linear feedback systems. Frequency response of a remote position control servomechanism with hard spring nonlinear characteristics and the transient behaviour of the system. The necessary torque requirements for a servomotor.

1955: Dual input describing function and its application to feedback systems. Application of analogue techniques to a continuously rotating drum. Frequency response of a servomechanism designed for optimum transient response.

1956: Integral control with torque limitation. The describing function analysis of a non-linear servomechanism subjected to stochastic signals and noise. Automatic control analysis. Use of frequency response analysis in non-linear control systems.

1957: Effect of non-linearity on the statistical behaviour of feedback systems. Drum store for analogue computing.

At the end of 1957 J.C. West, who had led his servomechanisms group very successfully for 11 years and had won several IEE Premiums in the process, as well as writing a successful text book on the subject, [28] left to take the Chair of Electrical Engineering at Queen's University, Belfast. Over the years he had supervised a number of very able research students, one of whom, J.L. Douce, who had graduated in Physics in 1952, went on to become Professor of Electrical Engineering at Warwick University. Another, J.L. Leonard, went into industry and enjoyed a very successful career.

Following the departure of John West, servomechanisms research was continued by M.J. Somerville (appointed 1954) and D.P. Atherton (appointed 1958 and resigned in 1962), though Somerville was as interested in measurements as servomechanisms. Some of the projects completed between 1958 and 1964 were as follows:

1958: Multi-gain representation for a single-valued non-linearity with several inputs and the evaluation of their equivalent gains.

1961: Stability of a feedback system containing a limited field-of-view error detector. Evaluation of the response of single-valued non-linearities to several inputs.

1962: Design of an accurate simulator for sample data systems. Evaluation of the response of single-valued non-linearities to several inputs. The rapid evaluation of the auto-correlation function of the output of single-valued non-linearities in response to sinusoidal and Gaussian signals.

1964: Double input describing functions for unrelated sinusoidal signals.

Electric breakdown in gases and solids

D.R. (Denis) Hardy was appointed in 1950 with the intention of starting high-voltage research and, eventually, being responsible for research in the proposed 1MV laboratory in the new building. However, at the time of his appointment the new building was still four years from completion and high-voltage facilities in the old department were limited to a 250kV multi-stage impulse generator donated by

Metropolitan-Vickers in the late 1930s, which had been designed by F.S. Edwards who had graduated from the department in 1921. Hardy, who had recently completed a London Ph.D. in high-voltage work, was very enthusiastic about research and scoured the sales of ex-service equipment to purchase stocks of high-voltage capacitors, transformers and anything else that might prove useful. In 1952 he enrolled his first research students; T.E. Broadbent, J.K.Wood and H. Wroe, all of whom had graduated from the department that summer, and B. Jackson who had graduated in Physics. A programme of research on various aspects of the electrical breakdown of air began, and more research students were taken on in 1953 and 1954. However, Hardy resigned at the end of 1954 to take up an appointment with Brush Electrical Engineering Co., and was succeeded in the summer of 1955 by R. Cooper, renewing his connection with the department after an absence of nine years. His interest at that time was in various aspects of the breakdown of solid dielectric materials and he instituted a vigorous programme of research. Broadbent gained his Ph.D. in 1955 and left to spend two years in industry, but returned in 1957 as a staff member and continued his researches into gas breakdown, mostly using the 1MV laboratory. These two lines of research, electrical breakdown in gases and in solids, thereafter ran in parallel. Some of the research projects carried out in the period covered in this chapter were as follows:

1953: Impulse calibration of 2 cm dia. sphere gaps.
1954: Effects of irradiation on the calibration of 2 cm dia. sphere gaps. Protective spark gaps with radioactive substances.
1955: Control of high-voltage impulse generators. A thermally-triggered spark gap.
1956: Mechanism of breakdown in triggered spark gaps. Time lags in the intrinsic breakdown of solid dielectrics. Electric strength and surface conditions of the alkali halides.
1957: The influence on electric strength of mechanical deformations caused by the application of voltage to KCl crystals. Directional effects in the electric breakdown of single crystals of KCl and NaCl.
1959: Surge diverters using trigatrons. Breakdown of alkali halide crystals.
1960: High-voltage triggered spark gaps. A new high-voltage multi-stage impulse generator circuit. The characteristics of the trigatron spark gap at very high voltages. Influence of ionic conductivity on the electric breakdown of KCl and NaCl.
1961: Tripping a single-stage impulse generator. Electric breakdown of sodium chloride.
1962: Spark track in air at very high voltages. The hot wire triggered spark gap at very high voltages. Anisotropy in the electric strength of alkali halide crystals.
1963: Development of the discharge in the trigatron spark gap at very high voltages. Intrinsic electric strength of polythene.
1964: A new high-voltage triggered spark gap. Spark initiation in small trigatrons. Electric strength of solid insulation.

In addition to the above, a lot of work was done between 1955 and 1964 in collaboration with British Insulated Callender's Cables on prototype submarine cables for the 600MW Cook Strait power link between the north and south islands of New Zealand and for submarine cables for the English Channel. Many tests at 400kV on special purpose capacitors were made for the same company. Work was also carried out in collaboration with Ferranti Ltd. on the calibration of high-voltage resistors and with the English Electric Co. in connection with the setting up of the new 2.5MV impulse generator at their transformer testing laboratory in Stafford.

Measurements

Research in this subject began in 1954 with the appointment of M.J. Somerville. Much of the work of this group was carried out in association with other groups and is

1950-1964: Golden years

therefore covered in the above lists. Some of the 'measurements' projects were:
1955:   A circuit for analogue formation of xy/z. An a.c. potentiometer for measurement of amplitude and phase of the fundamental component of a waveform.
1959:   Reduction of low-frequency noise in feedback integrators.
1960:   Work concerned with analogue and digital computers.
1962:   Design of an accurate simulator for sample-data systems. Filter synthesis using active RC networks.
1963:   A self-generating H.F. carrier feedback anemometer.

High-vacuum, Electronics and Computing

As might be expected, the work of various groups overlapped. Consequently several people who were not members of the computer design team nevertheless contributed to the design of computers, and conversely the work of the computer designers was often relevant to non-computing fields. The work of the computer group has been dealt with in detail earlier in this chapter and will not be repeated. The list below consequently includes project titles of general electronic interest but not those directly concerned with the design of a particular computer, nor computer programming work.

1951:   Various investigations relevant to computer storage systems.
1952:   Display of transistor characteristics in the cathode ray oscillograph. A method of designing trigger circuits. An automatically controlled Knudsen-type vacuum gauge.
1953:   A simple analogue divider.
1954:   The transistor regenerative amplifier as a computer element. Sealed off cathode ray tubes with very high writing speeds. A balanced electrometer amplifier. A method of measuring the emissive properties of insulators. The physics of cathode ray tube storage. A Pirani gauge circuit.
1955:   The reading of magnetic records by reluctance variation. Delay line circuits.
1956:   Quiescent core transistor counters. A new and simple type of digital technique using junction transistors and magnetic cores. A multi-input analogue adder for use in a fast binary multiplier. Design and operation of a parallel type CRT storage system. A feedback modulator. Noise and negative feedback in amplifiers.
1957:   An accurate electroluminescent graphical output unit for a digital computer. Graphical recording.
1959:   The reduction of low-frequency noise in feedback integrators. A new fast carry circuit. High speed digital storage using cylindrical magnetic films. A system for the automatic recognition of patterns. A d.c. tachometer generator. Experiments in machine learning and thinking.
1960:   A parallel arithmetic unit using a saturated transistor fast carry circuit. Ferrite magnetic core memory system with rapid cycle time. High-speed light output signals from electroluminescent storage equipment.
1961:   Character recognition by digital computer using a special flying-spot scanner. One-level storage. A cathode ray tube output for a digital computer. Electroluminescent matrix for pattern display and recording. The light waveforms emitted from electroluminescent cells energised by square waves and pulses of voltage. The physics of computer elements.
1962:   Ultra-high vacuum system. An electroluminescent fixed store for a digital computer.
1963:   Tunnel diode storage system with non-destructive read-out. Design of second harmonic detector heads and their application to the reading of digital information and measurement of low speed of rotation. Variation of coercivity of magnetic materials with driving field. A computer fixed store using light pulses for read-out. A high-speed large capacity fixed store for a digital computer. The shape variation of hysteresis loop of magnetic

materials with changes in the rate of rise of driving field. Investigation of computer-controlled traffic signals by simulation.
1964: High-speed ferrite core storage system. A method for the continuous measurement of thickness and deposition rate of conducting films during vacuum evaporation.

It is hoped that the above project titles give some idea of the research work being carried out in the Electrical Engineering Department between 1950 and 1964. It was a highly rewarding period in many different ways.

References: Chapter 12

1. Included in Professor Eric Laithwaite's tape-recorded recollections of the Electrical Engineering Department communicated in March 1991.
2. Faculty of Science *Syllabus of Classes*, various years.
3. Mr G.A.Cooper's tape recorded recollections, 29th April 1987.
4. Ibid.
5. As reference 1.
6. *Manchester University Extensions.* Article published in *Electrical Review*, 1st July 1955, pp. 25-27.
7. As reference 3.
8. S. H. Lavington. *A History of Manchester Computers*. NCC Publications, the National Computing Centre, Manchester, 1975, p. 26.
9. Ibid, p. 20.
10. Ibid, p. 22; also the following:
    Simon Lavington. *Early British Computers*. Manchester University Press, 1980, p. 42.
    R. A. Brooker. *An Attempt to Simplify Coding for the Manchester Electronic Computer*. British Journal of Applied Physics, **6**, 1955, pp. 307-311; and *The Programming Strategy used with the Manchester University Mark I Computer*. Proc. I.E.E., **103**, Part B, Supp. 1 - 3, 1956, pp. 151 - 157.
11. Electrical Engineering Department Annual Report, 1958-59.
12. E.R. Laithwaite: *An Automatically-Accessible Three-Dimensional Library for use in Digital Computers*. M.Sc. thesis, University of Manchester, 1950.
13. As reference 8, but pp. 24-25.
14. T. Kilburn, D. B. G. Edwards and G. E. Thomas. *The Manchester University Mark 2 Digital Computing Machine*. Proc. I. E. E., **103**, Part B, Supp. 1-3, 1956, pp. 247-268.
    Edwards, D. B. G. *The Design and Operation of a Parallel Type Cathode Ray Tube Storage System*. Proc. I.E.E., **103**, Part B, Supp. 1-3, 1956, pp. 319-326.
    F. C. Williams, T. Kilburn, C. N. W. Litting and G. R. Hoffman. *Recent Advances in Cathode Ray Tube Storage*. **100**, Part 2, No. 77, 1953, pp. 523-543.

G. E. Thomas. *The Use of Electromagnetic Delay Lines in the Manchester University Mark 2 Digital Computing Machine.* Proc. I. E. E., **103**, Part B, supp. 1-3, 1956, pp. 483-490.
15. K. Lonsdale and E.T. Warburton. *Mercury: A High-speed Digital Computer.* Proc. I.E.E., **103**, Part B, supp. 1956, 1-3, pp. 174-183.
16. S.H. Lavington (see ref. 8), pp. 25-26.
17. R. A. Brooker. *The Autocode Programmes developed for the Manchester (Mercury) Computer.* **1**, 1958, pp. 15-21; and R. A. Brooker. *Further Autocode Facilities for the Manchester (Mercury) Computer.*Ibid., pp.124-127.
18. As reference 8, but p. 30.
19. T. Kilburn, R. L. Grimsdale and D. C. Webb. *A Transistor Digital Computer with a Magnetic Drum Store.* Proc. I. E. E., **103**, Part B, Supp. 1-3, 1956, pp. 390-406.
20. As reference 8, but p. 20.
21. Ibid.
22. W. Renwick. *EDSAC 2.* Proc. I.E.E., **103**, Part B, Supp. 1-3, 1956, pp. 277-279.
23. Tape-recorded informal interview with Tom Kilburn, 1st July 1992, in which he recalled the early years of Manchester computers. The various quotes attributed to Tom Kilburn in this chapter are from this source except where stated otherwise.
24. As reference 8, but see the appropriate papers listed in the bibliography of Lavington's book and the further papers listed therein.
25. As reference 8, but pp. 34-35.
26. As reference 8, but p. 37.
27. Tom Kilburn's tape-recorded interview (see reference 23).
28. J.C. West. *Servomechanisms.* English Universities Press. London, 1953.

The author is indebted to Professor Simon Lavington of Essex University (formerly of Manchester University) for allowing him to quote extracts from the book, *A History of Manchester Computers.*

Chapter 13

1964-1977: Changing times, and the end of the Williams era

The years covered in this chapter were ones of great change, not only within the university but in society in general. The so-called 'swinging sixties', exemplified by The Beatles, the onset of the 'permissive society' and the challenging of authority, heralded major departures in outlook for the younger generation, for better or for worse according to one's viewpoint, from the norms of earlier years. A popular song of 1963 solemnly informed us that *The Times They Are A-Changin'*. It is unlikely that Bob Dylan had British universities in mind when he wrote it, but the universities were, nonetheless, not immune from the revolution affecting the rest of society.

Changes came about within the university for two main reasons. Firstly, it was a period of rapid expansion within the university system. Associated with this was a general increase in administrative bureaucracy and complexity. Secondly, because of "student troubles" throughout the country during the late 1960s, not unconnected with the revolution referred to in the opening paragraph, far-reaching changes were made in the way the universities operated and, in parallel with this, universities were subjected to ever-increasing financial stringency so that departments had to seek to "bring money in from outside" as never before.

The Electrical Engineering Department had its own particular problems. It was affected indirectly by the rapid expansion in university education that was taking place. Moreover, it was a period in which it had to accommodate itself to the loss of its thriving computing section. And finally, these years proved to be the last ones of the Williams era. His unique presence as leader of the department shaped its character for over thirty years, and it was therefore inevitable that life would be very different once he was no longer there.

University expansion

The main thrust of the Robbins Report on Higher Education, published in 1963,[1] was the need, as the committee saw it, for an enormous expansion in the university sector. This would be achieved both by increasing the number of students entering the existing universities and by creating a number of new universities. These changes would apply 'across the board' of academic disciplines but were seen as particularly important in subjects such as electronic and electrical engineering which were regarded, with justification, as crucial to the economic wellbeing of the nation.

1964-1977: Changing times, and the end of the Williams era

The recommendations of the Robbins Report came as no surprise, because large scale university expansion had been actively promoted by the main political parties as a desirable attainment for years. Even before the report was published new universities were being established in anticipation of its findings, for example Essex and Sussex (1961), Keele (previously the University College of North Staffordshire)(1962) and York (1963). After the report was published other institutions, whose creation or change of status was already being negotiated, became universities, including Kent and Lancaster (1964), Warwick (1965), Brunel, Loughborough and Surrey (1966) and Salford (1967). Some of these came about through the elevation of what had previously been Technical Colleges, or more recently, Colleges of Advanced Technology.

The Electrical Engineering Department at Manchester was somewhat immune to an increase in numbers because its expansion had already taken place in the years 1946 to 1964, during which the department had moved to the new purpose-designed building intended for an undergraduate intake of about 50 per year. In that period the number of full-time staff members had increased from six at the start of the 1946-47 session to 29 during the 1963-64 session, a figure which fell to 19 in the following session because of the departure of the Computer Science staff. During the period covered in this chapter the number of electrical engineering staff stabilised. The figures, taking three-yearly samples, were 23 (1967-68), 23 (1970-71), 24 (1973-74) and 24 (1976-77). In the same period the numbers of students graduating from the department increased but not dramatically so; in the three years 1964-1966 inclusive, 147 students graduated (B.Sc.) whereas in the three years 1975 to 1977 the number was 181, an increase of about 22%. [2]

So, if the number of staff in the department remained nearly constant during this period and if the number of students passing through did not increase much, in what way did university expansion elsewhere affect the department? The answer is that it became much more difficult than before to find the right numbers and quality of applicants, since so many places in engineering were now available elsewhere. Prior to about 1960, engineering had only been taught at Oxford, Cambridge, Durham and a handful of the so-called 'redbrick' universities which included Manchester. It had been sufficient, as far as Manchester was concerned, for word to spread, as though written on tablets of stone, that the Electrical Engineering Department at Manchester had a good reputation; applicants did not usually bother to ask about details of the course or compare it with the courses offered in other universities. Consequently there had always been an adequate supply of good applicants and the department had never needed to 'sell itself'.

Electrical Engineering at Manchester University

After the early sixties things were different. So many places were now available in engineering in the old and new universities that students became more selective and started to make careful comparisons between courses before making their choice. The result was that, by the mid 1960s, the number of students applying for admission to the department, whilst still reasonably high, was less satisfactory than before. So, for the first time in its history, the department had to pay attention to the business of attracting students to Manchester in preference to other universities. The need for the department to take positive action to attract students has continued as the years have gone by and is, of course, still with us at the present time. By the mid-sixties, the day of the glossy brochure was drawing closer.

The first measure that the department put in place to attract students was to change the title of its main course* from 'Electrical Engineering' to 'Electronic and Electrical Engineering'. This came about after listening to the views of applicants who were, in the main, attracted more by the word 'electronic' than the word 'electrical'. It was always made clear to applicants that the department's 'Electrical Engineering' course had just as much 'electronics' in it as courses in other universities which were called 'Electronic and Electrical Engineering' and that, indeed, the department had a proud recent history of electronic innovation. Nevertheless, nearly all prospective students said that they preferred to apply for courses which specifically included the word 'electronic' in the course title.

The introduction of the new course title of 'Electronic and Electrical Engineering' did not necessitate any major changes to the course syllabus, except for some reduction in the 'civil' and 'mechanical' content, and the department continued to be known as the Electrical Engineering Department. But an important difference between the old system and the new was that students would no longer graduate in 'Engineering', as they had done since the department was founded, but in 'Electronic and Electrical Engineering'. Under this new regime the 'electronic and electrical engineering' students would appear on their own degree results list, instead of being included with the civil, mechanical, aeronautical and nuclear engineers on the 'Engineering' results list as before. The change thus marked the end of a formal link with Engineering of more than 50 years' standing.

Certain changes had to be made in the regulations for the course and the first applicants for the newly-titled course entered in 1965, graduating in 1968. The effect of the new course title was dramatic; as soon as it was

---

* It will be remembered from the previous chapter that the department also gave half the teaching in the 'Physics and Electronic Engineering' course, given jointly with the Physics Department.

offered, applications doubled. This measure alone proved sufficient to ensure a ready supply of good applicants for the next few years, but by the mid seventies applications were once again drifting downwards because of the very large number of engineering places that had by then become available elsewhere through the proliferation of new universities and the expansion of existing ones. In 1975 the decision was taken to produce a departmental brochure for applicants which would be sent out free to all schools and other establishments on the Universities Central Council on Admissions (UCCA) list. The first issue (4,000 copies) was produced and issued in 1976 and, revised annually, it has been a regular feature of the recruitment strategy ever since.

Staff changes

New staff

Between 1964 and 1977, both years inclusive, 24 new young members of staff were appointed, including two Temporary Lecturers and one Honorary Lecturer (see Appendix 1). One (D.J. Kinniment) moved to Computer Science in 1964 when it opened. Five (D.G. Buchanan, J.L. Mudge, S.F. Gourley, P.A. Wade and N.P. Lutte), resigned before 1977 (names are given in chronological order of appointment) and the appointments of two of the Temporary Lecturers (R.A. Harris and D.A. Edwards) had expired before that date. Six resigned in later years (B.E. Jones, E.F. Taylor, W.W. Clegg, H.W. Thomas, F.G. Abbosh and J.R. Hawke) and three retired (P. Lilley, R.M. Pickard and I.E.D. Pickup), but seven are still on the staff at the time of writing (late 1997). They are (year of appointment in brackets) C.J. Hardy and B.R. Varlow (both 1965), D.W. Auckland and D. King (both 1968), M.J. Cunningham (1970), R.S. Quayle (1975) and R. Shuttleworth (1976).

Two new professorial appointments

In December 1966 R. Cooper and G.R. Hoffman were promoted from Reader to Professor. With F.C. Williams the undisputed leader, the professorial complement of the department was now three. It was not the first time the department had had more than one professor, as Tom Kilburn had been created Professor of Computer Engineering in 1960 and retained that title in the department until 1964 when the Department of Computer Science was established and he became Professor of Computer Science. Before that, Robert Beattie had been in solitary charge of electrical engineering from 1912 to 1938, Willis Jackson from 1938 to 1946, and F.C. Williams from 1946 to 1960.

Ron Cooper's original research interests had been on the use of very high frequencies in radar (see Chapter 10) and problems associated with discharges

Electrical Engineering at Manchester University

in gases and semi-conductors. Since joining the department from Liverpool University as a Senior Lecturer in 1955 (promoted to Reader in 1962), he had established himself as a leading international authority in the area of breakdown and pre-breakdown conditions in solid dielectrics.

"There's only one easy place to be in science and engineering, and that's in front."

"If you're there first you have nothing to read, you've got all your time to think."

This page and opposite: Four characteristic studies of Professor F.C. Williams, in which he explains his views on invention (as quoted) to a reporter from the magazine 'International Science and Technology' in 1964

1964-1977: Changing times, and the end of the Williams era

"I think it's a great mistake to learn too much, to be too good at any one thing, because this tends to become important in itself."

"The scientist is on an easy wicket compared with the engineer . . . research is more open to a method approach."

G.R. ('Pete') Hoffman joined the department as a research student in 1950 and passed through all the ranks from Assistant Lecturer to Reader before his promotion to Professor. His research interests were originally in high vacuum work but over the years they had widened to include high-speed storage systems for computing machines and problems relating to thin magnetic films and electro-luminescent and superconducting devices.

In September 1977 the retirement took place of Mr G.A. Cooper, known to everyone as Albert, who had been a member of the department's technical staff continuously from his appointment as a 14 year old in 1929. He was Laboratory Superintendent (originally called Laboratory Steward and later Chief Technician) for all except the first five years of this period. After nine years of retirement he died of a heart attack on 26th March 1988. He is remembered with affection by many generations of former students.

The undergraduate courses
Electronic and Electrical Engineering

With the introduction of the new course title of 'Electronic and Electrical Engineering' the opportunity was taken to reduce the 'civil' and 'mechanical' content, as stated earlier. The syllabus throughout the period described in this chapter evolved from year to year. Clearly it would be impracticable to record the detailed changes but the following account is representative: [3]

First year: In the earlier part of the period the course structure was as described in the previous chapter. From the late 1960s onwards a first-year 'Core plus Options' system operated, as follows:
Core: First half session: Fundamentals of Electricity and Magnetism (d.c., single-phase and three-phase a.c. circuits; electromagnetism; electrostatics; induced EMFs; waveform analysis).
Second half session: Electronics; Electrical Machinery (d.c. and a.c. machines); Electrical Measurements; Analogue Computation. Students also studied Mathematics throughout the year and in the later part of the period took a course in computer programming. The lecture courses were supported by related work in the Electrical Laboratory in which approximately 20 experiments were carried out during the year. Tutorials were a weekly feature of the course.
Options: Students took two subsidiary courses selected from General Engineering (civil and mechanical), Physics, Introduction to System Programming, and Metallurgy. Inorganic Chemistry featured on the list of options for several years but was dropped during the 1970s as it was rarely selected by students.

Second year: Students attended courses in Electrical Machinery, Measurement and Feedback, Electronics, Electric Circuit Theory, Transmission of Electrical Energy (which covered communications and power transmission), Magnetic Materials, and Mathematics. Laboratory classes were held in Electronics, Electrical Machinery, and Measurement and Control. There were also High-voltage and High-vacuum experiments.

Third year: The courses available (about 40 lectures each) were: Electrical Machinery; Electromagnetic Waves; Power Transmission; High-voltage Engineering; Control Engineering; Electronics; Physical Electronics; Communications and Noise; Computer Systems Design.
The third year laboratory work consisted of an experimental project which, from the mid 1960s, occupied the whole of the final year. Students generally worked in pairs, as had always been the case up to that time. A few students took projects offered and supervised by the Physics

1964-1977: Changing times, and the end of the Williams era

Department (usually by staff from Jodrell Bank), and some Physics students carried out Electrical Engineering projects.

In each year students were examined on the subjects studied during the year. Until 1964 the structure of the examination system was the same as it had been for many years; students took the Intermediate B.Sc. Examination at the end of the first year, Part I Honours at the end of the second year, and Part II Honours at the end of the third (final) year. In 1965 a new system was introduced in which students would take the 'First Examination' at the end of the first year, the 'Second Examination' at the end of the second year, and the 'Honours Examination' at the end of the final year. These changes were introduced for students entering in 1965, students already enrolled in the course completing their degrees under the 'old regulations'. The new regulations therefore worked through the system year by year and were in place for each year by the end of the 1967-1968 session. The main purpose of the change was to allow students a second chance to pass the examination held at the end of the second year. The university regulations had always stated that only one attempt to pass an Honours examination was allowed. Therefore, under the old regulations, in which students took Part I Honours at the end of the second year, students whose examination results at the end of that year were deemed not up to the necessary standard for an Honours degree had to be transferred to the Ordinary Degree course, and if their results were below Ordinary Degree standard they would have to come back as an external candidate another year to try again to pass the examination at Ordinary Degree level. Under the new scheme, students could try again in September to pass the Second Examination failed in June and, if successful, could proceed to the final year of the Honours Degree course. So, following the introduction of these changes, the number of Ordinary Degrees diminished. From then on, students whose final year examination results were not up to Honours standard, but of a standard equivalent to the Ordinary Degree, were awarded the Pass Degree of B.Sc.

Physics and Electronic Engineering

Throughout the period up to 1977 the syllabus of the course remained similar to that outlined on page 196; i.e. a combination of subjects relevant to Physics and to Electronic Engineering in roughly equal proportions. Projects were usually carried out in the Electrical Engineering Department but sometimes students undertook their projects in the Physics Department, usually under the supervision of a staff member from Jodrell Bank. In the earlier part of the period the examination structure followed that used in the Physics Department, in which students took the Preliminary Honours Examination in Physics at the end of first year, Part I Honours at the end of second year, and Part II Honours at the end of the final year. In 1968 the same system was adopted as had been introduced in the Electronic and Electrical Engineering course, in which students took the First Examination, Second Examination and Honours Examination at the end of the first year, second year and final year respectively.

'Student troubles' and their aftermath

The changes in social outlook occurring during the 1960s, especially amongst the young, manifested themselves in a period of prolonged student unrest and rebellion during the latter part of the decade, not only in Britain but worldwide. The departments comprising the Faculty of Science at Manchester, including Electrical Engineering and the other engineering departments, were not immune from the effects of the unrest but were less affected than many other departments in the university. The reasons are not hard to find. Firstly, science students in general, and possibly engineering students in particular, tend to be less politically-minded, and less militant, than those in departments such as history, economics and sociology where political matters are, by the nature of the courses, in the forefront of the students' minds. Secondly, engineering students are, in general, interested primarily in their subject of study. 'Engineering' is not a discipline normally studied at school, so a reasonably large measure of motivation and interest in the subject is needed for a student to want to study it at university. Finally, a glance at the students' timetables in departments such as Electrical Engineering will indicate that the students had little time for spending long hours in the university union plotting the downfall of the establishment. Consequently, the students' marches, the illegal occupation of various university buildings, the issue of inflammatory literature and similar pursuits, were not usually the province of electrical engineering students but were, in the main, the work of their more militant counterparts in other departments. This is not to say that the 'electrical' students took no interest in what was going on; they were rightly aware of all the issues being raised and had their share of grumbles, but they were content to pursue their studies normally and there were never any militant protests in the department.

Many of the complaints implicit in the 'troubles' were political and will not be considered further here. Those related to academic courses were twofold: firstly the lack of student representation on various university bodies at all levels from the University Council to departments, and secondly, inadequacies, as students saw it, in the university examination system. The year 1968 was marked by protests of many kinds, including marches and the occupation of the Whitworth Hall and the telephone exchange, the latter 'sit-in' making communication between the university and the outside world difficult. An Emergency General Meeting of the Manchester Students' Union was held on 4th December 1968 to air their complaints about examinations and assessment. Under heading (1) of a resulting document, a lengthy criticism of the existing examination system was put forward. It then continued: [4]

"This EGM mandates the Executive to demand of the University:
(a) The immediate investigation and implementation of a system of assessment on each student which depends on his actual ability rather than upon his ability to jump over the right hurdle. This investigation should consider particularly a mixed system of assessment which includes such methods as continuous assessment, 'open book' examinations, open ended examinations, dissertations and projects.
(b) The abolition in any future assessment of the three-hour closed examination.
(c) That systems of assessment be designed to conform with the requirements of individual subjects.
(d) That individual departments be required to publish
   (i) their purpose in setting examinations.
   (ii) the criteria for assessment agreed between the examiners within a department.
This information should be open to consideration by the students.
(e) That all scripts for written tests be returned to students.
(f) That an appeals procedure be set up. Such a procedure shall provide an opportunity for a student to bring to the attention of the department, social, medical, financial or other worries that may have affected his performance.
2. The EGM further mandated the Executive to demand that until the above is implemented:
(a) No student shall sit more than one Examination per day.
(b) There shall be automatic rights to re-sit.
(c) All marked scripts shall be returned to students.
3. This EGM further believes that any consideration of the examination system would be fruitless without more co-operation and cohesion between staff and students and that representation by all interested parties is an essential part of such co-operation.

This EGM therefore calls for the formation of staff/student committees in all departments and for the representation of students at a faculty level in order that the principles and implementation of methods of assessment may be discussed.
4. This EGM further believes that the University should make representations to Local Education Authorities to provide the necessary financial support in cases where the University recommends that a student should be given the opportunity to repeat a year or change his/her course.
5. This EGM finally mandates the Executive to report back to an EGM not later than the 24th January, and if the EGM considers the University's reaction to the above demands unsatisfactory, to organise a boycott of lectures and tutorials."

The university authorities, and the Vice-chancellor (Sir William Mansfield Cooper) in particular, were unhappy about the document, not so much about the proposals, which amounted only to the opening of a discussion, but to the tone, in which students expressed their views in terms of 'demands' backed up by threats of further action should these demands not be met to their satisfaction. However, the Vice-chancellor circulated a letter to departments requesting that the matters referred to should be discussed within departments. Lengthy papers were circulated, including *Examination by Assessment* [5] and *The Restless Student* [6], written by people from other institutions. Staff were asked to discuss the issues raised with students in their tutorial groups and staff meetings were held. Some of the students' 'demands' were thought by electrical engineering staff to be reasonable, others were not, and the views were communicated to the university authorities as requested. Similar discussions took place in all other departments, the students' deadline came and went, and eventually some of the students' demands regarding the examinations procedures were implemented, in particular, the wish that no more than one examination per day should be taken, and the request that different forms of examination be offered as an alternative to the standard 3-hour paper. The 3-hour paper was retained in electrical engineering as being by far the most reliable method of assessing a student's knowledge, but from that time onwards students have been given written statements of the exact criteria necessary to pass the examinations in each year. Continuous assessment has always been applied to laboratory work. However, in spite of the efforts of most departments to introduce new forms of assessment, the general unrest amongst sections of the student community continued for two more years. During that period work in the Electrical Engineering Department continued much as usual. None of its students was amongst the militant student leaders who by then had turned their attention to issues such as 'political files', but the department was affected, as was every department, by periodic student invasions of university buildings and their disruption of the university telephone service.

Following extensive discussions a staff/student committee was set up within the Electrical Engineering Department with two student members from each year, elected by the students, and about six staff members, the staff membership changing from meeting to meeting (except for the chairman and secretary) according to the nature of the items on the agenda. The Faculty of Science also set up a staff/student committee with one student representative and one staff representative from each course. These committees still exist and serve a useful consultative function, enabling many potential problems to be nipped in the bud. Further developments took place on 5th March 1970 when certain resolutions were approved by the University Senate which were a prelude to a revision of the University Charter. The new 'Supplemental Charter and Statutes' came into effect legally in February 1973.[7] One of its main provisions was that Departmental Boards be set up "which shall advise on courses of study within the several Departments and on other matters of concern to them". In accordance with the new legal requirement a Departmental Board was set up in electrical engineering. None of the three professors wished to be the Chairman. D. Tipping and T.E. Broadbent were nominated by Professor Williams and appointed Chairman and Secretary respectively, and the first meeting was held on 3rd April 1973. By the end of 1977, 24 meetings had been held and numerous matters had been discussed. The Board provided a useful forum for ideas to be exchanged, proposals put forward, and decisions made.

The student unrest of the late 1960s and early 1970s had been a bitter blow to the Vice-chancellor, Sir William Mansfield Cooper. Brought up in a more gentlemanly era and having served the university for over 30 years, his idea of a university was a place for the quiet furtherance of learning and scholarship. He was not equipped to deal with mobs of chanting students illegally occupying buildings. He had been due to retire in the autumn of 1970 and duly did so. It was unfortunate that the end of his distinguished service to the university was spent an atmosphere of aggressiveness, tension and unrest. He died in November 1992 at the age of 89.

After the measures described above were put in place and the new charter implemented the unrest gradually subsided, though occasional bursts of militancy arose from time to time throughout the 1970s.

'Atlas' is scrapped

When the staff of the newly-formed Computer Science Department moved out of the Electrical Engineering building in 1965 the Atlas computer which had come into service in 1962 remained there, the decision having been made that it would be pointless to move it to a new location. Consequently the top floor of the building remained a hive of activity for several more years as large numbers of people, from various departments of the university and from outside, continued to use it. The design of a successor to Atlas was initiated by the Computer Science staff around 1966 and by the early 1970s various pilot versions of the new computer, named MU5, were in service. Atlas was then almost at the end of what was considered to be its useful life and it was finally shut down, accompanied by what Simon Lavington described as "appropriate valedictory libation", on 30th September 1971. [8] Its physical removal from the building after its shut-down was rather less ceremonious. It was effected by the demolition men hurling the various units down from the third floor windows of the building to the ground below. How are the mighty fallen! Thus came about the rather ignominious end of a fine computer which, over a nine-year period, earned a lot of money for the university. The room that Atlas had occupied eventually became a micro-machines laboratory.

Electrical Engineering at Manchester University

University 'Open Day'

In the autumn of 1974 the university decided to hold an Open Day on a Saturday during the following summer. Its purpose was threefold:

1) To enable local citizens to observe some of the work going on and to learn of its practical relevance, in the hope that the myth of the university as an 'ivory tower' would be dispelled.

2) To present special demonstrations and exhibits of an entertaining kind that would be of attractive to the general public and to children.

3) To use the event as an opportunity to promote the university to any young people who might be thinking of applying for a course of study. Staff were expected to be available to discuss admissions policy and related matters and to have suitable recruiting literature on hand.

After months of meetings and preparation, the Open Day duly took place on 17th May 1975 and was generally considered to be a great success. Hundreds of people passed through the Electrical Engineering Department and many appreciative comments were received from visitors. Each research group in the department presented special displays or experiments. An appropriate 'tour of the building' was suggested and signposted. Special literature was produced as 'hand-outs', printed on 'Open Day' stationery, including an account of the research work being done in the department at the time. The department's demonstrations and exhibits were as follows: [9]

Ground floor - start of tour
1MV High-voltage laboratory: Demonstration, every 15 minutes, of the corona discharge, an electric mill, long sparks and the protection of buildings against lightning.
Sound recording demonstrations: Reduction of the "hiss" in audio recording systems. (Demonstration repeated every 15 minutes).

Children's electric go-kart

This page and opposite: Three scenes from the Open Day of 1975

1964-1977: Changing times, and the end of the Williams era

Children try a 'hands-on' experiment

Technician Arthur Gledson explains a point

Basement
Machines laboratories: Demonstrations of the interaction of permanent magnets and electromagnets with common materials. Working models of simple electric machines and commercial motors, including an 'electromagnetic gun'.

Third floor
Power supply laboratory: Micro-machines laboratory. Reaction of machines to short-circuits applied to their terminals.
PDP 11 computer: Description of the computer and its capabilities.
Digital computer games: The first game enabled the computer to measure the time it took the player to respond to the sudden switching on of a light. Three attempts were allowed and the

results were printed out by the computer. The second game tested the skill of the player to follow a moving band of light with a pointer. At the end the computer displayed an analysis of the player's performance. In another demonstration, a computer was able to punch the player's name on a length of paper tape. To participate, the player was asked to write his or her name on a piece of paper, and show it to the typist. The player was then given a card to 'post' in the computer room and the computer did the rest.

A simulated moon landing: This involved an analogue computer. Participants were invited to try their hands (both were required) at landing a lunar module on a flat part of the moon's surface. The hand controls were connected to the computer which simulated the landing and the results of control movements were displayed on a screen.

A floating ball demonstration: This demonstration was for amusement but the literature provided explained that it illustrated the principle of control systems using electric feedback. The system used was not a 'magnetic levitation' method, but the earlier method demonstrated many years ago by J.C. West. A ferrous ball was attracted upwards by a large electromagnet placed above it but if it progressed upwards beyond a certain point a light beam was interrupted which de-activated the electromagnet. It could be arranged, with appropriate system gain, for the ball to remain suspended in mid air.

Second floor
Watch timer: The equipment had been designed and built by third year students as their project and enabled a check to be made, quickly and accurately, whether a mechanical watch (of which there were still many in 1975) was keeping good time. Any gaining or losing was displayed.

Electronic table tennis: This was another third year student project and was a game of the type that is now very familiar. The display appeared on a standard television set with electronically generated bats, baselines, net and ball. The positions of the bats were under the control of the players and a point was scored when the opponent failed to return the ball. The effects of spin and bat profile could be incorporated.

Historical display: The development of valves, transistors, etc. through the years was illustrated.

First floor (in Lecture Room 'E', which later became the Student Common Room).
Electric go-kart: This was specially built for the Open Day as an entertainment feature for children. Boys and girls in the approximate age range of 6 to 11 years were invited to drive the go-kart, which worked from a car battery and was equipped with forward and reverse drive.

Outside the building (in the area between Electrical Engineering and Chemistry)
Electric bicycle: This was built initially as a student project to study the problems of battery traction. The modified pedal cycle used two standard lawn mower type batteries connected in series driving a 24 v motor. On a test track it had covered 20 miles at approximately 15 mph on one charge. It was equipped with electronic control and the cost of electricity was estimated at 0.04 pence/mile, but a more realistic operating cost was estimated at 0.5 pence/mile allowing for the replacement cost of batteries, the life of which was rather limited. People with reasonable cycling experience were allowed to try out the machine.

The Open Day was deemed by the authorities to have been so successful that the university decided to repeat the event in three years' time.

New M.Sc. course

In 1975 it was proposed that a new 'taught' M.Sc. course be set up with the title 'Instrument Design and Application', to be run jointly by the Electrical Engineering Department and the corresponding department at UMIST. It would combine an advanced course of instruction in which participants would be assessed by written examination, and a research project written up as a dissertation. Hitherto all M.Sc. degrees awarded in Electrical Engineering in the Faculty of Science had been on the results of a research project alone and the resulting thesis. The idea for the course, for which it was felt there was an urgent need, came from the measurements specialists Drs B.E. Jones and M.J. Cunningham of the department and their opposite number at UMIST, Dr J. Rawcliffe. During the autumn of 1975 the proposal was discussed at length by the Departmental Board and was approved in principle on 19th November. Further detailed discussion took place with UMIST about details of the syllabus and the establishment of the course was then approved by the Faculty of Science. The first students entered in October 1977. Over the years the course has proved successful with an entry of about ten students each year, and it still operates in 1997.

Research work in the department

The Computer Science Department, set up in 1964 from the Electrical Engineering Department's computing group, remained in the electrical engineering building, physically, for more than a year before suitable accommodation (though temporary) was found elsewhere. Once the computer staff had gone, the design of computers ceased to be a direct part of the electrical engineering department's interests but there were plenty of other areas of electronics to be investigated. The department's research interests developed extensively during the period covered in this chapter. In 1964 they comprised: electrical machinery; servomechanisms; measurements; electrical breakdown of solid and gaseous dielectrics; digital storage - magnetic and cryogenic; electro-luminescence, epitaxial growth and ion implantation; automated precision electron-beam machining.

As the 1960s progressed the word 'servomechanisms' went out of fashion and was replaced by the phrase 'control engineering' and later, 'control and systems engineering'. In the thirteen years after 1964 several new topics were added to the above list. These included vapour deposition (papers started to be published on the topic in 1971); digital memories using cylindrical magnetic domains (1973); computer-controlled electron-beam micro-pattern generators (1973); medical electronics (1974); control of power systems (1976); power electronics (1976); automation of textile machinery (1976); dedicated micro-processor systems (1977). The following is a list of research projects completed between 1964 and 1977:

Machines

Following the departure of Eric Laithwaite in 1964, research continued under the direction of F.C. Williams, D. Tipping, G.F. Nix, G.W. McLean, D.E. Hesmondhalgh and (after 1967) I.E.D. Pickup, assisted by research students. Professor Williams' interest in machines continued unabated throughout the period and extended to novel drive transmissions for motor vehicles. A go-kart was fitted with a transmission system invented by the group and was tested at the Harris Stadium in Fallowfield. An improved version of the system was later fitted into a Triumph Herald car and for several months Professor Williams commuted between his home in Prestbury and the university in the vehicle. Several IEE Premiums were won by Professor Williams and members of the group, and a Premium was also awarded by the Institution of Mechanical Engineers for the work on vehicle drives. Completed projects included:

1965:   Single-phase two-speed pole-changing motors using phase-mixing techniques. Brushless a.c. motor using silicon controlled rectifiers. General method for prediction of the characteristics of induction motors with discontinuous exciting windings.

1966:   Linear induction motors for low speed and standstill operations. Electromagnetic levitation of a conducting cylinder.

1967:   Design of the a.c. servomotor.

1969:   The induction-excited alternator. The performance and design of induction motors with square wave excitation.

1972:   A hybrid d.c. to a.c. polyphase variable frequency converter.

1973:   The influence of time and space harmonics on the performance of induction motors.

1974:   Methods of predicting the dynamic response of a variable-reluctance stepping motor. Prediction of the multi-phase response of a multi-stator variable-reluctance stepping motor. The dynamics of the sprag clutch.

1975:   Linear and non-linear theoretical representations of machine flux-linkages in a multi-stator variable-reluctance stepping motor and their use in the prediction of dynamic characteristics. Calculation of the lift/drag ratio for magnetic suspensions. Controlled linear induction actuator with integral position detector.

1976:   Prediction of the pull-in rate and settling time characteristics of variable-reluctance stepping motors and the effect of damping coils on these characteristics.

1977:   A mechanical torque converter and its use as an automobile transmission.

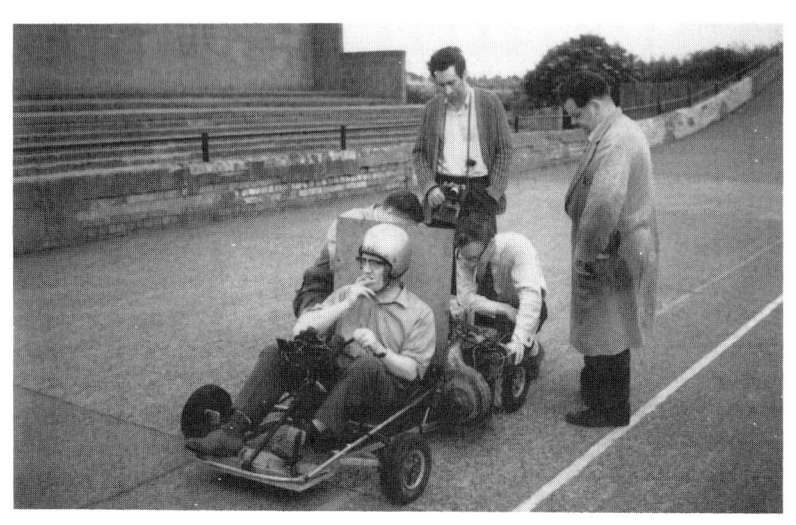

Professor Williams, cigarette in hand as usual, prepares for a test lap at the Harris Stadium, Fallowfield, early 1970s, on a go-kart equipped with a novel transmission system. The paper describing it won a Premium from the Institution of Mechanical Engineers.

1964-1977: Changing times, and the end of the Williams era

Electrical Measurements, and control and systems engineering

G.F. Turnbull carried out research in this subject until his resignation in 1967 and the work was continued by J.M. Townsend until 1970. Work was also done by B.E. Jones (whose particular speciality was measurements), P.A. Wade, N.P. Lutte, H.W. Thomas and F.G. Abbosh (all staff members) in the period under consideration, with help from their research students. Completed projects included the following, but the work overlapped with that of other groups. Some of the 'non-measurement' projects are quoted under the 'electronics' list below:

1965: A method for the analysis of relay amplifiers. Feedback pulse-width modulated audio amplifier.
1966: High input impedance clip-on a.c. voltmeter.
1967: Efficiency considerations in a class D amplifier.
1969: Electronic metronome using a novel frequency divider. Design of an accurate sampled-data simulator using field-effect transistors.
1970: Clip-on d.c. voltmeter. Capacitance transducer. Magneto-resistance and its application. Low-pass digital filters using least squared error design.
1971: Portable clip-on a.c. multimeter to measure voltage, current, active and reactive power and power factor.
1972: Effect of electrostatic screening on instrument accuracy.
1973: z-transform analysis of non-uniformly sampled digital filters.
1974: A new method for the study of nocturnal tooth contact using a passive technique of telemetry. A new multichannel optical data link with rotating machinery. An investigation of open and closed loop tilt monitors employing electrolytic spirit levels. Frequency domain approaches to MTI filters with staggered PRF. The gapped solenoid as a means of producing highly uniform magnetic field over an extended volume.
1975: Feedback measuring systems. Direct simultaneous displacement and velocity measurement by a single non-contacting capacitance transducer. The design of MTI filters with staggered PRF - a pole-zero approach.
1976: Stagger period selection for moving target radars. Tracking in a multi-radar environment. A study of the theoretical aspects of multi-radar tracking. Simultaneous direct measurement of displacement and velocity using a single capacitance transducer. A passive telemetry system applied in dentistry. An investigation into the use of a cone-jet sensor for clearance and eccentricity measurement in turbomachinery. Temperature-compensated precision solenoid.
1977: A general-purpose instrument for direct measurement and display of the dominant time-constant of decaying transients. The cascade realisation of MTI filters with staggered PRF and time variable weights. Radar tracking using adaptive Kalman filters.

Electrical breakdown in gases and solids; power systems and power electronics

R. Cooper worked with distinction on problems associated with solid dielectrics throughout the period, assisted by C.T. Elliott (until 1967), B.R. Varlow and D.W. Auckland (staff members from 1965 and 1968 respectively) and many research students. T.E. Broadbent continued his gas breakdown research, assisted by research students, until the early 1970s when it tailed off because of his transfer to other work. D.W. Auckland, who had initially worked on topics connected with dielectric breakdown, established a 'Power Systems and Power Electronics' group with Professor Cooper in the earlier part of the 1970s as an addition to his dielectric work, and a micro-machines laboratory was set up in the top-floor room previously occupied by the Atlas computer. From the mid-1970s onwards papers on power systems and power electronics were published at a steady rate. R. Shuttleworth was appointed to the staff in 1976 and assisted with the subsequent power systems work. Completed projects included:

1965: Impulse breakdown of perspex by treeing. Spark-gap switching with low voltage triggering. A direct-reading attracted disc absolute voltmeter. Characteristics and breakdown initiation of triggered spark-gaps with uniform applied field at very high voltages.

1966: Formative processes in the electric breakdown of KBr. Pre-breakdown light emission from alkali halide crystals. The electric strength of solid dielectrics.

1967: Investigation into a Kerr cell system for photographing the pre-breakdown phase of very short vacuum gaps. A study of the spark formation phenomena between short vacuum gaps using a Kerr cell camera system. A new design of Cockroft-Walton voltage multiplier using silicon diffused-junction rectifiers.

1968: Directional electric breakdown of KCl single crystals. Conduction of thin films of polyethylene at high stresses. Time lag of breakdown in solid dielectrics. Absolute measurement of high voltage by a null method using strain gauges.

1969: A generator of high voltage pulses with sub-nanosecond rise time and adjustable duration. A photographic study of electric breakdown of small gaps in vacuum. Development of the spark discharge in air with uniform applied field and long gaps. Stabilisation of a mains driven high voltage generator using silicon controlled rectifiers.

1971: Discharge propagation through single crystals of KBr. Conduction in polythene with strong electric fields, and the effects of pre-stressing on the electric strength. Measurement of the magnitude of impulse voltages using voltage comparators.

1972: The electrical breakdown of solid dielectrics in non-uniform fields.

1973: Dependence of conduction current in polythene on specimen thickness. The breakdown of composite insulation. The measurement of high voltage.

1974: The penetration of a gaseous discharge into a solid dielectric. A frequency-locked closed loop instrument for the measurement of transient changes in alternator rotor angle (jointly with the Control group).

1975: Investigation of water absorption in electrically-stressed polythene. A self-triggered high-speed camera for the photography of pre-breakdown luminescence of solid dielectrics.

1976: Electric strength of prestressed polythene following sudden cooling. The effect of ions on the absorption of water by electrically stressed polythene. The breakdown characteristics of air-filled tubules in solid insulation. The effect of prestressing on the electric strength of some polymers.

1977: A high speed photographic investigation of the alternating voltage breakdown of an air/solid composite. The intensification of under-exposed photographs of pre-breakdown luminescence in dielectrics (jointly with the Measurements group). The effect of heat treatment on some electrical properties of polythene. A multiplier wattmeter. Transistor time constant regulators for micro-synchronous alternators.

Electronics (includes high-vacuum, electronic circuits, computing circuits, electron beam machines, magnetics, microprocessors, etc.)

The staff involved in this work at various times between 1964 and 1977 included Professor G.R. Hoffman and, in order of appointment, C.N. W. Litting, D.C. Jeffreys, J.A. Turner, P.L. Jones, D.G. Buchanan, C.J. Hardy, P.Lilley, R.M. Pickard, J.L. Mudge, S.F. Gourley, E.F. Taylor, D. King, W.W. Clegg and R.S. Quayle. Numerous research students were also involved in the work. Completed projects included:

1965: A computer fixed store using light pulses for read-out. Problems of electron beam machining in miniature circuit design.

1966: Radio frequency switching of crossed film cryotrons. A small capacity high speed magnetic film memory.

1964-1977: Changing times, and the end of the Williams era

Back row, left to right: P. Lilley, G.W. McLean, B.E. Jones, M.J. Cunningham, H.W. Thomas, E.F. Taylor, D.W. Auckland, B.R. Varlow, D.E. Hesmondhalgh.
Middle row: D. King, R.S. Quayle, C.J. Hardy, G.F. Nix, D. Tipping, R.M. Pickard, P.L. Jones, W.W. Clegg, F.G. Abbosh.
Front row: I.E.D. Pickup, T.E. Broadbent, C.N.W. Litting, Prof. R. Cooper, Prof. F.C. Williams, Prof. G.R. Hoffman, G.A. Cooper (Lab. Superintendent), L.S. Piggott.

Staff of the Electrical Engineering Department, 12th December 1975. A photograph taken on the roof of the department with the Chemistry building in the background

Clockwise from top left: P.J. McGrory, D.E. Wells, R. Haxby, S.N. Iyer
The Electrical Engineering Department's team which won the IEE 'Student Challenge' in 1977, chaired by Nick Clarke.

1967: A high-precision electron beam machine. The suppression of the yellow emission of ZnS-In.
1968: Thin permalloy films: optimisation of properties for computer memories. Thermal electron beam machining of channels in steel surfaces. The epitaxial growth of zinc sulphide on silicon by vacuum evaporation.
1969: Laminated thin magnetic film memory elements.
1970: Domain configurations in laminated thin magnetic films.
1971: A selection system for a gas discharge display panel. A 500MHz thin film hybrid integrated circuit. The epitaxial growth of zinc sulphide on silicon by forced vapour transport in hydrogen flow. Simple capacitance transducer for instantaneous angular velocity measurements. A graphic fixed storage system for a remote terminal display.
1972: Observations on a quartz crystal deposition monitor. The epitaxial growth of zinc sulphide on silicon by forced vapour transport in argon flow. Electron beam damage: applications to bubble devices. Fabrication of bubble device overlay micro-patterns.
1973: Apparatus for the controlled co-deposition of MnBi thin films. Electron beam techniques for magnetic bubble device fabrication. Polymethyl methacrylate as an electron sensitive resist. The use of avalanche transistors for the generation of high voltage, variable width, rectangular pulses with fast rise and fall times. A novel electro-optic TV display.
1974: A technique for the growth of thick films of zinc sulphide. The influence of growth condition on the deposition of thick epitaxial (100) ZnS layers in $HCl:H_2$ gas flow. Magnetic bubble memory for video display. Thermo-magnetic writing on thin film MnBi using an electron beam.
1975: Electron beam induced conductivity in thin sputtered films of Al_2O_3. An adjustable pulse delay circuit. The use of electronic prediction to achieve fast response from a simple thermal mass flow meter. Electro-optic thin films of zinc sulphide. Substrate effects in autoepitaxial thick films of single-crystal zinc sulphide. Bubble dynamics in T-bar type circuits.
1976: Bubble device overlay fabrication using scanned electron beams. The epitaxial growth of thick smooth films of ZnS on GaAs. The effect of bypass flows on hetero-epitaxial growth of ZnS on GaP. A technique for the growth of single crystal films of zinc sulphide on (100) gallium arsenide by radio frequency sputtering. Optical waveguiding and modulation in zinc sulphide thin films. Flyback converter with a 555 timer. Theoretical analysis of bubble motion under T-bar tracks. A mathematical model for field-accessed bubble devices. Theoretically-computed bubble motion and experimental observations. Vector generators with 256 programmable steps.
1977: Comparison of experimental and theoretical results for high-frequency bubble propagation under T-bar tracks. Stroboscopic observations of bubble dynamics. Pilot-tone noise-reduction system using quantised control signals. Solid-state 1kV nanosecond pulse generator. The influence of source non-stoichiometry on UHV growth of crystalline films of zinc sulphide on silicon.

End of the Williams era

Professor Williams received a Knighthood in the Queen's Birthday Honours announced on 12th June 1976, a fact which gave great pleasure to his friends and colleagues.* Many people felt it should have been conferred about 25 years earlier. The high regard in which he was held in the

---

* By a remarkable coincidence G.H. Kenyon and J.A.L. Matheson, who each graduated with a 'First' in the same Honours School as Williams in 1932, were also knighted in the same Honours List.

1964-1977: Changing times, and the end of the Williams era

university world had brought him, over the years, Honorary Doctorates from the Universities of Durham, Liverpool, Sussex and Wales. Amongst his many awards were the Hughes Medal of the Royal Society, the first award of the Benjamin Franklin Medal of the Royal Society of Arts, the Faraday Medal of the IEE and the Pioneer Award of the IEEE.

During the hot summer of 1976 Professor Williams appeared to be in very good health and form, and was often to be found on the flat roof of the building conducting experiments on solar energy. However, on 25th August of that year it was announced that he was ill, the problem having been diagnosed as a tumour of the bowel. Operations were carried out on 1st and 8th September and initially he seemed to be making a good recovery. In the early spring of 1977, whilst not returning to work on a regular basis, he visited the department on several occasions and attended a few meetings. He called in my office one day for a chat and seemed to be in reasonably good health and spirits. However, to the sorrow of everyone, the disease reasserted itself and from about March onwards his condition deteriorated in ominous fashion. Sadly, the disease proved to be inoperable and he spent the last few weeks of his life in a Manchester hospital in declining health. He died on Thursday, 11th August 1977 aged 66, a year before he would have retired in normal circumstances. The named (Edward Stocks Massey) Chair of Electro-technics, founded in 1912, died with him. The funeral, held at Prestbury Parish Church the following Monday, was attended by a large number of present and past colleagues including Sir William Mansfield Cooper, Vice-chancellor from 1956 to 1970, who had been a personal friend of Professor Williams, as well as holding him in high regard professionally.

With the passing of Professor Williams a great sense of loss was felt by his staff, all of whom except one (Dr Litting) had been appointed during his 31 years of leadership. He had always been open and fair in his dealings with colleagues, and effortlessly created a happy, relaxed atmosphere of high morale in which his staff were free to do whatever research work they wished provided it was of good quality. His cheerfulness, enthusiasm and good humour was an inspiration, and his own technical brilliance was undisputed. It is no exaggeration to say that he commanded universal respect from anyone who had worked with him. Modest and unassuming, he was aware of his inborn gifts on technical matters but never flaunted them. He rarely gave lectures to undergraduates in later years, which is perhaps a pity because he was a brilliantly entertaining lecturer. Most of his lectures were given at the invitation of various learned societies and universities. On these occasions he would talk about computers, or machines, or invention in a most skilful and amusing manner.

There are plenty of 'Williams stories' and no-one who worked with him, or even worked in the same building, could fail to be aware of the experiments he was doing. After he turned his attention from computers to machines there were always interesting and unusual experiments going on with him at the centre. He was actively involved with the 'railway' in the basement and with the motor car and go-kart equipped with novel drives, mentioned earlier. At one time in the 1960s he arranged, as part of an experiment, for a large metal plate to fall vertically guillotine-like between two stretched wires running down the full height of the building's stairwell, from the top floor to the basement, a rather unnerving experience for anyone who saw the object suddenly whizz by as they climbed the stairs. Another time, he succeeded in setting the department on fire during the course of an experiment. The City Fire Brigade was called out and soon put out the conflagration. The chief fireman, surveying Williams' oil-soaked equipment scattered around in characteristic disorderliness, was heard to remark to someone: "Goodness knows what the Head of your Department would say if he saw this lot!" whilst the professor stood sheepishly in the corner.

When Williams was appointed he was 35 but looked much younger. Once, in his early years as a professor, he was away and not expected to be back for several hours, and his secretary had allowed a man from Metropolitan-Vickers to use his room to interview students for jobs. But Williams returned earlier than expected whilst his secretary was temporarily absent, and breezed into his office just as a candidate came out. "Oh, sit down, my boy!" said the man sitting at Williams' desk. Williams, realising what the situation was and entering into the spirit, dutifully sat down. The man was just about to start interviewing Williams for a job when the secretary walked in and said, "Oh, Mr So-and-So, have you met Professor Williams?"[10]

In similar vein, Williams in his early professorial years once turned up at the Whitworth Hall as one of the platform party due to officiate in a degree ceremony, with his academic robes rolled up under his arm. The Head Porter refused to let him in until someone confirmed that the young man was indeed Professor Williams. His sense of humour and total lack of pomposity ensured that he was always delighted by such incidents.

Albert Cooper, the department's Laboratory Superintendent who had known Williams since 1929 when they arrived as 'lab boy' and student respectively, said that he was the "world's best" at trouble shooting, whether the problem was electrical or mechanical. Williams was once leaving for the station to catch a train to London when he noticed Albert tinkering with his motor-cycle which was refusing to start. Forgetting all about his train, Williams joined in the investigation with great enthusiasm and within a few minutes he had sorted out the problem. "I'll miss my train now", said Williams, coming down to earth. "No you won't", said Albert. "Hop on the back of the bike and I'll take you to the station", which he did. [11]

Denis Hardy, a staff member of the department in the early 1950s, said that when Williams had after-lunch coffee in 'Staff House' with professors of other disciplines he could 'run rings' round them on technical matters, even on their own subject. When they told him of knotty problems they had encountered in their own particular discipline, Williams was often able to come up with a solution in minutes which they had failed to spot. [12]

On one July day in Staff House, Vincent Knowles, the former Registrar of the University, mentioned to Williams that so many students were now graduating that he had to arrive at the university very early in the morning to sign hundreds of Degree Certificates. Later he had a rubber stamp made, but even then stamping the certificates was a laborious process, and he asked Williams if he could invent something to do it automatically. He duly obliged. Knowles said he could not remember the details of the machine but it involved a large wheel and a bicycle chain! [13]

Williams' mind was very quick, not only on technical subjects but in all respects. Donald Camm, formerly Secretary to the Faculty of Science, said that when Williams was Dean of the Faculty (1955-58) he would turn up shortly before a meeting and say "What have we got today?" The Secretary to the Faculty would run through the agenda with him, pointing out potential problems, and all the salient facts were absorbed immediately, which enabled him to cope with any difficulty that arose as effortlessly as though he had been dealing with the subject all his life. [14] The same characteristics were evident when he was prevailed upon by the university to serve for several years as Dean of the Faculty of Music (from 1972 until his death), a role he was asked to undertake on the grounds that it was a very small Faculty and needed a leavening of wisdom from outside to make sure that it operated as it should.

If a criticism could be made of Williams' leadership it would be, as has sometimes been alleged, that in his later years he spent too much time on his own research interests and did not pay enough attention to the needs of the department. For instance, he did not travel around meeting important people unless it was absolutely necessary. Nor was he interested in encouraging his staff to become members of committees of outside bodies so that the department and its members were noticed by influential people. But that was not Williams' style; he did not seek favour. He believed in letting his own and his staff's work speak for itself, and many people would believe the criticism, even if valid, is counterbalanced by the immense benefits he brought to the department through his prestige, earned by the undisputed technical brilliance of his own work, to say nothing of the money he brought in to the university through his and Kilburn's computer work. It would be unrealistic to expect any man to possess every possible virtue. Williams' style was very different from that of his predecessor, Willis Jackson, who was not in the same class as Williams from the technical viewpoint but was an outstanding manager and leader of his department. Each was knighted and Jackson went on to the House of Lords. Ron Cooper, who knew them both well, once said that if the technical brilliance of F.C. Williams and the management skills of Willis Jackson could have been rolled up and put into one person, he would have been the greatest university professor the world has ever known.

References: Chapter 13

1. *Report of the Committee on Higher Education*, chaired by Lord Robbins. Cmnd 2154, HMSO, 1963.
2. Information on staff numbers is from university calendars, various years, and the student numbers are from departmental records.
3. Details of the course are from departmental information and the Faculty of Science publications *Syllabus of Classes* and *Prospectus* for various years.
4. Information from the Minutes of the Student Union E.G.M., 4th December 1968.
5. John G. Lang. *Examination by Assessment.* Paper circulated in November, 1968. The author was a Senior Lecturer in Education at the University of Bristol.
6. James A. Perkins. *The Restless Student.* Article from *Minerva*, 1969. The author had been President of Cornell University, USA, since 1963.
7. *Supplemental Charter and Statutes, together with General Ordinances and Regulations.* University of Manchester, 12th February 1973. 97 pages.
8. S.H. Lavington. *A History of Manchester Computers.* National Computing Centre Publications, Manchester, 1975, p. 36.
9. Some of the Open Day details quoted in the text are taken from the literature published by the Electrical Engineering Department for the occasion.
10. Anecdote included in Professor Eric Laithwaite's tape-recorded recollections of the Electrical Engineering Department communicated in March 1991.
11. Story told in Mr G.A.Cooper's tape recorded recollections, 29th April 1987.
12. Told to the author by Dr D.R. Hardy, c. 1952.
13. Told to the author in 1994 by Dr Vincent Knowles.
14. Told to the author by Mr Donald Camm, Secretary to the Faculty of Science and later Secretary to the Faculty of Medicine. Williams was Dean before Camm's years as Secretary, but the story is entrenched in Faculty folklore.

Chapter 14

1977-1994: 4-year courses, widening research, and a new School of Engineering

The years covered in the previous chapter had been ones of change, and between 1977 and 1994 the pace of change quickened. Particular features of the period in the Electrical Engineering Department were the introduction of 4-year courses, the abandonment of the B.Sc. degree in favour of B.Eng., the founding of two major new research groups and the opening of the Information Technology Laboratories. Nationally introduced changes also affected the department, in particular the accreditation of courses and the introduction of national 'research assessment' exercises. Money was extremely tight with constant clampdowns on spending, and the procurement of outside research contracts was therefore of vital importance. Finally, a reorganisation of engineering within the university culminated at the end of the period in the foundation of the Manchester School of Engineering and the requirement that Electrical Engineering would have to vacate the building it had occupied for over 40 years.

Accreditation

The introduction of accreditation of courses in Electrical and Electronic Engineering and related subjects by the Institution of Electrical Engineers came as no surprise to anyone. Much debate had been going on for years about various aspects of the engineering profession, one of which concerned the education of aspiring engineers. The introduction of many new courses in engineering by recently-created universities was one factor in the discussions. In 1978 the Merriman Report of the Institution of Electrical Engineers [1] proposed the introduction of accreditation of undergraduate courses and by mid 1980 the setting up of the system was in progress. On 4th November 1980 Professor Hoffman and I attended a meeting at the IEE's headquarters at Savoy Place, London, where the IEE's views on the subject were explained, including the fact that it was individual courses that were to be accredited, not the departments themselves. Within a year the scheme was in operation. In mid 1981 the department was asked to submit large amounts of documentation concerning its staff, courses, facilities and research, and the first visit to the department by the IEE's accreditation panel took place on 30th November and 1st December 1981. The procedures were so new that both sides were 'feeling their way', no one being exactly sure what was required. The panel members appeared to be reasonably favourably impressed by what they saw and heard, and the result was that the department's Electronic and Electrical Engineering courses were awarded accreditation for a period of three years from the middle of 1982 and the Physics and Electronic Engineering course, which was being assessed separately, achieved its accreditation soon afterwards. Parallel schemes were put into operation by the other engineering institutions for civil engineering, mechanical engineering, and other branches.

Further visits of the accreditation panel to the department took place on 13th and 14th March 1985, 7th and 8th December 1987, and 7th and 8th November 1994, the latter visit having been delayed. Each time a large amount of documentation had to be prepared and there were additional 'mini-accreditations' in between when the IEE panel wanted to know about changes that had taken place since their previous visit and more documents had to be produced. After each of the last two visits (1987 and 1994) all the department's courses were awarded accreditation for five years from the date of the verdict, which was the maximum period normally permitted.

Staff changes

Between 1977 and 1994, 24 new members of the academic staff were appointed of whom two were Temporary Lecturers and two were Honorary Lecturers (see Appendix 1).[2] Of these, 12 had left the department before 1994 by retirement, resignation or transfer to other departments. Notable amongst the new staff (with their dates of appointment in brackets) were D.J. Sandoz (1979) who left the department but later returned as a Visiting Professor and is now a Part-time Professor, P.G. Farrell (1981) who, on his arrival, established a new research group in Digital Communications, B.K. Middleton (1987), appointed as a Reader but promoted to Professor in 1993, who is in charge of the Information Storage research group, and Professor C.S. Xydeas (1987), who on his appointment founded the Speech and Image Processing research group. D.W. Auckland, who had been a Senior Lecturer since 1986, was promoted to Professor in 1991.

In 1982 Professor Cooper retired. He was a member of the department almost continuously from 1936 to 1946 as an undergraduate student and postgraduate research worker, and then from 1955 until his retirement as a staff member, of which the last 16 years were as a Professor.

In 1986 the department was saddened by the death at the age of 60 of Professor Hoffman who had been a staff member since 1951 and had been promoted to the rank of Professor in 1966 at the same time as Professor Cooper. He was taken ill in September 1985 with what appeared to be jaundice, but the condition turned out to be cancer. An operation was performed and he came back to the department a few times in October

1977-1994: 4-year courses, widening research, and a new Engineering School

though far from well. Sadly, the condition progressed and nothing could be done. He died on 21st January 1986. His funeral held the following week in Altrincham was attended by many colleagues, past and present.

Of the staff members appointed before 1977, 12 departed between 1977 and the end of 1994, in addition to those mentioned above. Long-serving staff members who retired, apart from Professor Cooper, were (date of retirement in brackets) C.N.W. Litting (1987), R.M. Pickard (1989), T.E. Broadbent (1990), G.F. Nix (1992), D.E. Hesmondhalgh and D. Tipping (1993). Most of them continued their link with the department, for varying periods, as Honorary Fellows. Another long-serving staff member, G.W. McLean resigned in 1990 to take up a post in industry.

Following the retirement of Albert Cooper as Laboratory Superintendent in 1977, P.G. Michaelson-Yeates was appointed to succeed him in 1978. He resigned in 1983 and W.N. Woodyatt was appointed in 1984. He remained in the post until his retirement in 1994, after which no further Laboratory Superintendent for Electrical Engineering was appointed.

In 1994, the end of the period covered in this chapter, there were four Professors of Electrical Engineering (Farrell, Xydeas, Auckland and Middleton), two Visiting Professors (Sandoz and Reece) and 21 other members of the full-time academic staff.

Until 1988 the university had never designated anyone within a department to be 'Head of Department'. It had always been assumed that the professors would share the necessary duties in an appropriate manner. However, for some years it had been usual for someone to be designated 'the person in administrative charge'. From 1988 departments were requested to elect someone, not necessarily a professor, to hold the title 'Head of Department' for a fixed term for administrative purposes. In the Electrical Engineering Department Professor Farrell was elected Head of Department in 1988 when the scheme was introduced, having previously been 'the person in administrative charge', and served for five years before resigning as Head. He was succeeded by Professor Auckland in 1993.

The Departmental Board of the Electrical Engineering Department, comprising the academic staff, existed from its establishment in 1973 until 1994 when it was disbanded on the creation of the Manchester School of Engineering. Initially the Chairman and Secretary were appointed, but a system of elections, which permitted re-elections, was introduced in 1977. No Chairman or Secretary was ever voted out of office but they resigned after varying periods of their own free will. Officials were:

Chairman: D. Tipping (1973-77); C.N.W. Litting (1977-86); G.W. McLean (1986-90); W.W. Clegg (1990-94).
Secretary: T.E. Broadbent (1973-88); W.W. Clegg (1988-90); M.J. Cunningham (1990-94).

Conversion of the 1MV high-voltage laboratory to other uses

The 1MV high voltage laboratory, commissioned in 1956, had been used for research work on long sparks throughout the late 1950s and the 1960s, leading to a long series of published papers on the fundamental mechanism of electrical breakdown in air. There had also been a lot of useful collaboration with local industrial firms such as BICC, as the laboratory provided testing facilities not available in their own factories. However, by the early 1970s most of the useful research work that could be done in a laboratory such as this had been completed. Also, the local firms for whom it had provided a valuable facility were no longer producing high-voltage equipment and therefore did not need it. Consequently the laboratory was little used after the early 1970s except for demonstration purposes; when applicants or other visitors were being shown round the department the high-voltage laboratory was usually included on the tour, as it enabled spectacular sparks several feet long to be displayed and the similarities with lightning explained.

By 1974 active consideration was being given to dismantling the equipment and using the space thus made available (40ft cube; 64,000cu ft) for purposes that would be more useful in the light of current needs. [3] Possible uses discussed at Departmental Board meetings in 1974 and 1975 included a new lecture theatre (for which there was a definite need), a student common room, more laboratory space for undergraduates, and more space for research. However, enquiries showed that no money was available for any of these purposes at that time.

Following the death of Professor Williams in 1977 application was made to the Vice-chancellor for the appointment of a third professor (to join Professors Cooper and Hoffman) and by February 1978 the request had been approved. Clearly the person appointed would need room for his or her research and, as space in the department was very much at a premium, the high-voltage laboratory was an obvious target. By the end of 1978 there was also a proposal that a microprocessor teaching laboratory be constructed, microprocessors having recently been introduced. Talks with various authorities had already taken place on this issue and it seemed that money to establish such a laboratory, to be operated jointly with the Computer Science Department, would be forthcoming provided the equipment was housed in a single room dedicated to the teaching of microprocessors and not as individual units scattered around the building.

These matters were discussed at a Departmental Board meeting on 12th January 1979; at which it was recommended that the high-voltage equipment in the 40ft cube be dismantled and removed, and the cube be converted into three rooms, one above the other, by inserting two floors at the existing first and second floor levels. The ground floor room would continue as part of the high-voltage group's research space, the first floor room would be made available to a new professor for his/her research needs, and the second floor room would become a microprocessor teaching laboratory.

Interestingly, even during the design stages of the building in the late 1940s and early 1950s it had been realised that the existence of the laboratory would add a reserve 'space bank' for other purposes should such space be required at some time in the future.

There was considerable urgency to construct the microprocessor teaching laboratory and it did not take long for all the necessary approvals to be obtained. Work on dismantling and removing the 1MV generator and installing the two floors proceeded rapidly in 1979 and by October of that year the Microprocessor Laboratory on the second floor was ready for use. A fire escape was installed running down the corner of each room from the second floor to an outside door at ground level, to satisfy current fire regulations. Brief details of the three rooms, each 40ft x 40ft in floor area, created as a result of the conversion of the 40ft cube, are as follows:

Ground floor

This remained in the hands of the high-voltage research group. High-voltage 'safety cages' housed a 300kV impulse generator built by a staff member, a 50kV discharge detector and an ampere-turn bridge for high-voltage measurements. A space of about 30ft x 25ft was left open for project and other work and in the remaining space there was an office for the laboratory supervisor with a large window overlooking the laboratory, this being a requirement of the current 'Health and Safety' regulations.

First floor

Initially this was left as a 40ft x 40ft open space for the use of the new professor's research. Following the appointment of Professor Farrell in 1981 it was divided into offices and laboratory space according to his wishes.

Second floor

Two offices were constructed, for the laboratory supervisor and a secretary, and most of the remainder of the 40ft x 40ft area was left 'open plan'. The cost of the laboratory, which was designed to train 20 people simultaneously, was about £100,000. It contained ten identical microprocessor teaching units; each included a Z80 single board with VDU and a floppy disk-based 38 OZ development system. The laboratory was used to augment the teaching of applications of microprocessors at under- and post-graduate level. Two ten-day courses providing "Professional Training in Microprocessors" for practising engineers were given during the 1979-80 session and the course continued at the rate of four courses per year in subsequent years.

New Microcircuit Design Laboratory

In the 1980-81 session it was announced that work was in progress to establish a Microcircuit Design Laboratory. The development was believed

Analogue and digital computers interacting to form a hybrid computer, late 1970s

Using an R.M. Nimbus computer in the new Microprocessor Laboratory, early 1980s

to be the first of its kind in Britain and was initiated in response to UGC encouragement to strengthen the teaching of microelectronics in universities. The laboratory was planned to contain a microfabrication facility using electron-beam lithography, and to have a small-volume production capability for package custom LSI microcircuits for internal needs in teaching and research. Designs would be implemented on part-processed wafers supplied by Ferranti Ltd. who would also provide technical support. The Faculty of Science allocated £50,000 for equipment and approval was given for providing the necessary clean environment for the satisfactory working of the fabrication facility. The staff involved in the project were Professor G.R. Hoffman assisted by C.J. Hardy, P.L. Jones, R.M. Pickard and R.S. Quayle. [4]

By the summer of 1982 the laboratory was nearing completion, and it became fully operational during the 1982-83 session. Following its success a wider collaboration began with Ferranti Electronics and the SERC Central Facility at the Rutherford and Appleton Laboratories to develop direct-write techniques for advanced types of uncommitted logic array. The group was also invited to join the UK Electron Beam Lithography Development Committee.

Two more University Open Days
1978:

The second University Open Day was held on 20th May 1978, three years after the first and on the equivalent Saturday. The Electrical Engineering Department's contribution followed the pattern of the first Open Day but there were several new exhibits and demonstrations. These included: [5]

Pendulum: An upside-down pendulum automatically kept vertical. This demonstrated feedback control.
Passenger lift: A demonstration showing how a model lift is controlled by a computer.
Signal Recovery: How an electrical signal 'lost' in a noisy background is recovered.
MSF clock: Exact time, direct by 'hot line' from the Post Office. This was a new technology at the time.
Crystal growth: Demonstration of crystal growth and the fabrication of electronic materials.
Computer games: Games not played on the previous Open Day included a tank battle and a football game.
Electromagnetic generation: How much power can you generate?
Stroboscope: How to make moving objects appear stationary.
Nail melting: Showing the heating effect of an electric current and the principle of the transformer.

Several demonstrations which had proved very popular on the previous Open Day were repeated, including the 1MV laboratory display of long sparks, the children's go-kart, and the electric moped. In the 1975 high-voltage laboratory demonstrations, which had taken about 15 minutes each,

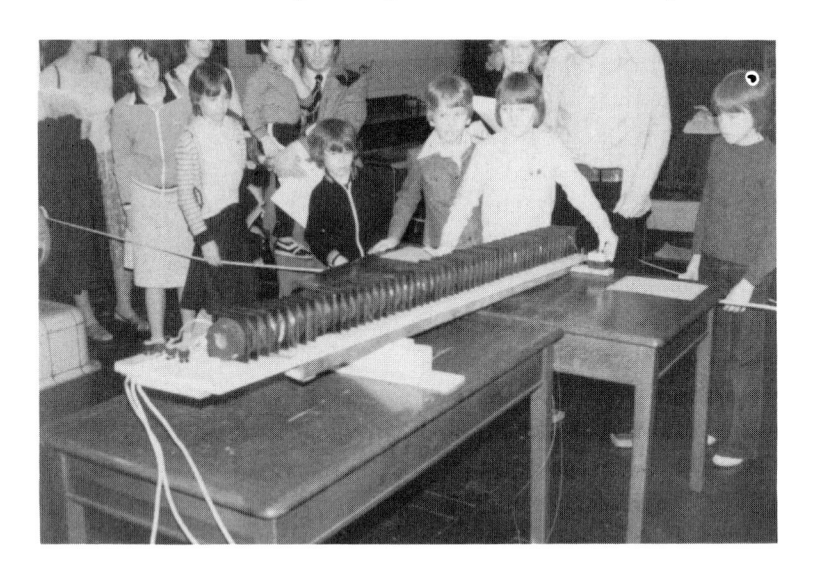

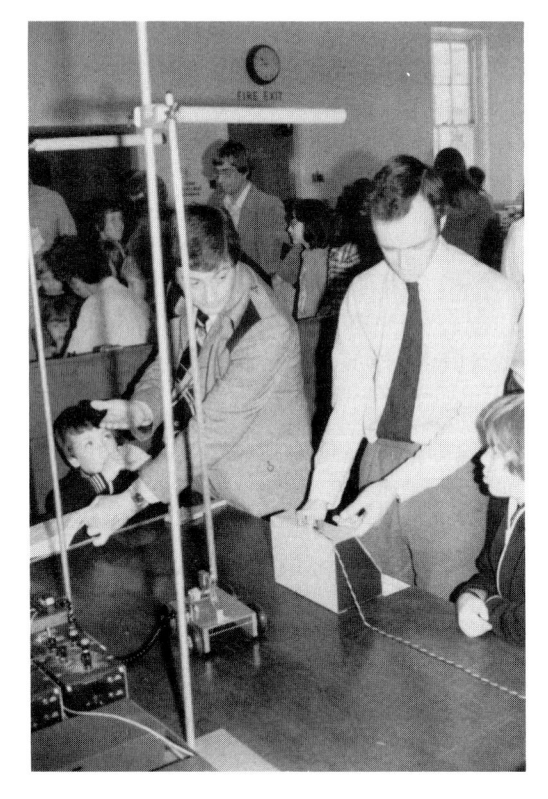

Above: An electromagnetic gun, which shot arrows very spectacularly
Below: The inverted pendulum, with Dr Thomas in attendance

Two scenes from the 1978 Open Day

Brian Varlow and I talked ourselves hoarse repeating the demonstration all day, so on this occasion a script was written, careful timings having been made to synchronise what was being described with what was being shown, and my wife, Marguerite, who was a very good reader, recorded it. The tape was played successfully at 15 minute intervals all day and was heard by the 1,000 or so visitors who passed through the control room to see the sparks. This event was effectively the swan-song of the 1MV high-voltage laboratory for in the following year the equipment was dismantled.

1981:

The third Open Day took place on Saturday, 23rd May 1981. Again the Department proved to be popular and was thronged by crowds of visitors all day. Most of the displays on this occasion were similar to those shown in 1978, with modifications, but the Microprocessor Teaching Laboratory (opened in 1979) was now available and special demonstrations were presented, including a microprocessor-controlled seismic recorder, and there were new games in which visitors took part. The children's electric go-kart had proved so popular on the previous two Open Days that another had been constructed, so that this time two children could drive around in the designated area at the same time, thereby reducing the queues. The 1MV laboratory was no longer available but the new 300kV impulse generator was demonstrated, and even that was capable of producing an impressive enough spark and bang to thoroughly startle the onlookers.

The 1981 Open Day, like the previous two, had proved very successful, but the University decided, after due reflection, not to hold any more because of the cost. This had proved very substantial because the university had to take out extra insurance, some loss of equipment on the day was inevitable, technicians had to be paid for coming to work on Saturday, copious literature had to be printed, and staff members spent a lot of time before the event preparing for it when they could have been doing other things. Consequently no more Open Days have been held, but those that did take place, each on a glorious summer's day, remain a pleasant memory.

Undergraduate courses

The 3-year course in Electronic and Electrical Engineering

At the outset of the period the course was virtually the same as that outlined on page 226, and the details will not be repeated here. Following the appointment of Professor Farrell it was decided to embark on a revision of the course, which had not changed radically for several years. After much deliberation a revised syllabus including more 'design' work was introduced in first year in 1982, second year in 1983 and third year in 1984. Further

Electrical Engineering at Manchester University

Above: Children discover how much power they can generate, watched by Dr Hesmondhalgh
Below: Dr Hardy helps children play a computer game

Two scenes from the 1981 Open Day

small changes took place year by year in the next three or four years. By 1987 the detailed syllabus was as follows:[6]

First year.

An introduction to the subject of Electrical and Electronic Engineering:
Fundamentals of electromagnetism: Units, electrostatics, electromagnetics, induced emfs, phasors.
Circuit theory I: d.c. and a.c. circuits, transients, symbolic method.
Electronic circuits I: Bipolar junction and field effect transistor circuits, amplifiers.
Signals, circuits and systems: Fourier series, Fourier and Laplace transforms, transmission lines.
Power engineering: Polyphase systems, d.c. machines, stepping motors, power transformers.
Dielectrics and their applications: Electrical properties of insulating materials.
Measurements: Theory of measurements and an introduction to the types of measuring instrument.
Digital processes I (including practical work): Introduction to microprocessors and programming. Assembly language.
Semiconductor devices I: Physical structure and theory of semiconductors.
General engineering: Thermodynamics, statics and dynamics. (Given by Engineering Department staff).
Mathematics: Complex numbers, differentiation, linear algebra, vectors, dynamics, differential equations.
The electrical engineering industry: A short course outlining the history and structure of the industry.
Computer programming (Pascal). Included lectures and laboratory.

In the electrical laboratory, 20 experiments were done (one per week) over a range of electronic and electrical subjects. There were also regular tutorials. There was also a practical course in Manufacturing Technology designed to satisfy the IEE requirements on 'Engineering Applications I'. At the end of the first year, students took examinations in the various subjects studied. Any subject failed could be retaken in September.

Second year.

Extension of the basic knowledge acquired in first year: detailed study of the following subjects:
Electronic circuits II: Amplifiers, various circuits, radio receivers.
Digital processes II: Logic gates and families, combinational and sequential systems, computer circuits.
Circuit theory II: Laplace transforms, two-port networks, filters, solution of problems.
Signal processing and communications: Signal processing and properties of signals, modulation, communication systems.
Electromagnetic waves: Transmission lines, electromagnetic theory, wave equations and propagation.
Control: Fundamentals of control systems, stability, practical control systems.
Power systems and electronic control of power: Transmission and distribution systems, power semiconductors for motor control.
Electrical machines: Transformers, single-phase and three-phase motors, alternators.
Semiconductor devices II: Metal-semiconductor and bipolar junctions, MOS structures, MOSFET, JFET.
Magnetic materials: Fundamental magnetism, technical and applied magnetism.
Instrumentation: Various types of measuring device including transducers.
Further Computer programming (Pascal). Included laboratory classes.
Mathematics: Functions of a complex variable, probability and statistics.
Design and product quality: Factors affecting the design, quality and reliability of industrial products.

Laboratory work was in the form of short projects covering the whole area of electronic and electrical engineering. There were also regular tutorials (introduced in the mid 1980s).

At the end of the second year students took examinations in the various subjects studied. Any subject failed could be retaken in September.

Third year.

Students specialised in the subjects which interested them most. Five subjects were selected for study from the following list:
Electronic systems: Amplifier systems, waveform generation, analogue/digital systems, TV systems.
Microcircuit systems design: An introduction to the principles of integrated systems design.
Semiconductor devices III: Semiconductor materials, applications and devices, special effects and

devices.
Communications: Information theory and coding, signal transmission, sequences and synchronisation, communications and networks.
Analogue and digital filters: Active and passive analogue filter design, digital signal processing.
Microwave and opto-electronic systems: Propagation and detection of microwaves, antennas, optical transmission and reception.
Control engineering: Theory and application of the digital control of dynamic systems.
Power systems: Generation, transmission and distribution of electrical power.
Power electronics, machines and drives: d.c. and a.c. drives, stepping motors, transients, synchronous, induction and linear machines.
High-voltage engineering: High-voltage technology, conduction and breakdown in solids, liquids and gases.
Digital speech and image processing: Signal theory, coding of speech signals, coding of images.
Information storage systems: Magnetic tape, disc and optical storage systems, solid state memories, storage hierarchy.
Computer systems design: Computer architecture and organisation - theory and practice.
Numerical analysis and computing: Practical numerical methods, including computer programming of specific examples.

There were also two compulsory courses, taken by all students, as follows:
Design for production: The various factors which relate to the design process (with case studies).
Management for production: General management, accounting and finance, management techniques (with case studies).

The final year laboratory work consisted of a project, undertaken by each student. Until the early 1980s students usually worked in pairs; after that time they always worked individually.

At the end of the third year students took the Honours Examination in the compulsory subjects and the chosen options. The 'general paper' was dropped in the mid 1980s. The mark awarded in the project also counted in determining the class of degree awarded. Until 1988 the degree was governed mainly by the final year work though earlier work could be taken into consideration. After that date a proportion of the second year mark counted directly towards the final degree assessment.

Physics and Electronic Engineering

The course continued into the 1980s with a syllabus similar to that outlined on page 196 but updated periodically to parallel the changes in the Electronic and Electrical Engineering course. During the 1980s various other new courses were proliferating at Manchester and it was found that the course was becoming less popular than in former days. In the mid 1980s the decision was made to discontinue it. The last entrants were admitted in 1988 for graduation in 1991, though two stragglers graduated in 1992. Since its creation in 1963 the course had fulfilled a useful purpose and 474 students graduated with a degree in Physics and Electronic Engineering.

B.Sc. degrees change to B.Eng.

The Finniston Report [7], published in 1980, proposed that engineering courses should lead to the degree of Bachelor of Engineering, not Bachelor of Science, and this was soon taken up as national policy, expressed through the professional institutions. Whatever the thoughts of the Electrical Engineering staff at Manchester were, there was no option but to comply, for the IEE let it be known that eventually engineering courses which did not lead to B.Eng. would not be accredited. Consequently the decision was made that the degree awarded at the end of the 3-year course in Electronic and Electrical Engineering would be B.Eng. instead of B.Sc. for students graduating in 1989 and subsequently. When the decision was made the reason for the change was explained to all the students already on the course, and they were given the option of graduating with a B.Sc. degree instead of B.Eng. if they wished. However, none did. The Physics and Electronic Engineering course, which was in the process of winding down, continued to award the B.Sc. degree.

New 4-year industrially-linked course

After much discussion in the late 1970s and early 1980s the department decided to offer a new 4-year industrially-linked degree course in Electronic and Electrical Engineering as an alternative to, but running in parallel with, the existing 3-year course. The course was created in response to the need identified by the Finniston Report, industrial organisations, professional institutions, and government for undergraduate teaching courses which were relevant to industrial design, development and manufacturing practice. The course led to the double degrees of B.Sc., M.Eng. About 18 students entered annually. The first were admitted in October 1983, graduating in 1987.

The course provided an integrated programme based on three main components: a full range of electronic and electrical engineering science subjects; university-based study of practical aspects and industrially-related problems; and industrially-based training and experience. The first two took place during four normal three-term academic years, and the third occurred during vacation periods by arrangement with the collaborating industrial organisations. All students registered for the 4-year course were required to obtain sponsorship by an industrial or other appropriate organisation by the end of the first academic year. The department assisted students in finding sponsorship by putting them in touch with companies known to be prepared to support the course. Some companies preferred to enter into a sponsorship with students at the outset of their studies, whilst others preferred to arrange it during the first academic year. Should a student fail to gain sponsorship during the first year, or decide not to continue in the 4-year course, he or she could transfer at the end of the first year to the second year of the 3-year course without detriment. Transfer was also possible from the 3-year to the 4-year course up to the end of the first year. The structure of the course at the time of its creation was as follows:

<blockquote>The course included all the technical content of the 3-year course but with the subjects spread out over four years instead of three. In addition, industrially orientated material was integrated into the course over the 4-year period. This covered production processes; engineering design; quality and reliability; management of manufacturing</blockquote>

units; sales and finance.

First year and second year laboratory work was similar to that of the 3-year course but with additional teaching in workshop practice. In the third year of the course students undertook a "team" project in which they designed and constructed an item of equipment with manufacturing and commercial considerations in mind. In their fourth year students carried out a "technical" project of the type done in the third year of the 3-year course.

The sponsoring company was responsible for providing its students with relevant work experience during the summer vacations. At the end of the vacations each student submitted a report on an industrially-orientated topic chosen jointly by the training officer and the academic supervisor. Throughout the four years of the course close liaison was maintained between the university and the sponsoring company regarding the progress of individual students.

The technical content of the course evolved over the years in the same way as that of the 3-year course.

Other new courses

Following the success of the industrially-linked 4-year course, the M.Sc. component of which was changed to M.Eng. in the 1990s, it was decided in the late 1980s to introduce a number of other new courses, 3-year and 4-year, with specific associations. These included, with the year of introduction and the degree awarded in brackets:

3-year courses:
Electronic Engineering Science (1989; B.Eng.)
Electronic Systems Engineering (1990; B.Eng.)
4-year courses:
Electronic and Electrical Engineering (Integrated European Programme) (1988, M.Eng.)
Electronic Engineering Science (Integrated European Programme) (1989; M.Eng.)
Electronic and Electrical Engineering (Industrial Programme) (1989; B.Eng.)
Electronic and Electrical Engineering (1990; M.Eng.)
Electronic and Electrical Engineering (Integrated Japanese Programme) (1990; M.Eng.)

The course structures were based on the Electronic and Electrical Engineering course, with appropriate variation as implied by the titles of the courses. Students enrolled on courses associated with foreign countries were required to undertake part of their course in those countries.

Modular courses

In the early 1990s the university, as a matter of policy, decided to move to a modular system of courses. This necessitated some rearrangement of material in the electrical engineering course structure outlined above though the total content of the various courses remained much as before. The revised system was introduced in 1993 and still operates in 1997 though, as always, minor changes are made from year to year. A general outline of the revised course is given in the next chapter.

1977-1994: 4-year courses, widening research, and a new Engineering School

The new Information Technology Laboratories

Under Phase 2 of the University Grants Committee's Engineering and Technology Programme, the Electrical Engineering Department was awarded a substantial increase in resources during the 1986-87 session. In addition, the Engineering Industry Training Board (EITB) provided funds to assist the department with the continued development of industrially-linked activities for both its extended (4-year) and 3-year undergraduate courses. Part of the extra UGC resource was a new building (the Information Technology Laboratories), to be shared with the Computer Science Department, planned to provide space additional to that available in the Dover Street building. Teaching would be expanded in the general area of Information Technology through the new facilities and resources. The Laboratories were opened, with due ceremony, by The Princess Royal on 21st June 1988.

The University of Manchester was one of only ten universities and polytechnics in the United Kingdom to be selected by the Commission of the European Communities under a recent ESPRIT action to develop skills within undergraduate courses for the design of very large scale integrated circuits. The initiative provided access to the industrial manufacture of microchips, CAD and test equipment, and the appointment of additional lecturers. The VLSI specialisation was focused on the 3-year honours course in Electronic Engineering Science, in which students would have the opportunity to design and verify silicon microchips of substantial complexity with applications in digital communications, signal processing, computer architectures or other areas where the department had particular strengths. Some of the teaching of VLSI would be shared with the Department of Computer Science who benefited similarly from the ESPRIT action. [8]

Research work in the department

The period covered in this chapter brought a considerable widening of the range of research done in the department. In 1977 the current work could be summarised briefly as:

Machinery; measurements and instrumentation; control and systems engineering; power electronics and control of power systems; medical electronics; luminescence, epitaxial growth and ion implantation; electrical breakdown of insulation; digital memories using cylindrical magnetic domains; computer-controlled electron-beam micro-pattern generators; vapour deposition; dedicated micro-processor systems.

With the appointment of Professor Farrell in 1981 an active research programme in Digital Communications was started. Then, in 1987, Dr B.K. Middleton was appointed to the staff with the status of Reader, bringing with him from Manchester Polytechnic active research work in Mass Data

Electrical Engineering at Manchester University

Information Technology building containing microcircuit fabrication facilities and ECAD suite

The Princess Royal being shown the laboratories by Professor Farrell before officially opening the IT building in June 1988

Storage Systems, thus enhancing the already flourishing work of the department's Information Storage (Magnetics) research group. Dr Middleton was promoted to Professor in 1993. Another new area of research began in 1987 with the appointment of Professor C.S. Xydeas. His subject was Signal Processing Techniques as applied to Speech and Image Processing, a continuation of the work he had been doing in his previous post at Loughborough University.

Through these appointments two completely new areas of research were opened up and an existing area was strengthened, leading to a large increase in the numbers of papers published and the winning of many new outside contracts. The latter point was particularly important for it was an era when collaboration with outside organisations who would help to finance the research was encouraged very vigorously by the university; indeed, the procurement of such financial support was crucial. All the research groups, old and new, sought research funding from outside bodies and in the period covered in this chapter many millions of pounds were awarded to the department in the form of grants, research contracts, etc.

Towards the end of the period some rationalisation of the research groupings took place, including the merging of the Machines, High-voltage and Power Systems groups to form the Electrical Energy Systems group, and the work in Measurement and Control was merged into that of other groups. By 1994 the research groupings in electrical engineering were classified under the following general headings, each group being led by a Professor:

Information Storage (including microelectronic design and fabrication); Signal Processing; Digital Communications; Electrical Energy Systems.

Lists of research projects completed by the various groups over the 17-year period are given in the following sections.[9] No attempt has been made to list all the work done; research work in the department seems to have increased almost exponentially in volume since 1977 and it would be unrealistic to list everything. The aim here is simply to quote a good representative selection of the work of each group so as to give an accurate picture to the discerning reader of the type of work that has been done. Even this limited review of research amounts to nearly eight pages of small print and I do not feel the inclusion of any more detail would be justified.

Machines

Following the death of Professor Williams in 1977, work in this area continued under the direction of D.E. Hesmondhalgh, G.W. McLean, G.F. Nix, I.E.D. Pickup and D. Tipping until the late 1980s. Between 1989 and 1993 all of these five left the full-time staff by resignation or retirement except Dr Pickup. By the end of 1994 machines research was in the hands of I.E.D. Pickup, assisted by research students, though D.E. Hesmondhalgh and D. Tipping were still involved on a part time basis as an Honorary Fellows.

Electrical Engineering at Manchester University

In October 1985 the 'Manchester Machines Research Group' (MMRG) consisting of the combined research forces from the department and from the corresponding department at UMIST was officially launched, thus making possible the pooling of resources. Research seminars were held, attended by many representatives from Industry. Completed projects over the period covered in the chapter included:

1978: Machines using rectified oscillatory motion. Analytical method for estimating electromagnetic damping in a multi-stator stepping motor. Parametric oscillatory motion in electro-magnetic devices.

1979: High-frequency instabilities in variable-reluctance stepping motors. Subharmonic resonance in stepping motors. Non-linear model for predicting settling time and pull-in rate in hybrid stepping motors. Development of a stabilisation scheme for the suppression of paramagnetic instability in stepping motors. A swing towards understanding electrical machine instability. Operational difficulties with stepping motors. Permanent magnet brushless drives using claw armatures. The zig-zag linear synchronous motor. Low-speed motorised dental handpiece.

1980: Principles of parametric instability (high-frequency resonance) in stepping motors and methods of stabilisation. A model for predicting the dynamic characteristics of hybrid (permanent magnet) synchronous/stepping motors. Parametric oscillatory motion in electro-mechanical devices. Fundamental design factors in rotary dental instrumentation. A multi-element magnetic gear. Brushless d.c. drives using claw type stators and disc rotors. A linear motor using claw stators and permanent magnet moving member.

1981: Translational forces on discrete metallic objects in travelling magnetic fields. Principles of operation of the parametric reluctance motor. A general theory of the parametric reluctance motor.

1982: Calculations of the pull-out torque of hybrid stepping motors in the half-step mode. Non-linear resonance phenomena in stepping motors. Power conditioning hardware for a.c. traction based on the utilisation of TLVR hardware and technology.

1983: Slotless construction for small synchronous motors using samarium cobalt magnets. Miniature sheet-rotor induction motors for extremely high speeds. A comparison of locomotive drives using current-forced a.c. traction motors. Performance of the zig-zag linear synchronous motor in urban vehicle drives. Calculation of the lift-drag ratio for a three-limb axial flux suspension magnet. An analysis of the induction motor: steady-speed operation; damping and synchronising torque production; power/frequency relationships. A circuital method for the production of pull-out torque characteristics of hybrid stepping motors. An alternative energisation mode for the 4-phase, variable-reluctance stepping motor.

1984: Effect of stator slotting on the performance of sheet-rotor induction motors. Equivalent rotor impedance of sheet-rotor induction motors. Some problems associated with designing motors for speeds up to 150,000 rpm. Torque pulsations and losses in inverter-fed synchronous motors. Combined lift and thrust for maglev vehicles using the zig-zag synchronous motor.

1985: Inverter drives for a.c. locomotives. Analysis of asymmetric rotor faults in turbogenerators.

1986: Torque availability from small synchronous motors using high coercivity magnets. A miniature transformer-coupled high-speed synchronous motor. Investigation of the Delco permanent magnet actuator motor.

1987: Relating instability in synchronous motors to steady-state theory using the Hurwitz-Routh criterion. Design and construction of a high-speed, high-performance, direct-drive handpiece. Analogue and digital controllers for magnetic suspension systems. Analysis and design of consequential pole-change induction motor drives. Aspects of the dynamics of single-phase reluctance machines. Stabilisation of hybrid permanent magnet stepping motors. Calculation of pull-out characteristics of hybrid stepping motors with current-regulating drive circuits. Dynamic instability in permanent-magnet synchronous/stepping motors.

1988: Design of a high-speed permanent magnet motor. A new magnetic rotor position sensing unit for brushless d.c. motors. Finite-element analysis of synchronous machines. Design of rotor windings to improve the performance of converter-fed synchronous motors.

1989: Torque calculations for hybrid stepping motors under dynamic conditions. Representation of damper bars in computer models of synchronous motors supplied from variable-frequency inverters. The use of finite elements in the analysis of electrical machine performance.

1977-1994: 4-year courses, widening research, and a new Engineering School

1990: Performance and design of an electromagnetic sensor for brushless d.c. motors. Design of a permanent magnet linear synchronous motor drive for a position/velocity actuator. Analysis of current waveforms in hybrid permanent-magnet stepping motors exhibiting dynamic instability.

1992: Analysis of frequency and amplitude modulation in the stabilisation of permanent-magnet synchronous/stepping motors. Analysis of current waveforms in permanent-magnet synchronous/stepping motors and synthesis of a stabilising signal.

1993: The development of a linear d.c. actuator for valve control in aircraft.

Electrical measurements, and control and systems engineering

B.E. Jones, a specialist in measurements and related matters, resigned in 1981 but work in measurements was continued by M.J. Cunningham. Staff contributors to research in control systems between 1977 and 1994 included M.J. Cunningham, H.W. Thomas, F.G. Abbosh, A.M. Solomons, D.J. Sandoz, P. Bridge, J.R. Beebe, D.J. Cole, and P.J. Turner. In 1990 D.J. Sandoz transferred to Vuman but later returned as a Visiting Professor. Completed projects included:

1978: A cathode ray tube actuator in a non-contact analogue tachometer. An electrostatic wattmeter employing automatic voltage feedback to the quadrants to achieve torque balance. Feedback electrostatic wattmeter using a non-contact moving-coil actuator. Pressure transducers for measuring underwater explosions. The detection and handling of aircraft manoeuvres.

1979: Feedback in instruments and its applications. Phase-nulling telemetry system incorporating a remote passive transponder. Manoeuvre handling in multi-radar ATC system. Applications of aircraft-derived data in automatic radar tracking systems.

1980: Continuous monitor of the safety limits of vehicle tyre pressure using a non-contact method of passive telemetry. Capacitance transducer using a thin dielectric and variable-area electrode. Remote passive transponder incorporating a reactive transducer. An application of computer-aided control system design to milk-drying plant.

1981: Computer aided design of industrial control systems. Linear induction motors controlling carriages moving round a circular track. Low-distortion feedback voltage-current conversion technique. Precision constant current source. Low-distortion high-output class B current converter using error feedforward. Digital displacement transducer using pseudo-random binary sequences and a microprocessor. Miniature capacitance transducers using thin films and having high sensitivities. Optical shaft followers using CRT and LED-array actuators. Optical fibres and electro-optics for instrumentation.

1982: Measurement errors and instrument inaccuracies. Direct method of contact identification. Design current output amplifiers using current mirror circuits. Current mirrors, amplifiers and dumpers. Low distortion feedback bipolar current amplifier technique. Optically isolated phase-angle power controller. Feedforward temperature compensation for Hall effect devices. Identification of a 270MW turbo generator. CAD for commissioning and maintaining process control systems.

1983: New desaturation strategy for digital PID controllers. Air-cored solenoid with a very small temperature coefficient of magnetic field.

1984: CAD for the design and evaluation of industrial control systems. Speed and acceleration measurement. Systems of units. Investigation of underwater transient pressure measurement using manganin foils. The expression of uncertainties.

1985: Fibre optics in aircraft utility systems. Transducers for measuring underwater shock pressures.

1987: Investigation of microvoids in voided polyvinylidenefluoride. Optimal tracking controller design for invariant dynamics direct-drive arms. Reducing aliasing artifacts in images by matched resampling. Piezoelectric polymer pressure sensors.

1988: Investigation of voided PVdf after exposure to high-pressure pulses. PVdf shock sensors.

1989: PVdf gauge and header amplifier design. A formula for least-squares projection and its application in image reconstruction.

1990: A function space model for digital image sampling and its applications in image reconstruction. A Fourier-domain formula for the least squares projection of a function onto a repetitive basis in N-dimensional space.

N.B. From 1991 the work done by this section is included under the headings of other sections.

Electrical breakdown of insulation; power systems and power electronics

After Professor Cooper's retirement in 1982, work on insulation breakdown was maintained by D.W. Auckland and B.R. Varlow, and the power systems work was continued by D.W. Auckland and R. Shuttleworth, with assistance from J.R. Hawke, a Temporary Lecturer, until 1980 and numerous research students throughout the period. Dr Auckland was promoted to Professor in 1991.

In the late 1980s a joint research group was set up between the department's high-voltage group and the corresponding one at UMIST. Named the 'Manchester High-voltage Research Group', it followed the setting up of a similar group in machines research.

Completed projects included:

1978: Photographic investigation of breakdown in composite insulation. The development of electrical discharges in simulated 'tree' channels. A thread-tension transducer using torque-balance about the axis of a motor. The stabilisation of mains-driven high voltage generators using transistors.

1979: Kerr-type electro-optical effects in solid dielectrics.

1980: Intrinsic breakdown in diverging fields - formative stage. The measurement of shaft torque in microalternators. The parallel operation of power transistors. A voltage-sensitive protection system using a microprocessor.

1981: Photographic investigation of formative stage of electric breakdown in diverging fields. Effect of morphology on some electrical properties of P.E. film. The measurement of transient changes in rotor angle of a microsynchronous alternator. Compensation system for a micromachine model.

1982: Micromachine systems for teaching and research.

1983: Control method for compensation of switching delays in transistors. A digital controller for the synchronisation of parallel-connected power transistors operating in switched mode. The measurement of shaft torque in synchronous alternators using an optical shaft encoder.

1985: Thermally stimulated current and morphology in low density polyethylene. Electrical prestressing, trapping centres and morphology in polyethylene.

1987: The influence of vibration on the initiation of trees in dielectrics. Micromachine model for the simulation of turbogenerators. Electrostatic discharges from the human body.

1988: Mechanical degradation of solid insulation. Digital synchronisation of switching transistors.

1989: Dependence of electrical tree inception and growth on mechanical properties. Investigation of large actuators. Ageing in high-voltage insulation. Improvement in power system stability by cross coupling generator control systems. The accurate modelling of large synchronous generators.

1990: Light emission in solid dielectrics during the early stages of tree initiation. Ultrasonic detection of insulation degradation. Mechanical aspects of electrical treeing in polyester resin. Rare-earth permanent magnets in switchgear actuators. Thermo-electric generators for undersea use.

1991: Dependence of electric tree inception and growth on mechanical properties. Mechanical aspects of water treeing. A permanent magnet actuated solid-state control auto-reclose circuit breaker. The feasibility of kilowatt combined heat and power systems for domestic use. The modelling of brushless exciters.

1992: Effects of barriers on the growth of trees in solid insulation. Factors affecting electrical tree testing. The electrical breakdown of insulating oil.

1993: Charge deposition in gas filled channels with insulating walls. Use of a medical ultrasonic scanner for the inspection of high voltage insulation. The role of particles in discharge initiation in liquids. Correlation of mechanical properties with electric treeing behaviour at elevated temperatures. Artificial neural network (ANN) -based machine model for transient stability studies and for the detection of rotor inter-turn faults. Detection of tracking carbon paths using visual and thermal images.

1994: Application of ultrasound to the NDT of solid insulation. A new test method for measurements of surface tracking resistance. Measurement of electric field-induced crack growth in polymeric insulation using laser interferometry. Image recognition for reading housing estate layouts in the design of distribution systems. Water penetration via fibres and the formation of water trees in insulating liquids. Stress relief by non-linear fillers in insulating solids.

1977-1994: 4-year courses, widening research, and a new Engineering School

Electronics, microelectronics and information storage *(includes medical electronics; luminescence, epitaxial growth and ion implantation; digital storage systems using magnetic bubble domains; computer-controlled electron-beam micro-pattern generation; ferromagnetic thin-film magnetometry; vacuum deposition of thin dielectric, ferromagnetic and magnetic films; dedicated microprocessor systems; magnetic and magneto-optic recording, solid state memories).*

Professor Hoffman was in charge of electronics research until his death in 1986. The work of the large group then continued on similar lines as before and was strengthened in 1987 by the appointment as Reader of Dr B.K. Middleton, as a result of which the magnetics research personnel were re-formed as the Information Storage group. Other staff involved in the work between 1977 and 1994 included (in order of their appointment as full-time staff members) C.J. Hardy, R.M. Pickard, P. Lilley, E.F. Taylor, D. King, W.W. Clegg, R.S. Quayle, E.W. Hill, C.D. Wright, J.J. Miles and C.G.M. Harrison. Completed projects included:

1978: Analysis of the step and ramp response of a feedback gain-control system. Aspects of electron resist exposure. Detection and processing of secondary electron signals for pattern re-registration in scanning electron beam exposure. Distortion in low noise amplifiers. A multiple device interfacing system for mini-computers. Measurement of the magnetic properties of thin ferromagnetic films. Optimisation of overlay properties for use in magnetic bubble devices. Theoretical considerations in the design of magnetic bubble detectors. Limitations of zinc sulphide leaky waveguide electro-optic modulators. Electro-optic films on semi-conductors. Transport kinetics in horizontal ZnS epitaxial growth systems.

1979: Minimising the effects of filter overlap in multi-band signal/noise improvement systems. Stroboscopic measurement of strip domain expansion in chevron stacks. Teaching and research requirements of microprocessor development equipment. Mass transport processes associated with epitaxial growth of ZnS in H_2: the effects of a by-pass flow. Electro-chemical characterisation of carrier concentration profiles in silicon. The effects of finite contact energy barriers and traps on space charge limited currents in thin insulators.

1980: Use of microprocessors in reactor guard line circuits. Electro-chemical carrier concentration profiling in silicon. The electrolyte-silicon interface: anodic dissolution and carrier concentration profiling. A video-sync separator. Application of a pulsed solid-state laser diode array as an illuminator in high-speed photography. Investigation of thin-film ferromagnetic materials for use as a magnetic field sensor and the design of electronic circuits to energise the sensor. Thin film magnetometer: the output signal in the presence of an orthogonal field; the effect of magnetic film skew on the output signal; response to large fields. Evaluation of charge transfer devices in image filtering of TV signals.

1981: Minimising the noise in a thin film inductance variation magnetometer. The use of polymer dispersions to improve the sensitivity of electron resists.

1982: The measurement of small magnetic fields using thin ferromagnetic fields. Customisation of uncommitted logic arrays using electron beam lithography. Microprocessor control of thin film alloy deposition. Optical properties of zinc sulphide thin films prepared by sublimation in ultrahigh vacuum and by R.F. sputtering in argon. VPE ZnSe on GaAs: photoluminescence and conductivity. Control of optoelectronic properties of ZnSe films grown in GeAs by VPE. Characteristics of magnetic bubble detector performance. Performance characteristics of a chevron expander. The control of ignition timing to achieve maximum fuel economy.

1983: Microprocessor control of ignition advance angle. Performance of magnetoresistive vector magnetometers with optimised conductor and anisotropic axis angles. The magnetisation distribution in magnetoresistive stripe arrays. Factors affecting the performance of a thin film magnetoresistive magnetometer. Characteristics of magnetic bubble detector performance. A microprocessor-based cursor measuring system. Microprocessor control of thin film alloy deposition. The use of poly(ethylene oxide) to improve the exposure characteristics of poly(methyl methacrylate) electron resists. Optical films of single crystal zinc selenide.

1984: Electron beam direct-write customisation of uncommitted logic arrays. Microprocessor monitoring of the incident flux during the MBE growth of zinc selenide on gallium arsenide. Microcomputer controlled growth of zinc selenide on gallium arsenide using molecular beam epitaxy.

Electrical Engineering at Manchester University

Thin film magnetoresistive vector sensors with submicron gap widths. The performance of magnetoresistive vector magnetometers with optimised conductor and anisotropy axis angles. A position-controlled rectangular-coordinate table. Microprocessor control system for a high precision specimen stage. The growth and photoluminescence of ZnSe on GaAs by VPE in the temperature range 300-500°C. Electrical properties of zinc selenide crystals.

1985: Novel magnetoresistive devices with sub-micron features. Rapid prototyping of ULA designs by direct-write electron beam lithography. Formation by E-beam lithography and atom milling of accurately dimensioned sub-micron slots in nickel-iron thin films. A theoretical study of Bloch curve motion and punch-through in garnet films by computer simulation. Some aspects of thin film magnetoresistive sensors. Photoluminescence and electrical properties of VPS SnSe grown on GaAs.

1986: Platinum cobalt magnets produced by D C triode sputtering. A three-dimensional computer simulation of a translating magnetic bubble. Material considerations for vertical Bloch line memory. A three-dimensional computer model of domain wall motion in magnetic bubble materials. An integrated design to fabrication route for gate-array customisation by direct write.

1987: The role of E-beam in quick turn around and low volume production. Analogue-to-digital converter macro cell design. Observation on vapour species concentration during the growth of single crystal zinc selenide films by molecular beam epitaxy. False alarm rate and its implication in a range ambiguous radar. Sputtered permanent magnet arrays for MR sensor bias. VBL memory: projections of performance. Study of VBL punch-through thresholds by pulse in-plane fields. Compensating temperature induced sensitivity changes in thin film NiFeCo magnetoresistive magnetometers. Microstructural and magnetic properties of amorphous TbFe films. Spectral determination of transitional length in digital magnetic recording. Stability phenomena in amorphous rare earth-transition metal films.

1988: Customisation of high-performance uncommitted logic arrays using electron-beam direct-write. A low-power bipolar gate array. Fast prototyping for gate arrays. Assessment of strategy for optimum control of ignition angle advance. Heteroepitaxial ZnSe on GaAs: high-growth rates at low temperatures by conventional VPE. Stability and microstructural phenomena in RE-TM films for thermo-magnetic-opto-recording. Stroboscopic observation of high-speed VBL motion using an optical sampling system. Stray field interaction in PtCo/SiO$_2$/NiFe multilayer thin film structures. Velocity measurements on vertical Bloch lines. Field distributions produced by heads of different geometries. Field-induced ordering in parametric suspensions. Predicting the performance of thin film recording media. A new theoretical approach to digital magnetic recording. Monte Carlo simulations of light transmission in dispersions of paramagnetic particles.

1989: Digital control of ZnS and ZnSe vapour sources in an MBE system. A breathing monitoring device. Minimisation technique for series gated emitter coupled logic. Monte Carlo modelling applied to linewidth control for electron beam direct-write over device topography. A scanning laser magneto-optic microscope. Magneto-optic recording studies using a scanning laser microscope. An imaging system to investigate the dynamics of the write process in ICI digital paper. The digital recording properties of thin film media having sawtooth magnetisation transitions. Comparison of various methods for characterising the head-medium interface in digital magnetic recording. MR vector sensor with trimmable sense axis direction. Observations of magnetisation rotation in narrow permalloy strips using a scanning laser microscope. Modelling multiple MR elements for compound read heads. Development of magnetoresistive sensors for low power magnetometer applications.

1990: Short wavelength recording studies in amorphous TbFeCo films for magneto-optic recording applications. Observations of magnetisation distribution in narrow permalloy strips. Modelling multiple MR elements for compound read heads. Micromagnetic model for the study of magnetoresisitive devices. Design transfer to semicustom ASIC technology. Preparation of zinc sulphoselenide alloys using MBE. Magnetic properties of materials, transition shapes and digital recording properties. Magnetic transition structure in longitudinal thin film media.

1991: Development of a scanning laser microscope for magneto-optic studies of magnetic films. The effect of bulk hysteresis loop on the recording performance of magneto-optic films. Modelling the effects of boundary pinning in magnetoresistive elements. Magnetisation distributions in thin film permalloy strips. Integrated analogue multiplier combining MOS and bipolar transistors. Modulated beam mass spectrometry applied to monitoring the evaporation of zinc sulphoselenide alloys. An

1977-1994: 4-year courses, widening research, and a new Engineering School

instrument for recording drop-outs from magnetic tape. A simple high field hysteresis loop plotter for magnetic recording tapes. Experimental study of a write equaliser for unbiased digital magnetic recording. Micromagnetic simulations of transverse recordings. Remanence curves of longitudinal thin films. Observations of magneto-optic phase contrast using a scanning laser microscope. Comparison of various readout techniques in magneto-optic recording. Measurements of the temperature and spatial variations of coercivity in magneto-optic recording media using a scanning laser microscope.

1992: Time of flight and particle density measurement using a modulated beam mass spectrometer. Analogue voltage multiplier/divider with stabilised scale factor. Curve fitting to Monte Carlo data for the determination of proximity effect correction parameters. Bipolar VLSI - an application for a high-performance microcontroller. Crystal structure and photoluminescence spectra of applied layers of ZnSe on GaAs (100) substrates. Code merit in digital magnetic recording channels. Characterisation of magnetic recording heads. Representation of a digital recording channel. Recording properties of thin magnetic films at high densities. Crystallisation studies of electron-beam deposited telluride films. Micro-annealing studies of PtCo films for magneto-optic recording applications.

1993: A combined scanning optical and force microscope using interferometric detection. A two-stage model for the transition process in perpendicular magnetic recording. The effect of ultrasonic frequency on the accuracy of gas flowmeters. Effects of pinning magnetisation with geometrical features in thin permalloy films. Effect of the deposition process on the magnetic properties of coupled permalloy films. The role of dipole coupling in multilayers. Photo-assisted vapour phase epitaxy of ZnSe/GaAs (100). Theory of recording on media with arbitrary easy axis orientation. Theory of longitudinal digital magnetic recording on thick media. Modelling of recording system performance and its relationship to recording media properties. Effect of track width on transition noise in longitudinal thin film media. Simulations of the magnetisation reversal of Co-Cr and Fe particles.

1994: A differential interferometer for scanning force microscopy. Barkhausan transitions in single and bi-layer thin permalloy films. Energy intensity distributions of 30keV electrons in indium phosphide. Wavelength, frequency and interface effects in the replay process in digital magnetic recording. A qualitative multilayer recording model for media with arbitrary easy axis orientation. Scanning laser microscopy of magneto-optic storage media. Design of a combined optical disc tester and R-θ scanning laser microscope. Study of the relationship between temperature gradient and noise caused by irregular bit marks and internal sub-domains in magneto-optic recording.

Digital communications *(includes error detection and correction codes and coding systems; signal processing in communication systems; information theory and source coding).*

The group was instituted by Professor Farrell on his appointment in 1981. Staff members who have assisted in the work at various times include M. Beale, D.J. Tait, C.A. Boyd, Violet F. Leavers and T. O'Farrell.

In the late 1980s a joint research group, the 'Manchester Communications Research Group', was set up, linking the departmental research work with that in the corresponding group at UMIST.

Completed projects included:

1982: Error-correction methods for HF data channels. Digitally implemented HF data transmission modem. Study of EDC codes. Decoding algorithms for a class of burst-error-correcting array codes. An upper bound on the minimum distance of binary cyclic codes, and a conjecture. Array codes for correcting cluster-array patterns. Pros and cons of spread-spectrum for mobile communication.

1983: Array codes for correcting cluster-error patterns. Noise and false-lock performance of the PSK-Tanlock loop. The influence of LSI and VLSI technology on the design of error-correction coding systems. Practical applications of channel coding techniques. Composite sequences for SAWE matched filters in spread-spectrum receivers. A class of composite sequences and its implications for matched filter signal processing.

1984: Finite field transforms and symmetry groups. Code structure and decoding complexity. Coding for project UNIVERSE. Codecs for CERS.

1985: Soft decision decoding for the HF channel. Measurement and verifications of an HF channel model. Soft decision techniques for HF channels. Further developments in burst-error correcting array codes. Minimum weight decoding of cyclic codes. Burst-b characterisation of a mobile radio channel. A class of composite sequences for matched filter signal processing.

1986: Bandpass adder channel for multiuser (collaborative) coding schemes. An investigation of forward error control schemes for high error rate channels. T-SAT payload: Codec subsystem. Synchronisation of error control codecs for mobile communications via satellite.

1987: A class of burst-error-correcting array codes. Complexity and autocorrelation properties of a class of de Bruijn sequences. Low signal multiplexing system for pyroelectric linear arrays. Noise processes in a data acquisition system for pyroelectric linear arrays used in low resolution thermal imagers.

1988: Error performance of maximum likelihood trellis decoding of (n, n-1) convolution codes: a simulation study. Soft minimum weight decoding of Reed-Solomon codes. A soft decision Reed-Solomon decoder. Division algorithms for hard and soft-decision decoders. Correlation properties of dual-BCH, Kasami and other sequences for spread-spectrum multiple-access systems.

1989: A new multiple key cipher and an improved voting scheme. Low complexity decoding algorithms for general linear codes. Multiple burst-correcting array codes. Land mobile satellites using the highly elliptic orbits. Error control techniques applicable to HF channels. Calculation of error probabilities for distance-invariant error control codes. Low complexity concatenated coding schemes for digital satellite communications. An ARQ/FEC coding scheme for land mobile communication. Simulation of communication systems using the transputer.

1990: Hidden assumptions in cryptographic protocols. Joint source-channel coding for raster document transmission over mobile radio. Multilevel pseudocyclic codes. Coding as a cure for communication calamities: the successes and failures of error control. Soft-decision syndrome based node synchronisation. Punctured convolutional codes for mobile communications. Soft-decision syndrome based node synchronisation. Extension of the Smith construction for self-orthogonal array codes.

1991: New approaches to reduced complexity decoding. New signature code sequence design techniques for CDMA systems. Performance of phase invariant trellis coding for differentially-encoded coherent M-PSK signals on Rayleigh fading channels. Burst error correction capability of square array codes. Concatenated coding and interleaving for half-rate GSM digital mobile radio systems. Cyclic square and its coding characteristics.

1992: A survey of array error control codes. Ring-TCM codes for QAM. Node synchronisation for high rate convolutional codes. Information set decoding of maximum rank array codes. Implementation of a coded m-PSK modulation scheme based on rings of integers modulo-m. An adaptive coding scheme and its VLSI implementation. Applying compression coding in cryptography. A formal framework for authentication.

1993: Security architectures using formal methods. Generalised array codes and their trellis structure. Asymptotic shaping gains from non-Gaussian distributions. An efficient soft-decision decoding algorithm for array codes. A maximum likelihood decoding algorithm for variable-length error-correcting codes. Construction and decoding of a class of parity check-based codes. Soft-decision decoding algorithm for cyclic block codes. Combined DC-free, runlength limited, and burst error correcting code for a four-track magnetic tape. New multilevel (3-ary) optimum error-control codes derived from anticodes. A maximum A-Posteriori (MAP) decoding algorithm for variable-length error-correcting codes. Split syndrome soft-decision decoding for block codes.

1994: Ring-TCM codes for QAM on fading channels. Coded modulation based on rings of integers modulo-q: block codes and convolutional codes. Multilevel block subcodes for coded phase modulation. Array codes for cluster-error correction. Performance analysis of an optical correlator receiver for SIK DS-CDMA communication systems. Channel sense spread spectrum with overload detection versus FDDI. Sequential decoding of variable-length error-correcting codes. Error codes for cluster-error correction. Note on new digital signature scheme based on discrete logarithm. Designing secure key exchange protocols. Strengthening authentication protocols to foil cryptoanalysis. Partitioning algorithms for soft-decision decoding of block codes. A new block shaping scheme. Optical correlator receivers for SIK DS CDMA communication systems. Fast optimal decoding of product codes based on optimization of subcodes.

1977-1994: 4-year courses, widening research, and a new Engineering School

Signal processing techniques, as applied to speech and image processing (includes coding of speech signals; speech recognition; noise cancellation; modelling of the excitation source; high-compression coding of facsimile images).

The group was founded by Professor Xydeas on his appointment in 1987. Staff members who have assisted in the research include A.F. Erwood, C. Daskalakis, C.G.M. Harrison and J.P. Oakley. Completed projects included:

1988: Advances in analysis by synthesis LPL speech coding. Theory and real time implementation of a CELP coder at 4.8 and 6 Kbits/sec. A new approach to speech coding.

1989: Speech processing in mobile radio systems. Advances in low bit rate speech coding. Improving vector excitation codes through the use of spherical lattice codebooks. A novel approach to boundary finding. New method for images interpolation and its applications in image analysis.

1990: Overview of speech coding techniques. A single DSP high speech quality CELP codec. A progressive encoding technique for binary images. Advances in the processing and management of multimedia systems. Advances in transform coding of speech signals.

1991: An efficient method for finding the position of object boundaries to sub-pixel precision range. Multimedia information systems: the management and semantic retrieval of all electronic data types. Local methods for curved edges and corners. A novel graph theoretic texture segmentation algorithm. Classification of shape for content retrieval of images in multimedia databases. A comparison of acoustic noise cancellation techniques for telephone speech.

1992: Detection of the extreme points of closed contours. A hybrid bandsplitting acoustic noise canceller. A hierarchical classification method and its application in shape representation. Comparison of excitation methods for electrical capacitance tomography. A simulation study of sensitivity in stirred vessels. Digital signal processor based architecture for EIT data acquisition.

1993: Microfossil image characterisation within a content addressable image database. Detection and characterisation of carboniferous foraminifers for content-base retrieval from an image database. Measurement of tooth positions on a spur gear with an accuracy of 10µm from a single front-lit image. A long history quantisation approach to scalar and vector quantisation of LSP coefficients. Document skew correction system. Speech coding at 2.4kbits and below. High-quality low bit rate speech coding system. Scrambling of speech signals.

1994: Detection of circular arcs for content-based retrieval from an image database. Study of robust isolated word speech recognition based on fuzzy methods. Combining fuzzy vector quantisation and neural network classification for robust isolated word speech recognition. Channel noise detection and suppression in G 721 decoded speech for CAI CT2 applications. Low bit-rate coding using an interpolated ZINC excitation model. A neuroethological approach to automatic language identification. Digital EIT data acquisition. A database management system for vision applications. Evaluation of an application-independent image information system. Conversion of scanned documents to the open document architecture. Detecting the skew angle in document images.

Debate on the future of Engineering in the university

During the 1970s, and increasingly in the 1980s, the character of universities changed considerably, due in part to the fact that money had become the all-important policy criterion, which meant that the university was run as a business. Academic staff members were being encouraged to retire early, irrespective of the value of their experience and expertise, so that money could be saved; research assessment exercises were putting every departmental activity under a financial microscope; and the view that 'big is beautiful' as far as departmental size was concerned was becoming increasingly the prevailing philosophy of the day. In short, the smaller departments, of which Electrical Engineering could be considered one, were being put under ever-increasing financial scrutiny, amounting virtually to their having to justify their very existence. It was not enough for the professors to say to the Vice-chancellor, "This is a good department with a fine history going back nearly seventy years which has produced a wealth of fine graduates and first-rate research, and intends to go on doing so." Different criteria were now applied in a changed university world.

This was the situation the Electrical Engineering Department found itself in towards the end of the 1980s. In spite of all that had been achieved in recent years, and the acquisition of a half share in the new Information Technology Laboratories, the two departmental professors at that time, Professors Farrell and Xydeas, judged that the department faced immense difficulties. Matters came to a head when, at the end of a Departmental Board meeting on 29th November 1989, Professor Farrell made a statement proposing major changes to the structure of the department. The proposals, put forward in the name of both professors, suggested four possibilities for the future of the Electrical Engineering Department, as follows:

a) Remain a single Department but with internal changes.
b) Consider joining Simon Engineering as part of a Faculty or Department of Engineering.
c) As (b) but join with UMIST - either just our own Department or together with the Simon Engineering Department.
d) Restructure both internally and externally in that parts of the present Department would be distributed between Simon Engineering, Computer Science, and UMIST.

Announced without warning, these radical proposals came as a bombshell to most of the electrical engineering staff who greeted options (b), (c) and (d) with some hostility. In the aftermath of the proposals the matter was subsequently debated at length within the department. On 8th December 1989 a meeting of the department's non-professorial staff was held. A number of people had submitted documents beforehand outlining their own views, supported by appropriate written evidence. Although it was accepted that some degree of change was essential in order for the department to remain viable, the general feeling of those present was doubt as to whether proposals as revolutionary as (b), (c) and (d) above, which amounted to a dismembering of the department, were necessary. There was widespread rejection of the idea of splitting the department between those of Engineering and Computer Science on the grounds that this would mark the end of Electrical and Electronic Engineering as a subject; some people felt that if a major change had to be made it would be better for electrical engineering to become part of a wider engineering group, thus renewing a former association and producing a structure which would be large enough to exist in a competitive world.

1977-1994: 4-year courses, widening research, and a new Engineering School

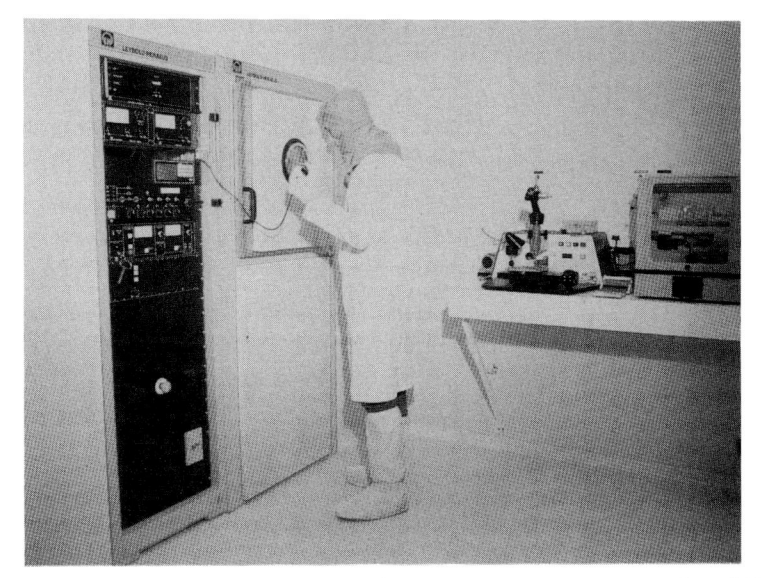

Clean room facility in the Electrical Engineering Department, late 1980s

Back row, left to right: D.J. Tait, J.J. Miles, M.J. Cunningham, C.D. Wright, C.A. Boyd, R. Shuttleworth, B.R. Varlow, C.J. Hardy, C.G.M. Harrison.
Middle row: D.E. Hesmondhalgh, R.S. Quayle, D. King, A.F. Erwood, J.P. Oakley, D. Tipping, Margaret E. Clarke, D.W. Auckland, W.W. Clegg.
Front Row: I.E.D. Pickup, E.W. Hill, P.L. Jones, G.W. McLean, Prof. P.G. Farrell, Prof C.S. Xydeas, B.K. Middleton, T.E. Broadbent, G.F. Nix, R.M. Pickard (Honorary Fellow). Not on picture: P.Lilley.

Staff of the Electrical Engineering Department, 2nd July 1990.
A picture taken outside the new IT Laboratories

Predictably, it was not long before news of the turmoil within the Electrical Engineering Department reached the ears of those outside, and surprise was expressed in some quarters that a department which was widely thought to be successful now wished to dismantle itself. On 21st December 1989 a meeting was held between Professors Jackson, Montague, Stanley and Tomlinson of the Engineering Department and Professor Xydeas and Drs Auckland and McLean from Electrical Engineering. The various alternative proposals that had been put forward by the electrical engineering professors, and variants of the proposals, were discussed, largely from the viewpoint of how the Engineering Department would be affected. The discussion then moved to the Faculty of Science. On 10th January 1990 a meeting was held between the Vice-chancellor, the Dean of the Faculty of Science, Professors Donnachie (Physics), Montague (Engineering), Walsh (Mathematics) and Farrell (Electrical Engineering). In his report of the meeting Professor Farrell stated that the need for restructuring, not only in Electrical Engineering but also in Engineering, had been re-affirmed.

The options were discussed further at the Electrical Engineering Board meeting of 24th January 1990. In a brief summing-up the chairman (Dr McLean) expressed his view that staff seemed to favour the idea of forming a School embracing the the activities of the departments of Electrical Engineering, Engineering, Computer Science and Materials Science.

Talks were still continuing outside the department. A meeting was held on 1st February 1990 attended by the Vice-chancellor, two Pro-vice-chancellors (Sir Francis Graham Smith and Joan Walsh), the Dean (Professor Max Irvine), Professor Peter Montague (Head of Engineering), John Gurd (Head of Computer Science), Professor Frank Sale (Head of Metallurgy and Materials Science), Professor A. Donnachie (Head of Physics) and Professor Farrell (Electrical Engineering). Of all the options for restructuring, the one involving two-thirds of the Department joining Computer Science to form a new department of (say) Information Engineering, and one-third joining a re-structured Department of Engineering, was thought to be the most effective. However, the matter was still open to further discussion.

The genie was by now out of the bottle and could not be pushed back inside. Some kind of re-structuring would have to occur, but what? It had become clear in the many discussions which had taken place that it was not just Electrical Engineering that would be affected; other departments also had their problems and would have to be included in any discussion about the best way of facing the future. The attitudes of UMIST would also have to considered if there was any possibility of further association with them. The future of Engineering in the university was in the melting pot.

1977-1994: 4-year courses, widening research, and a new Engineering School

A means of deciding on the best way forward was established in the early summer of 1990 when the Vice-chancellor decided to set up a "Working Party on Engineering". An interim report issued by the Working Party in January 1991 stated the membership and outlined the terms of reference: [10]

The University of Manchester
WORKING PARTY ON ENGINEERING

1. Membership and Terms of Reference.

The working party was established by the Vice-Chancellor in June 1990 to review engineering and related subjects in the Faculty of Science. The purpose of the review was to provide the Vice-Chancellor with advice on the future of engineering in the University.

The membership of the working party is as follows:

University members: Professor K.M. Entwistle (Chairman)*, Professor Michael Hart (Physics), Professor B.T. Robson (Geography), Professor A.P.J. Trinci (Biological Sciences), Emeritus Professor Jack Zussman (Geology).

External Assessors:
Professor Colin Andrews, Department of Engineering, University of Cambridge
Emeritus Professor Sir Bernard Crossland, Queen's University of Belfast
Professor H.R. Evans, School of Engineering, University of Wales, Cardiff
Professor W.A. Gambling, Department of Electronics and Computer Science, University of Southampton
Professor R.M. Needham, Computer Laboratory, University of Cambridge
Sir Robin Nicholson, Director, Pilkington plc.

The working party was not given formally stated terms of reference, but essentially they see it their task to gauge the standing of the engineering departments, both nationally and internationally; to identify their strengths and weaknesses; to make recommendations that would enhance their standing.

At their first meeting, the internal members of the Working Party agreed that, although UMIST was not formally part of the review, they would wish to ensure that all their recommendations should take account of the engineering activities in UMIST. To this end, the Principal of UMIST agreed to provide some information about UMIST activity and agreed that a number of UMIST engineering professors should meet the Working Party.

The Working Party met separately the following groups:

(i) The Professors of Engineering
(ii) The Professors of Electrical Engineering
(iii) Representatives of the non-professorial staff in the Department of Engineering
(iv) Representatives of the non-professorial staff in the Department of Electrical Engineering
(v) Professor Gurd and senior colleagues in the Department of Computer Science
(vi) Professor Garner and senior colleagues in the Department of Chemistry
(vii) Professor Curtis and senior colleagues in the Department of Geology
(viii) Professor Gregory and senior colleagues in the Department of Mathematics
(ix) Professor Hall and senior colleagues in the Department of Physics
(x) Professors Sale and Young, Materials Science Centre
(xi) Professors Chalmers, Launder, Reid and Singer, UMIST.

2. The Working Party received the following documents:

(i) Relevant extracts from the University's Planning Statement for 1994-95.
(ii) Session 1989-90 statistics for relevant departments.
(iii) Statements from Professor Montague on how Engineering see the future.
(iv) Statements from Professor Farrell on how Electrical Engineering see the future.
(v) Submissions from the non-professorial staff in the Department of Electrical Engineering.
(vi) Letters from individual members of staff.
(vii) Views from Heads of related departments.
(viii) Relevant extracts from UMIST Research Selectivity submission to the UFC.

---
* The Chairman, Professor Entwistle, graduated in Electrical Engineering from the department in 1945.

Re-housing of the Electrical Engineering Department

Whilst the above committee was still deliberating another unsettling event occurred. In the early 1990s the university conducted a survey on the use of space within the campus, taking into account student numbers and various other factors according to a specified formula. One of the conclusions arising from this was that the Simon Engineering Laboratories possessed more space, in the opinion of the enquiry panel, than was justified, a fact which led to the proposal that part at least of the Electrical Engineering Department should leave the building it had occupied since 1954 and move into the Simon building to join the other engineering departments. The proposal soon became a definite plan approved by the university and it was not long before certain areas of the Electrical Engineering Department's Dover Street building were earmarked for the new 'Campus Ventures' project (a commercial enterprise unrelated to the Electrical Engineering Department). Initially it was intended that the whole department would move during one long vacation, but in fact the transfer to other premises has been phased over several years and is still not complete at the time of writing.

Establishment of the Manchester School of Engineering

The general view of Professor Entwistle's 'Working Party on Engineering', as its discussions progressed, favoured a closer liaison between the various departments of Engineering at the University and the corresponding departments at UMIST. This proposal was formally set out in the "interim report" of the Working Party issued in January 1991 which stated:

" . . . the advice of the Working Party comprises the single and vital proposal that a Manchester School of Engineering be created by bringing together the engineering resources of Owens and UMIST into a co-ordinated whole."

However, whilst recognising the laudable objectives of the proposal this idea was not looked upon favourably by UMIST for a variety of reasons including the following:

1) The logistical difficulties of combining or having close liaison at undergraduate level between two large institutions on different sites;
2) A difference in approach in the teaching programmes in the two Institutions, Owens' general approach to Engineering being more orientated to Engineering Science;
3) Differences in the academic organisation of Engineering in the two institutions.

Moreover, UMIST at the time was considering its own future, and it decided to break away from the university, in effect, as from 1st August 1994 and become virtually a separate and autonomous institution instead of continuing as a Faculty of the University. This would mean a loosening, not a tightening, of links between the engineering departments of the two institutions, and it also meant that professors from UMIST would no longer be members of the University Senate. Any proposed scheme for closer links at undergraduate level was therefore doomed to failure. For these reasons the possibility of close links between the two institutions, or amalgamation, as proposed by the Entwistle Working Party, was in due course abandoned. The Working Party took no further part in the course of events from that stage.

A hurried re-think was thus required. Discussions took place within the university and a new Working Party was set up by the Vice-chancellor, with Mr S.A. Moore (Deputy Vice-chancellor) in the chair. The other members were:

Professor Peter Montague (Dean of the Faculty of Science)
Professor F.H. Read ((Research Dean of the Faculty of Science)
Professor J.S. Metcalfe (Dean of Economics and Social Studies)
Professor D.I.A. Poll (Head of the Department of Engineering)
Professor I.M. Smith (Department of Engineering)
Professor G.R. Tomlinson (Department of Engineering)
Dr C.S. Merrifield (Chairman of the Departmental Board in Engineering)
Professor D.W. Auckland (Head of the Department of Electrical Engineering)
Professor P.G. Farrell (Department of Electrical Engineering)
Professor Duncan Dowson, F. Eng., F.R.S., University of Leeds (as external assessor)

Several meetings of the Working Party took place during 1993 and by the latter part of the year it had come to a decision. It proposed that the best way of producing a large, viable, teaching and research institution was to amalgamate the various Departments of Engineering (including Electrical Engineering) within the Faculty of Science to form "The Manchester School of Engineering". A draft report of the Working party outlining the proposal was circulated widely towards the end of 1993 and the early part of 1994 for consultation. The separate departments of Electrical Engineering, Civil Engineering, etc. would, under this new scheme, be dis-established as departments in their own right and would become Divisions of the new Engineering School. The recently-approved physical transfer of much of the Electrical Engineering Department into the Simon building fitted in neatly with the plan. The scheme had the great advantage of 'safety in numbers' by providing a large organisation for teaching and research which would compare favourably, it was believed, with any in Britain. The report of the Working Party was considered by the Departmental Boards of Engineering and Electrical Engineering early in 1994 and met with almost universal approval. The proposal was considered by the University Senate on 25th May 1994 and by Council on 6th July 1994, both of which approved it.

Almost co-incident with the establishment of the new School, the Faculty of Science was renamed the 'Faculty of Science and Engineering'. The change of title was formally approved by the University Court at its meeting of 11th May 1994 and provided formal rectification of the anomaly whereby the Departments of Engineering within the Faculty had been awarding Bachelor of Engineering degrees to their graduating students instead of Bachelor of Science since 1989, to satisfy the requirements of the professional engineering institutions.

Arrangements for the creation of the new School were quickly put in hand and the Manchester School of Engineering was formally established on 1st August 1994, the same day that UMIST broke away from the University. An official opening was held on Wednesday, 12th October 1994 at 5.00pm to 7.00pm, when there were speeches by Professor Farrell (Head of the new School), Professor Harris (Vice-chancellor), Professor Donnachie (Dean of the Faculty of Science) and Professor Montague (Pro-vice-chancellor). Professor Derek Jackson gave a lecture on the history of Engineering at the University and there was an exhibition of Osborne Reynolds artefacts. The founding of the School marked a new chapter in the long history of Electrical Engineering at the university.

References: Chapter 14

1. *Qualifying as a Chartered Engineer.* Report of a working group chaired by Mr J.H.H. Merriman. Institution of Electrical Engineers, May 1978.
2. Details of staff changes are from departmental sources and university calendars.
3. Information from the Minutes of Electrical Engineering Departmental Board meetings and other departmental meetings.
4. Electrical Engineering Department Annual Report, 1980-81.
5. Details of the department's contribution to the Open Days are from literature produced by the department at the time.
6. Information about the courses is from departmentally produced literature: *Notes for Students*; Student Brochure; accreditation documents, etc.
7. *Engineering our Future.* Report of the Committee of Inquiry into the Engineering Profession. Chairman, Sir Montague Finniston. HMSO Cmnd 7794, London, January, 1980.
8. Electrical Engineering Department's Annual Reports, 1986-87, 1987-88.
9. Information on research projects is from the Electrical Engineering Department's Annual Reports, various years, and other departmental sources.
10. The information quoted is from the Working Party's report of January 1991.

Chapter 15

1994 onwards: The present and future

This chapter brings us to the present day. We are approaching the turn of the century and, coincidentally, it is just over one hundred years since Robert Beattie was appointed to teach electrical engineering. Naturally there have been enormous changes during that period, probably more so in the second fifty years than the first fifty. In four years' time the university will be celebrating the 150th anniversary of the foundation of Owens College. One wonders what the early luminaries of that institution, Scott, Greenwood, Williamson, Frankland and Roscoe, would think, were they able to look down on the scene on Oxford Road today. One can only speculate, but it is fairly safe to assume they would have been well satisfied with the success and growth of Owens College, and subsequently Manchester University, in the intervening years. Another landmark will be with us in 1998 - the 50th anniversary of the creation in the Electrical Engineering Department of the world's first stored program electronic computer. It is an appropriate time for reflection, but for the moment let us look at the present state of electrical engineering in the university.

Move from the Dover Street building

As explained in the previous chapter, the university's investigation in the early 1990s into the use of available space had led to the conclusion that the Engineering Department (Simon Laboratories) possessed more space than was appropriate to its needs. In the light of the decision in 1993-94 to create a Manchester School of Engineering of which Electrical Engineering would form part, the university made the obvious and perhaps logical decision that some of the Electrical Engineering staff should move into the Simon building, with their laboratories. It was not part of the plan that the whole department should transfer there, as several of the Electrical Engineering staff and their associated specialist laboratories and facilities had been located in the Information Technology Laboratories since 1988 and it was appropriate that they should stay.

It was originally intended that the move from the Dover Street building should be effected at one fell swoop during a long vacation, but in fact this has not happened.* Instead, it has occurred in stages over a number of years. By the autumn of 1993 there was pressure to locate part of the new 'Campus Ventures' project on the top floor of the building and the process of

---

* When the department moved from Coupland Street to the new building in Dover Street in 1954 the whole transfer of staff and equipment was completed over a period of a few weeks.

moving the electrical engineering staff and equipment began. At the time of writing (Autumn, 1997) only the ground floor, first floor and part of the basement of the Dover Street building are still occupied by electrical engineering staff and equipment; the remaining areas have been allocated to non-electrical engineering occupants. The electrical engineering staff are currently located as follows:

Dover Street Elect. Eng. building	*Simon building*	*Information Technology Labs.*
Prof. D.W. Auckland	Prof. P.G. Farrell	Prof. B.K. Middleton
B.Al-Zahawi	C.G.M. Harrison	Prof. C.S. Xydeas
Violet F. Leavers	P.L. Jones†	M.J. Cunningham*
P. Lilley†	R.S. Quayle	C.J. Hardy
J.P. Oakley		E.W. Hill
I.E.D. Pickup†		D. King
R. Shuttleworth		J.J. Miles
D.J. Tait		C.D. Wright
B.R. Varlow		

Electrical Engineering as part of the Manchester School of Engineering

The Electrical Engineering Department ceased to exist as a department in its own right on 1st August 1994 when the Manchester School of Engineering was founded. Consequently the Departmental Board of Electrical Engineering and the Boards of the other engineering disciplines were abolished but a larger Board representing the whole School was created in their place. The founding of the new School made little difference to the day-to-day running of teaching and research. The undergraduate courses and research work continued as before and the degrees continued to be in 'Electronic and Electrical Engineering' or related titles. However, the existence of the large School, bringing together all the engineering disciplines and presenting a large and coherent front to the outside world, undoubtedly strengthened the position of the whole of engineering in the university, from viewpoints both within the university and outside.

Electrical Engineering staff in the late 1990s

In the 1996-97 session the lecturing staff of the Electrical Engineering Division of the Manchester School of Engineering comprised the 21 members listed above. (See also the University Calendar list on the last page of Appendix 2). The most recently appointed members of staff are Drs Al-Zahawi and Leavers who both took up their posts in 1994. Professor Farrell was appointed 'Head of the Manchester School of Engineering' when it was established in 1994 and retains the post at the time of writing. The

---

* As 'Chairman of the Electrical Engineering Division' of the newly-formed School, Dr Cunningham also has an office in the Simon building.

† Retired from the full-time staff; now 'External Lecturers'.

title 'Head of Department' in Electrical Engineering no longer exists as Electrical Engineering is not now a department but a Division of the Manchester School of Engineering. However, Dr Cunningham currently holds the title 'Chairman of Division' which is the present equivalent of the former title.

The undergraduate courses

Since 1994 the number of courses available has continued to proliferate. The 'standard' 3-year courses are shown in block letters below, and, by 1997, each had six 4-year variants, making 21 courses in all, as follows:[1]

Honours School	Degree	Duration	UCAS code
Electronic and Electrical Engineering	B.Eng.	3 years	**HH 56**
Extended and Enhanced	M.Eng.	4 years	HHN6
Industrial Programme	B.Eng.	4 years	HHNP
European Programme	M.Eng.	4 years	HHMP
Japanese Programme	M.Eng.	4 years	H580
Singapore Programme	M.Eng.	4 years	HH5Q
USA Programme	M.Eng.	4 years	HHMQ
Electronic Engineering	B.Eng.	3 years	**H600**
Extended and Enhanced	M.Eng.	4 years	H608
Industrial Programme	B.Eng.	4 years	H604
European Programme	M.Eng.	4 years	H602
Japanese Programme	M.Eng.	4 years	H601
Singapore Programme	M.Eng.	4 years	H612
USA Programme	M.Eng.	4 years	H613
Electronic Systems Engineering	B.Eng.	3 years	**H611**
Extended and Enhanced	M.Eng.	4 years	H614
Industrial Programme	B.Eng.	4 years	H605
European Programme	M.Eng.	4 years	H603
Japanese Programme	M.Eng.	4 years	H606
Singapore Programme	M.Eng.	4 years	H607
USA Programme	M.Eng.	4 years	H609

'Electronic and Electrical Engineering' is a comprehensive and balanced programme covering light current and heavy current topics, thus opening up employment prospects in any branch of the profession.

'Electronic Engineering' concentrates on Electronics and Information Technology, thereby providing a fundamental understanding of circuits and devices as the basis for the development of future technologies.

'Electronic Systems Engineering' is a specialised variant of electronics, developing the ability to implement complex systems from specification and validation, through the stages of design, construction and testing.

Syllabuses

In accordance with university policy introduced in the early 1990s, all courses now follow a modular pattern in which students must complete satisfactorily a certain number of 'units' in each year. With so many different degree courses available it is obviously not possible to give a detailed account of the content of each one, but briefly, the syllabuses are as follows. All courses comprise one unit in the modular system unless stated otherwise. Twelve units are studied in each year, and each year comprises compulsory and optional units. Restrictions operate within the choice of options to ensure an appropriate subject balance:

Electrical Engineering at Manchester University

Electronic and Electrical Engineering Degrees (Codes HH56, HHN6, H580, HH5Q, HHMP, HHMQ, HHNP)

First year (all programmes in Electronic and Electrical Engineering): Engineering skills and computer programming I; design and manufacture; electronic and electrical systems and applications; introduction to electronic circuits; electronic circuit design I; introduction to digital design; introduction to microprocessors; measurements and instrumentation; power engineering I; engineering mathematics I and II*; materials and devices; foreign language.

* In these lists, where combinations of 'I', 'II' or 'III' of any subject are grouped together to save space, they count as 1 unit each.

Second year (all programmes in Electronic and Electrical Engineering): Design and quality; computers in engineering; engineering group project; electronic circuit design II; signals and circuits I and II; introduction to control; electromagnetism; power engineering II; electrical machines; engineering mathematics III; digital design methods; microprocessors; mechanical engineering I; foreign language.

Third year (HHNP): Year in relevant industry, gaining experience and undertaking training.

Third year (HHMP, H580, HH50, HHMQ): Year in a university abroad taking relevant lecture courses, doing project work and reinforcing cultural and linguistic experience.

Third year (HHN6) and final year (all other programmes in Electronic and Electrical Engineering): Project laboratory (3 units); pulse and radio frequency circuits; electronic circuits; design for production; introduction to communications; digital transmission; introduction to digital signal processing; digital image processing; control engineering; power systems; high-voltage engineering; power electronics; a.c. variable speed drives; engineering management; mechanical engineering II; numerical methods; computer systems design.

Final year (HHN6); each course counts 1.5 units except the project: Project (3 units); microcircuit systems design; data communications; mobile communications; information coding and encryption; image processing; speech and image compression; predictive control; electromagnetic fields; power systems analysis; high-voltage technology; information storage systems; optical data storage; magnetic design and modelling; magnetic measurements; engineering management.

Electronic Engineering Degrees (Codes H600, H608, H601, H602, H604, H612, H613)

First year (all programmes in Electronic Engineering): Engineering skills and computer programming I; CADMAT; introduction to electronic circuits; electronic circuit design I; introduction to digital design; introduction to microprocessors; measurements and instrumentation; software engineering I; materials and devices; engineering mathematics I; engineering mathematics II and III; electronic and electrical systems and applications; foreign language.

Second year (all programmes in Electronic Engineering): Design and quality; computers in engineering; engineering group project; electronic circuit design II; electronic systems; digital design methods; microprocessors; signals and circuits I and II; introduction to control; electromagnetism; semiconductors and devices; software engineering II; foreign language.

Third year (H604): Year in relevant industry, gaining experience and undertaking training.

Third year (H601, H602, H612, H613): Year in a university abroad taking relevant courses, doing project work and reinforcing cultural and linguistic experience.

Third year (H608) and final year (all other programmes in Electronic Engineering): Project laboratory (3 units); pulse and radio frequency circuits; electronic circuits; design for production; introduction to communications; digital transmission; introduction to digital signal processing; digital image processing; control engineering; engineering management; numerical methods; computer systems design.

Final year (H608); each course counts 1.5 units except the project: Project (3 units); microcircuit systems design; data communications; mobile communications; information coding and encryption; image processing; speech and image compression; information storage systems; optical data storage; magnetic design and modelling; magnetic measurements; engineering management.

Electronic Systems Engineering Degrees (Codes H611, H614, H603, H605, H606, H607, H609)

First year (all programmes in Electronic Systems Engineering): Engineering skills and computer programming I; CADMAT; introduction to electronic circuits; electronic circuit design I; introduction to digital design; digital design methods; introduction to microprocessors; measurements and instrumentation; software engineering I; engineering mathematics I and II; electronic and electrical systems and applications; foreign language.

Second year (all programmes in Electronic Systems Engineering): Design and quality; computers in engineering; VLSI group project; electronic circuit design II; electronic systems; microprocessors; signals and circuits I and II; electromagnetism; engineering mathematics III; VLSI design; software engineering II; semiconductors and devices; foreign language.

Third year (H605): Year in relevant industry, gaining experience and undertaking training.

Third year (H603, H606, H607, H609): Year in a university abroad taking relevant lecture courses, doing project work and reinforcing cultural and linguistic experience.

Third year (H614) and final year (all other programmes in Electronic Systems Engineering): Project laboratory (3 units); pulse and radio frequency circuits; electronic circuits; design for production; introduction to communications; digital transmission; introduction to digital signal processing; digital image processing; engineering management; computer systems design; compilers; integrated systems.

Final year (H614); each course counts 1.5 units except the project: Project (3 units); microcircuit systems design; data communications; mobile communications; information coding and encryption; image processing; speech and image compression; information storage systems; optical data storage; system and machine architecture; advanced architecture; neural networks; engineering management.

The total annual intake to the various courses in the second half of the 1990s is about 100 students.

Current research work in electrical engineering

At the time of writing there are four research groups in electronic and electrical engineering, each headed by a professor. The groups, with a brief summary of their areas of speciality, are as follows:[2]

Communications (Professor Farrell)

The group is concerned with all aspects of digital communication systems theory and practice. These include error control coding for all types of communication and information storage systems, multifunctional coding, multiple access techniques applied to personal and mobile communications and broadband networks, security and cryptographic protocols, performance analysis and simulation of communication systems, and autonomous systems. Non-professorial staff members of the group since 1994 are D.J. Tait, Violet F. Leavers, C.A. Boyd (to 1995) and T. O'Farrell (to 1996).

Information Storage (Professor Middleton)

This is currently the largest of the research groups, as measured by the numbers of staff contributing to the research, and is concerned with studies of magnetic and magneto-optic recordings for high density digital applications. Interests range from magnetic materials for recording media and heads through to recording systems and applications. Related instrumentation, scanning laser and force microscopes are being developed and used. Specialist facilities also include magnetic film deposition equipment, lithographic facilities and recording decks and spin stands. Non-professorial staff members are W.W. Clegg (to 1996), M.J. Cunningham, C.J. Hardy, E.W. Hill, P.L. Jones, D. King, J.J. Miles, P. Lilley, R.S. Quayle and C.D. Wright.

Signal processing (Professor Xydeas)

The group is concerned with digital signal processing theory and information systems. Current work is focused on efficient speech coding systems for a variety of applications including digital mobile radio and satellite systems, packet-based voice systems and voice-store and forward systems, acoustic noise reduction schemes, robust speech recognition, real-time implementation of algorithms/DSP architectures, image analysis, shape and texture analysis, object classification, image reconstruction for electrical impedance tomography, fusion of multi-sensor information and directed model based motion analysis, content addressable multimedia database systems, generic visual query languages, automatic analysis and representation of document images. Non-professorial staff members are A.F. Erwood (to 1996), C.G.M. Harrison and J.P. Oakley.

Electrical Energy Systems (Professor Auckland)

Activities span the subject area from electrical machines and drives to high-voltage technology. Interests are centred on generation, power systems, drives, and high voltage. Current work involves the production of electrical energy from waste recovery, the modelling of aircraft power systems, solid polymeric insulation and the design of distribution networks using image processing methods. Non-professorial staff members are B. Al-Zahawi, I.E.D. Pickup, R. Shuttleworth and B.R. Varlow. D.E. Hesmondhalgh and D. Tipping were involved part-time until 1996.

Examples of the current research work of the four groups

This is best illustrated by listing some of the projects which were completed during the calendar years 1995 and 1996. The various research groups are not compartmentalised in a watertight way and there is considerable liaison and overlap between the work of different groups and some staff members have done work for several groups. Consequently, some of the work listed below is the result of co-operation and could have been listed in the work of groups other than the one in which it has been placed. Examples of completed research projects which led to publications are as follows:[3]

Communications:

1995: Reducing the complexity of trellises for block codes. Adaptive forward error control schemes in channels with side information at the transmitter. Combined diversity detection and soft decoding for fading channels. Low complexity soft-decision decoding for RAC codes. The general threshold decoding rule and related codes. Modified minimum weight decoding for Reed-Solomon codes. Adaptive error control schemes in broadcast slotted DS/SSMA. Variable length error correcting codes.

1996: Effective trellis decoding techniques for block codes. Efficient ring-TCM codes on over fading channels. Performance of an OFDM system in frequency selective channels using Reed-Solomon coding schemes. Information set decoding of generalised concatenated codes. Scarce-state transition vertibi decoding of punctured convolutional codes for VLSI implementation. Optimised threshold switched diversity with soft-decision decoding for multipath fading channels. Minimum function-equivalent trellis of linear block codes. Class of TCM codes over rings of integers modulo-q. Ring-TCM for M-PSK modulation AWGN channels and DSP implementation. Ring TCM for fixed and fading channels - land mobile satellite fading channels with QAM. Trellis decoding of combined diversity-coding schemes (MSLD) for fading channels. Efficient permutation criterion for obtaining minimum trellis of a block code.

1994 onwards: The present and future

Information storage (including measurements):
1995:   A study of PtCo films used for recording studies. Pit formation in dye-polymer optical storage materials. Piezo-ceramic polymer films for micro-actuation and control applications. Magneto-optic media testing using a scanning laser imaging system. Conductor stress-induced anisotropy on vertical Bloch lines. Stripe-domain chopping in the presence of in-plane fields. Active vibration control and actuation of a small cantilever using deposited piezo-electric films. Design considerations for an ASIC to implement a maximum likelihood decoder for CMI and Manchester line codes. Integrated circuit design and implementation for a new optimum Manchester decoder. Imaging of laser modulated written domains in magneto-optical films for direct overwrite. Active vibration control of a small cantilever. Modelling the perturbative effect of MFM tips on soft magnetic thin films. The limit of fluxgate sensitivity due to Barkhausen noise for single layer and bi-layer permalloy thin film cores. Modelling damaged MFM tips using triangular charge sheets. Measurement of the model shapes of inhomogenous cantilevers using optimum beam deflection. Output waveforms in the replay process in digital magnetic recording. Head on domain boundary and magnetization transition motion in thin magnetic layers. The effect of cluster size on thin film media noise. Micromagnetic simulation of texture induced orientation in thin film media. A mathematical model for the multi-electrode capacitance sensor. Reciprocity in magnetic forced microscopy. A depth discrimination based focus error detection scheme for optical storage applications.
1996:   A linearised electrostatic actuator. An investigation into the optimum design of piezo-actuators. A study of the initial growth of PtCo thin films on silicon nitride. Xilinx FPGA design in a group environment using VHDL and synthesis tools. A versatile vibrating reed magneto-optic magnetometer. A GMR magnetometer. Active vibration control using ZnO thin films. The use of thin sputtered films for moving micro-mechanical structures with nanometric precision. A 1D model of spin-valve read heads. Micromagnetic modelling of rigid disk media. Optical readout systems for high density magneto-optic recording. A study of PtCo films used for longitudinal recording. Digital record theory for a thin medium with easy magnetisation direction tilted out of its plane. Wavelength, frequency and interface effects in magnetic recording system replay noise spectra. Models of metal evaporated tape. Identification of noise and ISI limited performance in digital magnetic tape recording systems. A simulation of the role of physical structure on feature sizes in exchange coupled longitudinal thin films. Spectral response nulls for single pole heads in perpendicular recording. Reciprocity based transfer function analysis in magnetic force microscopy.

Signal processing:
1995:   Calibration of an airborne camera. Codesign and testing of voice operated embedded systems. Design for voice operated wheelchair systems. Model matching in intelligent document understanding. Real time speech processing to eliminate slamdowns in digital voice systems. Slamdown detection: a practical application of neural networks. Pre-attentive computer visions: a new model. A low cost quantitative reconstruction algorithm for ECT and EIT in process tomography. Implementation of oriented filters for arc detection. Efficient coding of LSP parameters using split matrix quantisations. Methodology for detecting man-made structures in sequences of airport aerial images. Robust speech recognition in a car environment. A new shape representation for fast 3-D reconstruction from multiple 2-D perspective images. Application of electrical impedance tomography to a plant-scale mixing vessel.
1996:   Automatic language identification using supra-segmental prosodic features. Automatic language identification using a linguistically motivated multi-parameter model. Enhancement of low level feature extraction using a probabilistic Hough transform in conjunction with an inhibit-surround mask in image space. Simulation of photon-limited images using video data. Advances in prototype interpolation coding at 1.5 and 2.4kb/sec. Source driven variable bit rate prototype interpolation coding. A new approach to robust speech recognition using fuzzy matrix quantisation, neural networks and hidden mirror models. Comparison of system architectures for document image understanding. A curvature sensitive filter and its application in microfossil image characterisation. Pitch synchronous multi-band coding of speech signals.

Electrical Engineering at Manchester University

Electrical energy systems:

1995: Design of a thermoelectric generator for remote subsea well heads. Novel approach to alternator field winding interturn fault detection. Artificial neural network-based method for transient response prediction. The use of neural networks for discrimination of partial discharges in transformer oil. Smart insulation for tree resistance and surge absorption. Automatic map reading for distribution system design. Water treeing in insulating liquids. Effects of barriers on tree growth in solid insulation. A surface tracking model for failure prediction. The use of ultrasound for the detection of water trees in XLPE cable. Water absorption in composite dielectrics. A high speed alternator for a small scale gas turbine CHP unit. A 1kW high speed alternator for a turbocharger based micro CHP unit.

1996: Reducing variability in inclined plane tracking test results. A case-based approach to alternator modelling. Mechanical interaction of electric trees and barriers in insulating resins. Zinc oxide filled polyester resin for non uniform behaviour at high fields. Degradation of oil-paper systems due to the electric field enhanced absorption of water. A 250,000 rpm drilling spindle using a permanent magnet motor. Matrix converter application for direct-drive gas turbine generator sets. Development of a new magnetic actuator for a vacuum switch. Steady-state analysis and simulation of a sinusoidal sub-synchronous converter cascade. Electricity in the home: an evaluation of single-house combined heat and power. Analysis of single-step damping in a multistack variable reluctance stepping motor. New tap changing scheme. Development of a novel snubber circuit for an on-load GTO assisted tap changer.

The Manchester School of Engineering in 1997

The Simon Laboratories, the home of much of the new School, occupy the centre to right of the picture, with the Electrical Engineering Dover Street building, home of the Electrical Engineering Department for 40 years, on the left. In the background (left) is the Church of the Holy Name, and (right) the Moberley Tower and part of Alfred Waterhouse's building.

1994 onwards: The present and future

Brief summary of electrical engineering over the years

The university can look back with pride on the achievements in Electrical Engineering. From the earliest days of Owens College the subject was taught at a time when Frankland and Roscoe were establishing the reputation of the college as a leading centre for scientific teaching and research. The good work was continued in the 1870s, 1880s, 1890s and early 1900s by Arthur Schuster, a fine physicist who recognised the importance of electrical technology. He was well acquainted with the theory of electrical machines, transformers and related technical devices, and lectured on the subject himself. It was Schuster who was responsible for putting the teaching of electrical technology on a formal basis not only in the lecture room but also in the laboratory. The dynamo house he set up in 1888 for teaching and research purposes marked a formal recognition of the importance of electrical engineering as an academic discipline.

During the 1890s Physics students were able to specialise in electrical technology ("electrotechnics") and in 1896 Robert Beattie was appointed to give lectures, supervise laboratories and do research in the subject. Electronics did not exist as a subject at that time so the work was naturally concentrated on measurements, machines, power transmission and the other 'heavy' subjects which were beginning to change people's lives. Beattie's efforts were successful, the number of students opting to study electrical technology grew, and the Department of Electrotechnics was founded by Rutherford in 1912 with Beattie as professor in charge. However, it had hardly found its feet when the Great War began in 1914 and the department's academic staff of two were diverted into war work, to which they contributed with distinction.

After the war Beattie remained in charge of electrical engineering for 20 more years. In the early part of this period little research work was attempted, due to a variety of causes, but the standard of teaching was high, leading to an output of graduates who were much sought after. In the late 1920s and early 1930s research work became an increasingly important aspect of the department's activities and in the years leading up to World War II, when the youthful F.C. Williams was an Assistant Lecturer, much original and high-quality work was done and papers were published at a prolific rate.

Following the retirement of Robert Beattie in 1938, Willis Jackson took up the reins and led the department most capably until 1946. During all but the first and last years of his tenure the country was at war. In that period the team of young research students he led, "Jackson's Circus", made a significant, but generally unacknowledged, contribution to the scientific war effort.

Willis Jackson was succeeded as Professor of Electrotechnics by F.C. Williams, universally regarded as one of the shining stars of electronics research and invention. As soon as he arrived at the beginning of 1947 he and his young assistant, Tom Kilburn, set about building a computer store and within 18 months (by June 1948) had achieved the notable distinction of inventing the world's first successful stored program computer. It is doubtful whether anyone at the time realised the tremendously significant part that computers would play in the lives of future generations; it was just part of the interesting work they wanted to do at the time. No one had cause to give a thought to the possible historical significance of what was being achieved. Such is usually the case when major innovations are introduced, whatever the particular field.

Within a short time the computing work made rapid advances. Between 1948 and 1962 a series of highly successful computers was produced including commercial versions designed in the department and built by Ferranti; Mark I, Mark II ('Meg' and 'Mercury'), 'Muse' and 'Atlas'. It is perhaps invidious to compare the relative merits of different kinds of research, but by any standards the computing work carried out in the Electrical Engineering Department at Manchester between 1947 and 1964 must rank as highly as work done anywhere in the world.

In 1954 the Electrical Engineering Department moved from Coupland Street to its new purpose-built Dover Street building. Electrical Engineering had occupied its old building, as a department, since 1912 and before that it had been located in Coupland Street since 1900 as part of the Physics Department. Going back further still, the earlier teaching and research had taken place in the basement of the old Waterhouse building, the original dynamo house of 1888 being situated in what was called 'the courtyard' (which still exists) adjacent to the back quadrangle. The new Dover Street building, the first structure to occupy the university's new science area, was regarded at the time as a 'no expense spared' development. It served the department well for 40 years.

With the departure of the computing staff to form the Department of Computer Science in 1964, the Electrical Engineering Department lost one of its most successful areas of expertise but there were other areas which, since the war, have put and kept the department in the forefront of research, notably electronics (various aspects), electrical machines, servomechanisms and control systems, and high-voltage and dielectric breakdown. One can think of many notable people who contributed to the research work in the 20 or so years after World War II apart from Williams and Kilburn. J.C. West, Eric Laithwaite and Ron Cooper spring to mind, and there were others.

1994 onwards: The present and future

The death in 1977 of Professor Williams undoubtedly marked the end of an era, but the strong tradition of research work in the Electrical Engineering Department has been maintained. During the last two decades the appointment of Professors Farrell and Xydeas has resulted in the establishment of major new areas of research, and in recent years Drs Auckland and Middleton have been promoted to chairs, resulting in each case in the strengthening of their research groups.

Research has always had a more exciting image than teaching, but the work of the electrical engineering staff in teaching generations of students must not be forgotten. Even during the 1920s when little research work was done, the electrical engineering graduates left the department with such a sound base of technical knowledge that many of them soon occupied important positions in their chosen fields. This state of affairs has prevailed throughout the period covered in this book. At least thirty of the department's first-degree graduates have been appointed to Chairs in British Universities and a number of others in universities overseas. Numerous graduates have occupied or still occupy high positions in industry, and some are known to have become millionaires. It is no accident that since the second world war three Presidents of the Institution of Electrical Engineers are men who obtained their first degree in the Electrical Engineering Department; Willis Jackson (graduated 1925; President 1959-60; J.C. West (1943; 1984-85) and P.J. Lawrenson (1954; 1992-93). J.M. Meek, who was President in 1968-69, was not a graduate of the department but was a staff member during the war. Willis Jackson, who was then 'Professor the Rt. Hon. Lord Jackson of Burnley', was elected an Honorary Fellow of the IEE in 1968. The IEE awarded its Faraday Medal, which is given for notable scientific or industrial achievement in electrical engineering, to Professor F.C. Williams in 1972, to Professor J.M. Meek in 1975 and to Professor P.J. Lawrenson in 1990. Professor Tom Kilburn has also been the recipient of numerous honours including the Computer Pioneer Award in 1982, and several honorary degrees. These are just a few examples of the achievements of people from the department.

At the present time the Electrical Engineering Division of the new Engineering School is active in four main areas of research, numerous papers are being published, a lot of money is being brought in through research contracts, etc., and 21 different undergraduate courses are available. As always in its long history, electrical engineering is playing its full part in the work of the university, through its research and by educating the technologists whose expertise will be vital to the country in the coming years.

Prospects for the future

It is difficult to forsee the future of electrical engineering at Manchester University without the aid of a crystal ball. Most predictions turn out to be wrong. It is interesting to note that Sir John Stopford, the then Vice-chancellor and a very astute and wise man, noted in his Annual Statement in 1945 that, in accommodating the expected post-war expansion, a continuous increase in size could not be tolerated. He remarked:[4]

> "It is worth stating that when Owens College was transferred to the present site, about 1870, I do not imagine that the trustees in their wildest dreams expected the numbers at any time would exceed 1,000, and the accommodation was planned accordingly.
>
> After the first world war there was a tremendous increase in numbers, and our figures ultimately settled down to approximately 2,500. When the position was reviewed about fifteen years ago it was generally agreed that we should not exceed 3,000, and the expansion of departments, unions and refectories was planned to meet such a figure.
>
> Today we are facing an immediate increase to 4,000, which may be temporary, but I doubt it; and in view of the demands which we know are to be made upon the university and the need for a considerable increase in the number of teachers, doctors, scientists, technologists and dentists, it is clearly our duty to review the situation once again.
>
> The moral of this, as I see it, is that we should endeavour to make this the final review. To continue piecemeal this extension of the university would be uneconomic, inefficient and foolish. All will, I think, agree that we cannot afford, for academic and other reasons, to continue indefinitely to extend our commitments. Therefore, I believe we should try now to discover what is the optimum number of students for this particular civic university, paying the fullest regard to standards and all the educational requirements as well as the national and regional needs."

This example of Professor Stopford's glimpse into the future shows just how wrong predictions can be, even when they are based on the best available information. At the present time the population of the university comprises more than 18,000 students and 2,800 academic staff.

It is not just in size that the university has changed. When I joined the staff of the Electrical Engineering Department forty years ago nearly all my time was spent in teaching and research. Staff meetings were few and far between; the department ran on an informal basis. Professor Williams' policy was to appoint people whose abilities and integrity he trusted and let them get on with their job. If anyone wished to introduce a new course or other innovation, all that was required was a brief discussion with 'The Prof' and a few colleagues, and it was soon arranged. Nowadays the university system is very different. Arrangements are more formalised; there are rigid committee structures and, as noted in the previous chapter, external influences affect university departments so they are no longer able to proceed in the way they please. One example that comes to mind is the process of accreditation by the professional institutions, which to some degree governs the syllabuses of the courses offered, and which takes up much staff time and resources in preparing for periodic visits of the accreditation panel; one of the reasons why it was introduced was the need to monitor the courses of the large number of new universities and colleges teaching electrical engineering. * Other examples are the periodic national assessment exercises in teaching and research, which result in the publication of 'league tables' of the performance of departments based on criteria which are sometimes in dispute but against which there is no appeal. Staff appraisal also comes to mind, in which every staff member's 'performance' has to be assessed by someone half a millimetre further up the scale.

Above all, the universities in general are in such a parlous financial position at the present time that there is always a desperate need to 'bring money in', and it is only natural that the type of work which attracts the attention of sponsors is that which will provide a relatively short-term "answer to a problem" rather than fundamental studies which offer no obvious commercial benefit. This is not the fault of the university or any particular professor; they have no option but to look more favourably on the efforts of those with good industrial or similar contacts than those whose research does not lead to such financial benefits, for the university has to maintain its buildings, buy equipment, pay its wages, and remain solvent. But one does wonder how the efforts of some of the famous scientists, discoverers and inventors of the past might have fared under the present system, when they carried out research offering no clear profit to mankind. One thinks of Faraday's famous answer to a question about what use his research was; he replied "What use is a baby?" Or how would Millikan have fared in the present system of assessment where a constant stream of publications is necessary to show that one is doing something useful and thus have one's contract renewed or achieve promotion. Millikan spent years working on his project to measure the charge on the electron. Eventually he published a paper, the gist of which was, "I have measured the charge on the electron. Its value is 4.774×10^{-10} e.s.u."[7] Not a lot to show in terms of the number of publications - but work of outstanding merit.

Because the performance of departments and individuals is scrutinised in such great detail it is necessary to record all sorts of things that used never to be recorded; conferences attended, factories and research establishments visited, grants, etc. It is an interesting commentary on the way that bureaucracy governs the conduct of departments, through no fault of their

---

* In 1923 there were only 10 full English universities; Oxford, Cambridge, London, Durham, Manchester, Birmingham, Liverpool, Leeds, Sheffield and Bristol.[5] In 1997 the number is 115, of which nearly half are institutions granted university status in 1992.[6]

own, that the Electrical Engineering Annual Reports presented to the University Council are currently more than 20 pages long, including appendices; in the early 1960s, when the Atlas computer had recently been designed and put into operation and a lot of other research of the highest quality was being done in the department, the reports were about three pages long. No-one would argue that seven times as much is being achieved now than was the case 35 years ago when the number of staff was 28 compared with the present figure of 21, but the university authorities now demand that so much more detailed paperwork has to be recorded and presented.

The factors outlined above have resulted, I feel, in a certain loss of morale amongst university staff. At one time it was commonplace for staff approaching their 65th birthday to seek permission to stay on for a further two years before retiring at 67, the maximum age allowed. Now many staff are retiring in their 50s, feeling that the job does not offer the same enjoyment as a few years ago. Students are not directly affected. They are in the university for a short time and are not concerned with how the present system compares with that of yesteryear. What is of prime interest to them is the quality of the course they are taking and the probable job prospects to which their degree will lead. However, the financial strictures affecting the whole educational system affect students indirectly, and the problem of maintaining themselves financially whilst at university is of direct concern.

The present rigid management structures, the various forms of assessment of staff and courses, and the financial stringency of the present day, seem set to continue well into the next century. Similar conditions are prevalent not just within the universities, but in all professional occupations where every activity is scrutinised by accountants so that money can be saved, and job security is a thing of the past.

I hope the picture portrayed here is not too gloomy; it is not meant to be, for I believe there are grounds for cautious optimism. The changes noted are more apparent to older people who knew the university in an earlier era than to the youngsters of today who have entered the university system recently and seen it only as it is at the present time. The future of the university, and more particularly of electrical engineering, is in their hands. Given a talented and dedicated younger generation, which I am sure we have, it is certain that the future holds prospects of achievement just as great as those of the past. I like to think of the old Waterhouse building, whose corridors were trod by Osborne Reynolds and Arthur Schuster and which looks as splendid today as it ever did, as a symbol of stability in a changing world. We have seen 125 years of achievement in Electrical Engineering at Manchester. Let us look forward to even greater achievements in the future.

References: Chapter 15

1. The information on courses is from brochures published by the Electrical Engineering Division and similar sources.
2. The description of the work done by the research groups is from recent and current post-graduate research brochures published by the Electrical Engineering Division.
3. List of research projects is from Electrical Engineering Division Annual Reports for 1995 and 1996.
4. Vice-chancellor's Annual Statement dated 26th June 1946, quoted in the 1946-1947 University Calendar.
5. Whitaker's Almanack, 1923, pp. 286-293.
6. Whitaker's Almanack, 1997, p. 441 and a list published in 1997 by the Higher Education Information Services Trust, quoting 115 institutions currently regarded as having university status.
7. R.A. Millikan. Philosophical Magazine, **34**, p. 1, 1917.

Appendix 1. Members of Staff in Electrical Engineering, 1896-1997.

Note: Between 1896 and 1912 lectures in Electrical Engineering were given by Dr Beattie in the Department of Physics. In 1912 the Department of Electro-technics was founded with Dr Beattie as the first Professor, assisted by one other staff member. The Department continued to be known as 'Electro-technics' until 1948 when its title was changed to Electrical Engineering. This name was retained until 1994 when the Department became the Electrical Engineering Division of the newly-formed Manchester School of Engineering.

Key: AL = Assistant Lecturer, L = Lecturer, SL = Senior Lecturer,
R = Reader, P= Professor, T = Tutor, Sp. L = Special Lecturer (part-time members of staff who came into the Department to give lectures),
JD = Junior Demonstrator, D = Demonstrator,
SD = Senior Demonstrator, HL = Honorary Lecturer.
* Indicates transfer to the staff of another Department
The stated ranks are the ones held whilst serving in the Electrical Engineering Department.

Academic staff:

Robert Beattie	1896-1938 (retired). D, AL, L (in the Dept. of Physics). P (in the Dept. of Electro-technics).
Harold Gerrard	1912-1953 (retired). JD, SD, SL. Continued as Sp.L from 1953-1959.
Joseph Higham	1920-1952 (retired). L and Assistant Director of the Labs.
Frank Roberts	1930-1935 (deceased). AL.
William Makinson	1935-1936 (1-year appointment). AL.
F.C. Williams	1936-1939 (resigned). AL. 1946-1977 (deceased). P.
Willis Jackson	1938-1946 (resigned). P.
Raphael Feinberg	1939-1950 (appointment expired). L, Sp.L.
Thomas Havekin	1942-1955 (retired). Sp. L.
J.M. Meek	1944-1946 (resigned). Sp. L.
C.N.W. Litting	1946-1987 (retired). AL, L, SL, R, Director of Studies in Physics and Electronic Engineering.
J.C. West	1946-1957 (resigned). AL, L, SL.
L.S. Piggott	1947-1980 (retired). L, SL, SL & T.
Tom Kilburn	1948-1964* (transferred to the newly-formed Computer Science Department). Dr Kilburn had been in the Electro-technics Department since 1946 in another capacity. L, SL, R, P.
D.R. Hardy	1950-1954 (resigned). L.
E.R. Laithwaite	1950-1964 (resigned). AL, L, SL.
G.B.B. Chaplin	1951-1954 (resigned). AL, L.

Appendix 1: Members of staff in Electrical Engineering, 1896-1997

R.L. Grimsdale	1951-1961 (resigned). AL, L.
G.R. Hoffman	1951-1986 (deceased). AL, L, SL, R, P.
R.A. Brooker	1951-1964* (transferred to new Comp. Sci. Dept.). L, SL.
D.B.G. Edwards	1952-1964* (transferred to new Comp. Sci. Dept.). (on staff of Comp. Mach. Lab. from 1948-1952). AL, L, SL.
G.E. Thomas	1952 -1955 (resigned). (on staff of Computing Machine Lab. from 1948-52). AL, L.
M.J. Somerville	1954-1962 (resigned). AL, L.
Ronald Cooper	1955-1982 (retired). SL, R, P.
Michael Barraclough	1956-1959 (resigned). AL.
M.J. Lanigan	1957-1963 (resigned). AL, L.
Dennis Tipping	1957-1993 (retired). AL, L, SL.
T.E. Broadbent	1957-1990 (retired). L, SL, SL & T.
D.P. Atherton	1958-1962 (resigned). AL, L.
J.F. Eastham	1958-1965 (resigned). AL, L.
Cicely M. Popplewell	1958-1964*. (transferred to new Comp. Sci. Dept.). (on staff of Computing Machine Lab; 1949-58). L.
David Aspinall	1958-1964* (transferred to new Comp. Sci. Dept.). L.
D.C. Jeffreys	1958-1965 (resigned). L.
J.A. Turner	1960-1967 (resigned). AL, L.
G.F. Nix	1960-1992 (retired). AL, L, L & T.
R.B. Payne	1960-1964* (transferred to new Comp. Sci. Dept.). L.
G.F. Turnbull	1961-1967 (resigned). AL, L.
E.M. Dunstan	1961-1964* (transferred to new Comp. Sci. Dept.). L.
F.H. Sumner	1961-1964* (transferred to new Comp. Sci. Dept.). L.
Derrick Morris	1962-1964* (transferred to new Comp. Sci. Dept.). L.
G.W. McLean	1962-1990 (resigned). AL, L, SL.
J.M. Townsend	1962-1970 (resigned). AL, L.
K.F. Bowden	1962-1964* (transferred to new Comp. Sci. Dept.). L.
D. E. Hesmondhalgh	1962-1993 (retired). L, SL.
C.T. Elliott	1962-1967 (resigned). AL, L.
P.L. Jones	1963-1997 (retired). L, SL, SL & T.
D.G. Buchanan	1964-1967 (resigned). AL, L.
D.J. Kinniment	1964-1964* (transferred to new Comp. Sci. Dept.). AL.
B.E. Jones	1964-1981 (resigned). AL, L.
C.J. Hardy	1965-present. AL, L.
B.R. Varlow	1965-present. AL, L, SL, R.
Peter Lilley	1965-1995 (retired). L, SL.
R.M. Pickard	1966-1989 (retired). L, SL.
J.L. Mudge	1966-1967 (resigned). AL, L.
S.F. Gourley	1967-1974 (resigned). L.
I.E.D. Pickup	1967-1997 (retired). AL, L, SL.
E.F. Taylor	1967-1983 (resigned). AL, L.

P.A. Wade	1967-1970 (resigned). AL, L.
D.W. Auckland	1968-present. AL, L, SL, P.
David King	1968-present. AL, L.
N.P. Lutte	1969-1974 (resigned). AL, L.
W.W. Clegg	1969-1996 (resigned). TL, L, SL, R.
M.J. Cunningham	1970-present. L, SL.
R.A. Harris	1971-1973 (appointment expired). TL.
D.A. Edwards	1971-1972 (appointment expired). TL.
H.W. Thomas	1973-1984 (resigned). L.
F.G. Abbosh	1974-1978 (resigned). L.
R.S. Quayle	1975-present. L, SL.
J.R. Hawke	1975-1980 (resigned). HL.
R. Shuttleworth	1976-present. L.
A.M. Solomons	1978-1980 (resigned). HL.
D.J. Sandoz	1979-1990*(transferred to Vuman). L, SL. Later Visiting Professor and then Part-time Professor.
Brett Wilson	1980-1982 (resigned). HL.
E.W. Hill	1981-present. L, SL.
P.G. Farrell	1981-present. P.
Maurice Beale	1982-1991 (retired). L.
Philip Bridge	1984-1986 (resigned). L.
J.R. Beebe	1984-1986 (resigned). TL.
D.J. Cole	1984-1986 (resigned). TL.
C.D. Wright	1986-present. L, SL.
D.J. Tait	1986-present. L.
C.S. Xydeas	1987-present. P.
B.K. Middleton	1987-present. R, P.
Margaret E. Clarke	1987-1990* (transferred to Comp. Sci. Dept.). L.
P.J. Turner	1987-1989 (resigned). L.
A.F. Erwood	1987-1996 (resigned). L.
J.J. Miles	1987-present. L
Costas Daskalakis	1988-1991 (resigned). (transferred to the Dept. from the Computer Science Dept.) L.
C.A. Boyd	1989-1995 (resigned). L.
J.P. Oakley	1989-present. L.
C.G.M. Harrison	1990-present. L (previously Experimental Officer).
B. Al Zahawi	1994-present. L.
Violet F. Leavers	1994-present. L.
T. O'Farrell	1994-1996 (resigned). L.

Note that the post of Assistant Lecturer was abolished in 1970 so all Assistant Lecturers became Lecturers and no further appointments as Assistant Lecturer were made. Many of the people in the above list were in the Department as research students, research assistants, etc., prior to their appointment to the academic staff. Members of staff who retired after 1982 usually became Honorary Fellows for at least three years.

Appendix 1: Members of staff in Electrical Engineering, 1896-1997

Academic-related staff:

G.A. Cooper	1929-1978 (retired).	Chief Technician, Laboratory Superintendent.
P.G. Michaelson-Yeates	1978-1983 (resigned).	Laboratory Superintendent.
W.N. Woodyatt	1984-1994 (retired).	Laboratory Superintendent.

Experimental Officers, Computer Officers, Training Officers:

Key: EO = Experimental Officer, SEO = Senior Experimental Officer, CO = Computer Officer, TO = Training Officer.

J.K. Birtwistle	1962-1996 (retired). SEO.
Derek Moore	1968-1995 (retired). EO, SEO.
I.W. Stutt	1975-present. EO.
B.P. Garner	1980-1985 (resigned). EO.
M.B. Stephens	1985-present. EO.
C.G.M. Harrison	1987-1990 (appointed to academic staff). EO.
David Leonard	1990-1994 (resigned). TO.
C.M. Dente	1992-1994* (transferred to Computer Science Dept.). CO.

Appendix 2. Staff lists at invervals of a few years, 1897-1997

Physics,
1897-98

Langworthy Professor and Director of the Physical Laboratory,
    Arthur Schuster, Ph.D. (Heidelberg), F.R.S.
Professor, T.H. Core, M.A. (Edinburgh).
Demonstrators and Assistant Lecturers,
    Charles H. Lees, D.Sc. (Vict.).
    Albert Griffiths, M.Sc. (Vict.).
    Robert Beattie, B.Sc. (Durham).

Physics,
1904-05

Langworthy Professor and Director of the Physical Laboratory,
    Arthur Schuster, Sc.D. (Cambridge), Ph.D. (Heidelberg), F.R.S.
Professor, T.H. Core, M.A. (Edinburgh).
Lecturer and Assistant Director of the Laboratories,
    Charles H. Lees, D.Sc. (Vict.), Late Bishop Berkeley Fellow of Owens College.
Demonstrator and Lecturer in Electro-Technics, Robert Beattie, D.Sc. (Durham).
Demonstrator and Lecturer in Electro-Chemistry, R.S. Hutton, M.Sc., (Manchester).
Demonstrators, H.E. Wood, B.Sc. (Manchester).
    H. Morris-Airey, B.Sc. (Manchester).

Physics,
1910-11

Langworthy Professor and Director of the Physical Laboratory,
    Ernest Rutherford, B.A. (Cambridge), M.A., D.Sc. (New Zealand),
    Ph.D. (Giessen), F.R.S., Nobel Laureate.
Honorary Professor,
    Arthur Schuster, Sc.D. (Cambridge), Ph.D. (Heidelberg),
    D.Sc. (Manchester) F.R.S.
Reader in Mathematical Physics, C.G. Darwin, B.A. (Cambridge).
Lecturer and Demonstrator in Electro-Technics,
    Robert Beattie, D.Sc. (Durham), M.I.E.E.
Assistant Lecturers and Demonstrators,
    Walter Makower, M.A. (Cambridge), D.Sc. (London), Late John Harling Fellow.
    Herbert Stansfield, D.Sc. (London).
    E.J. Evans, B.Sc. (Wales and London).
    H.G.J. Moseley, B.A. (Oxford).
Special Lecturer in Electro-Chemistry, R.S. Hutton, D.Sc. (Manchester).
Lecturer in Electro-Chemistry, J.N. Pring, D.Sc. (Manchester).

Electro-technics,
1914-15

Edward Stocks Massey Professor and Director of the Electro-Technical Laboratories, Robert Beattie, D.Sc. (Durham), M.I.E.E.
Senior Demonstrator, Harold Gerrard, M.Sc. (Manchester).

Electro-technics,
1920-1921 (this list
remained unchanged
up to and including
1929-30)

Edward Stocks Massey Professor and Director of the Electro-Technical Laboratories, Robert Beattie, D.Sc. (Durham), M.I.E.E.
Senior Lecturer, Harold Gerrard, M.Sc. (Manchester).
Lecturer, Joseph Higham, B.Sc. (Manchester).

Electro-technics,
1930-31

Edward Stocks Massey Professor and Director of the Electro-Technical Laboratories, Robert Beattie, D.Sc. (Durham).
Senior Lecturer, Harold Gerrard, M.Sc. (Manchester), A.M.I.E.E.
Lecturer, Joseph Higham, M.Sc. (Manchester), A.M.I.E.E.
Assistant Lecturer, Frank Roberts, M.Sc. (Manchester).

Electro-technics,
1936-37

Edward Stocks Massey Professor and Director of the Electro-Technical Laboratories, Robert Beattie, D.Sc. (Durham).
Senior Lecturer, Harold Gerrard, M.Sc. (Manchester), A.M.I.E.E., F.Inst.P.
Lecturer, Joseph Higham, M.Sc. (Manchester), A.M.I.E.E.
Assistant Lecturer, F.C. Williams, D.Phil. (Oxford), M.Sc. (Manchester).

Appendix 2: Staff lists at intervals of a few years, 1897-1997

Electro-technics, 1939-40

Edward Stocks Massey Professor and Director of the Electro-Technical Laboratories, Willis Jackson, D.Sc.(Manchester), D.Phil.(Oxford), A.M.I.E.E., F.Inst.P.
Senior Lecturer, Harold Gerrard, M.Sc. (Manchester), A.M.I.E.E., F.Inst.P.
Lecturer, Joseph Higham, M.Sc. (Manchester), A.M.I.E.E.
Assistant Lecturer, Raphael Feinberg, Dr. Ing. (Kahlsruhe).

Electro-technics 1942-43

Edward Stocks Massey Professor and Director of the Electro-Technical Laboratories, Willis Jackson, D.Sc.(Manchester), D.Phil.(Oxford), A.M.I.E.E., F.Inst.P.
Senior Lecturer, Harold Gerrard, M.Sc. (Manchester), A.M.I.E.E., F.Inst.P.
Lecturer, Joseph Higham, M.Sc. (Manchester), A.M.I.E.E.
Special Lecturer, Thomas Havekin, B.Sc. (Manchester), Ph.D. (Birmingham).
Assistant Lecturer, Raphael Feinberg, Dr. Ing. (Kahlsruhe).
Demonstrators,
    Peter Dénes, B.Sc. (Manchester).
    A.C. Normington, B.Sc. (Eng.) (London).
    Waldemar Rosenberg

Electrical Engineering, 1947-48

Edward Stocks Massey Professor and Director of the Electro-Technical Laboratories, F.C. Williams, O.B.E., D.Sc. (Manchester), D.Phil. (Oxford), A.M.I.E.E.
Senior Lecturer, Harold Gerrard, M.Sc. (Manchester), A.M.I.E.E., F.Inst.P.
Lecturer and Assistant Director of the Laboratories,
    Joseph Higham, M.Sc. (Manchester), A.M.I.E.E.
Special Lecturer, Thomas Havekin, B.Sc.,(Manchester),Ph.D.(Birmingham),A.M.I.E.E.
Assistant Lecturers,
    Raphael Feinberg, Dr. Ing. (Kahlsruhe), M.Sc. (Manchester).
    C.N.W. Litting, B.Sc. (Manchester).
    J.C. West, B.Sc. (Manchester).
Demonstrators,
    A.C. Normington, B.Sc. (Eng.) (London).
    G.C. Collins, B.Sc. (Eng.) (London).
    C.T. Baldwin, M.A. (Cambridge).
    Dorothy E.M. Garfitt, M.Sc. Tech. (Manchester).

Electrical Engineering, 1951-52

Edward Stocks Massey Professor and Director of the Electro-Technical Laboratories, F.C. Williams, O.B.E., D.Sc. (Manchester), D.Phil. (Oxford), M.I.E.E., F.R.S.
Senior Lecturers,
    Harold Gerrard, M.Sc. (Manchester), A.M.I.E.E., F.Inst.P.
    Tom Kilburn, M.A. Cambridge, Ph.D. (Manchester).
Lecturers,
    Joseph Higham, M.Sc. (Manchester), A.M.I.E.E., Assistant Director of the Electrical Engineering Laboratories.
    L.S. Piggott, M.Sc. (Eng.) (London), A.M.I.E.E.
    C.N.W. Litting, B.Sc. (Manchester), A.Inst.P.
    J.C. West, B.Sc. (Manchester), A.M.I.E.E.
    D.R. Hardy, M.Sc. (Eng.), Ph.D. (London), A.M.I.E.E.
Special Lecturer, Thomas Havekin,B.Sc.,(Manchester),Ph.D.(Birmingham),A.M.I.E.E.
Assistant Lecturer, E.R. Laithwaite, M.Sc. (Manchester).
Demonstrators,
    Eric Rawlinson, B.Sc. (Manchester).
    Frank Harlen, B.Sc. (Sheffield).

Electrical Engineering, 1958-59

Edward Stocks Massey Professor and Director of the Electro-Technical Laboratories, F.C. Williams, O.B.E., D.Sc. (Manchester), D.Phil. (Oxford), M.I.E.E., F.R.S.
Reader in Electronics, Tom Kilburn, M.A. (Cambridge), Ph.D., D.Sc. (Manchester), A.M.I.E.E.
Senior Lecturers in Electrical Engineering,
    Ronald Cooper, M.Sc., Ph.D. (Manchester), A.M.I.E.E.
    C.N.W. Litting, B.Sc., Ph.D. (Manchester), A.Inst.P., A.M.I.E.E.
    E.R. Laithwaite, M.Sc., Ph.D. (Manchester), A.M.I.E.E.

Electrical Engineering at Manchester University

Senior Lecturer in the Theory of Computing,
    R.A. Brooker, B.Sc. (London), M.A. (Cambridge).
Lecturer and Tutor, L.S. Piggott, M.Sc. (Eng.) (London), A.M.I.E.E.
Lecturers in Electrical Engineering,
    D.B.G. Edwards, M.Sc., Ph.D. (Manchester).
    R.L. Grimsdale, M.Sc., Ph.D. (Manchester).
    G.R. Hoffman, B.Sc. (St. Andrews), Ph.D. (Manchester).
    M.J. Somerville, B.Sc., (Manchester).
    T.E. Broadbent, M.Sc., Ph.D. (Manchester).
Special Lecturer in Electrical Engineering,
    Harold Gerrard, M.Sc.(Manchester), A.M.I.E.E., F.Inst.P.
Assistant Lecturers in Electrical Engineering,
    Michael Barraclough, B.Sc. (Manchester).
    M.J. Lanigan, B.Sc. (Manchester).
    Dennis Tipping, B.Sc. (Manchester).
    D.P. Atherton, B.Eng. (Sheffield).
    J.F. Eastham, M.Sc. (Manchester).

Electrical Engineering, 1966-67

Edward Stocks Massey Professor and Director of the Electro-Technical Laboratories, F.C. Williams, C.B.E., D.Sc. (Manchester), D.Phil. (Oxford), D.Sc. (Durham), D.Eng. (Liverpool), M.I.E.E., F.R.S.
Readers,
    Ronald Cooper, Ph.D., D.Sc. (Manchester), A.M.I.E.E.
    C.N.W. Litting, B.Sc., Ph.D. (Manchester), A.M.I.E.E., A.Inst.P.
    G.R. Hoffman, B.Sc. (St. Andrews), Ph.D. (Manchester).
Senior Lecturer and Tutor, L.S. Piggott, M.Sc. (Eng.) (London), A.M.I.E.E.
Senior Lecturers,
    T.E. Broadbent, M.Sc., Ph.D. (Manchester), A.M.I.E.E.
    Dennis Tipping, B.Sc., Ph.D. (Manchester).
Lecturers,
    G.F. Turnbull, M.Sc. (Manchester).
    J.A. Turner, M.Sc., Ph.D. (Manchester).
    P.L. Jones, M.Sc., Ph.D. (Manchester).
    G.F. Nix, M.Sc., Ph.D. (Manchester).
    D.E. Hesmondhalgh, M.Sc. (Manchester), Ph.D. (Queen's University, Belfast).
    J.M. Townsend, B.Sc. (Manchester).
    C.T. Elliott, B.Sc., Ph.D. (Manchester).
    D.G. Buchanan, B.Sc. (Manitoba), M.Sc. (Manchester).
    B.E. Jones, M.Sc. (Manchester).
    G.W. McLean, M.Sc. (Manchester).
    R.M. Pickard, M.Sc., Ph.D. (Manchester).
    J.L. Mudge, B.Sc. (London), M.Sc., Ph.D. (Manchester).
Assistant Lecturers,
    Peter Lilley, B.Sc. (Wales).
    C.J. Hardy, B.Sc. (Manchester).
    B.R. Varlow, B.Sc. (Manchester).

Electrical Engineering, 1974-75

Edward Stocks Massey Professor and Director of the Electro-Technical Laboratories, F.C. Williams, C.B.E., D.Sc. (Manchester, Durham, Wales and Sussex), D.Phil. (Oxford), D.Eng. (Liverpool), C.Eng.,F.I.E.E., F.R.S.
Professors of Electrical Engineering,
    Ronald Cooper, Ph.D., D.Sc. (Manchester), C.Eng., M.I.E.E.
    G.R. Hoffman, B.Sc. (St. Andrews), Ph.D. (Manchester).
Reader in Electrical Engineering and Director of Studies in Physics and Electronic Engineering,
    C.N.W. Litting, B.Sc., Ph.D. (Manchester), C.Eng., M.I.E.E., F.Inst.P.
Senior Lecturers and Tutors in Electrical Engineering,
    L.S. Piggott, M.Sc. (Eng.) (London), C.Eng., M.I.E.E.
    T.E. Broadbent, M.Sc., Ph.D. (Manchester), C.Eng., M.I.E.E.

Appendix 2: Staff lists at intervals of a few years, 1897-1997

    Senior Lecturers in Electrical Engineering,
        Dennis Tipping, B.Sc., Ph.D. (Manchester).
        R.M. Pickard, M.Sc., Ph.D. (Manchester).
    Lecturers in Electrical Engineering,
        D.E. Hesmondhalgh, M.Sc. (Manchester), Ph.D. (Belfast).
        G.F. Nix, M.Sc., Ph.D. (Manchester), C.Eng., M.I.E.E.
        P.L. Jones, M.Sc., Ph.D. (Manchester).
        G.W. McLean, M.Sc., Ph.D. (Manchester).
        B.E. Jones, M.Sc., Ph.D. (Manchester), C.Eng., M.I.E.E.
        Peter Lilley, M.Sc. (Wales), Ph.D. (Manchester).
        S.F. Gourley, M.Sc. (New Brunswick), Ph.D. (Manchester).
        C.J. Hardy, B.Sc., Ph.D. (Manchester).
        B.R. Varlow, B.Sc., Ph.D. (Manchester).
        E.F. Taylor, M.Sc., Ph.D. (Manchester).
        D.W. Auckland, M.Sc., Ph.D. (Manchester).
        I.E.D. Pickup, M.Sc. (Manchester).
        David King, M.Sc., Ph.D. (Manchester).
        M.J. Cunningham, M.Sc., Ph.D. (Wales).
        H.W. Thomas, B.Sc., Ph.D. (Manchester).
    Laboratory Superintendent, G.A. Cooper, A.M.C.T.
    Senior Experimental Officers,
        J.K. Birtwistle, M.Sc. (Manchester), M.Inst.P.
        Derek Moore, B.Sc, (Durham), C.Eng., M.I.E.E.

Electrical Engineering, 1981-82

    Professors of Electrical Engineering,
        Ronald Cooper, Ph.D., D.Sc. (Manchester), C.Eng., F.I.E.E.
        G.R. Hoffman, B.Sc. (St. Andrews), Ph.D. (Manchester), C.Eng., F.I.E.E.
        P.G. Farrell, B.Sc. (City), Ph.D.(Cambridge),C.Eng., M.I.E.E., M.I.E.E.E.
    Reader in Electrical Engineering and Director of Studies in Physics and Electronic Engineering,
        C.N.W. Litting, B.Sc., Ph.D. (Manchester), C.Eng., M.I.E.E., F.Inst.P.
    Senior Lecturers and Tutors in Electrical Engineering,
        T.E. Broadbent, M.Sc., Ph.D. (Manchester), C.Eng., F.I.E.E.
        P.L. Jones, M.Sc., Ph.D. (Manchester).
    Senior Lecturers in Electrical Engineering,
        Dennis Tipping, B.Sc., Ph.D. (Manchester).
        R.M. Pickard, M.Sc., Ph.D. (Manchester).
        G.W. McLean, M.Sc., Ph.D. (Manchester), C.Eng., M.I.E.E.
        D.W. Auckland, M.Sc., Ph.D. (Manchester), C.Eng., M.I.E.E.
    Lecturer and Tutor,
        G.F. Nix, M.Sc., Ph.D. (Manchester), C.Eng., M.I.E.E.
    Lecturers in Electrical Engineering,
        D.E. Hesmondhalgh, M.Sc. (Manchester), Ph.D. (Belfast).
        Peter Lilley, M.Sc. (Wales), Ph.D. (Manchester).
        C.J. Hardy, B.Sc., Ph.D. (Manchester), C.Eng., M.I.E.E.
        B.R. Varlow, B.Sc., Ph.D. (Manchester).
        E.F. Taylor, M.Sc., Ph.D. (Manchester).
        I.E.D. Pickup, M.Sc., Ph.D. (Manchester), C.Eng., M.I.E.E.
        David King, M.Sc., Ph.D. (Manchester), C.Eng., M.I.E.E.
        M.J. Cunningham, M.Sc., Ph.D. (Wales), C.Eng., M.I.E.E.
        H.W. Thomas, B.Sc., Ph.D. (Manchester), C.Eng., M.I.E.E.
        R.S. Quayle, M.Sc. (Manchester), C.Eng., M.I.E.E.
        Roger Shuttleworth, B.Sc., Ph.D. (Manchester).
        D.J. Sandoz, B.Eng., Ph.D. (Liverpool), C.Eng., M.I.E.E.
        W.W. Clegg, B.Sc. (Liverpool), M.Sc., Ph.D. (Manchester), C.Eng., M.I.E.E., M.Inst.P.
        E.W. Hill, B.Sc., Ph.D. (Nottingham), M.Inst.P.
    Honorary Lecturer,
        Brett Wilson, B.Sc., Ph.D. (Manchester).

Laboratory Superintendent, P.G. Michaelson-Yeates, B.Sc. (London).
Senior Experimental Officers,
    J.K. Birtwistle, M.Sc. (Manchester), M.Inst.P.
    Derek Moore, B.Sc. (Durham), C.Eng., M.I.E.E.
Experimental Officers,
    B.P. Garner, B.Sc. (Salford).
    I.W. Stutt.

Electrical Engineering, 1989-90

Head of Department and Professor of Electrical Engineering,
    P.G. Farrell, B.Sc. (City), M.Sc. (Manchester), Ph.D. (Cambridge), C.Eng., F.I.E.E., M.I.E.E.E.
Professor of Electrical Engineering,
    Costas Xydeas, M.Sc., Ph.D. (Loughborough), C.Eng., M.I.E.E., M.I.O.A.
Visiting Professor,
    A.B.J. Reece, M.Sc. (Birmingham), C.Eng., F.I.E.E.
Reader,
    B.K. Middleton, B.Sc. (Sheffield), Ph.D. (Salford), C.Eng., F.I.E.E., M.Inst.P., C.Phys.
Senior Lecturers and Tutors in Electrical Engineering,
    T.E. Broadbent, M.Sc., Ph.D. (Manchester), C.Eng., F.I.E.E.
    P.L. Jones, M.Sc., Ph.D. (Manchester).
Senior Lecturers in Electrical Engineering,
    Dennis Tipping, B.Sc., Ph.D. (Manchester).
    G.W. McLean, M.Sc., Ph.D. (Manchester), C.Eng., M.I.E.E.
    D.W. Auckland, M.Sc., Ph.D. (Manchester), C.Eng., M.I.E.E.
    D.J. Sandoz, B.Eng., Ph.D. (Liverpool), C.Eng., M.I.E.E.
    W.W. Clegg, B.Sc. (Liverpool), M.Sc., Ph.D. (Manchester), C.Eng., M.I.E.E., C.Phys., M.Inst.P.
    D.E. Hesmondhalgh, M.Sc. (Manchester), Ph.D. (Belfast), C.Eng., M.I.E.E.
    R.S. Quayle, M.Sc., Ph.D. (Manchester), C.Eng., M.I.E.E.
    M.J. Cunningham, M.Sc., Ph.D. (Wales), C.Eng., M.I.E.E.
Lecturers in Electrical Engineering,
    G.F. Nix, M.Sc., Ph.D. (Manchester), C.Eng., M.I.E.E.
    Peter Lilley, M.Sc. (Wales), Ph.D. (Manchester).
    C.J. Hardy, B.Sc., Ph.D. (Manchester), C.Eng., M.I.E.E.
    B.R. Varlow, B.Sc., Ph.D. (Manchester).
    I.E.D. Pickup, M.Sc., Ph.D. (Manchester), C.Eng., M.I.E.E.
    David King, M.Sc., Ph.D. (Manchester), C.Eng., M.I.E.E.
    Roger Shuttleworth, M.Sc., Ph.D. (Manchester).
    E.W. Hill, B.Sc., Ph.D. (Nottingham), M.Inst.P.
    Maurice Beale, B.Sc., Ph.D. (Kent).
    C.N. Daskalakis, M.Sc., Ph.D. (Manchester).
    C.D. Wright, B.Sc. (London), M.Sc. (Sheffield), Ph.D. (C.N.A.A.).
    D.J. Tait, B.Sc. (Kent).
    Margaret E. Clarke, B.A., Ph.D. (Cambridge), M.Sc. (Manchester).
    J.J. Miles, B.Sc. (Liverpool).
    A.F. Erwood, B.Sc. (C.N.A.A.), M.Sc. (Loughborough).
    C.A. Boyd, B.Sc., Ph.D. (Warwick).
Laboratory Superintendent,
    W.N. Woodyatt, B.Sc. (Manchester), C.Eng., M.I.E.E., C.Phys., M.Inst.P.
Senior Experimental Officers,
    J.K. Birtwistle, M.Sc. (Manchester), C.Phys., M.Inst.P.
    Derek Moore, B.Sc, (Durham), C.Eng., M.I.E.E.
    C.G.M. Harrison, M.Sc., Ph.D. (Manchester).
Experimental Officers,
    I.W. Stutt.
    M.B. Stephens, B.Sc. (Manchester).
Honorary Fellow,
    R.M. Pickard, M.Sc., Ph.D. (Manchester).

Appendix 2: Staff lists at intervals of a few years, 1897-1997

Electrical Engineering, 1996-97

Head of the Manchester School of Engineering and Professor of Electrical Engineering,
> P.G. Farrell, B.Sc. (City), M.Sc. (Manchester), Ph.D.(Cambridge), F.Eng., F.I.E.E., M.I.E.E.E., F.I.M.A.

Chairman of Division and Senior Lecturer,
> M.J. Cunningham, M.Sc., Ph.D. (Wales), C.Eng., M.I.E.E., C.Phys., M.Inst.P.

Professor of Electrical Engineering and Graduate Dean in the Faculty of Science and Engineering,
> D.W. Auckland, M.Sc., Ph.D. (Manchester), C.Eng., M.I.E.E.

Professors of Electrical Engineering,
> B.K. Middleton, B.Sc. (Sheffield), Ph.D. (Salford), C.Eng., F.I.E.E., C.Phys., M.Inst.P.,
> C.S. Xydeas, M.Sc., Ph.D. (Loughborough), C.Eng., M.I.E.E., M.I.O.A.

Royal Academy Visiting Professor in Engineering Design,
> G. Birchby, B.Sc., C.Eng., M.I.E.E.

Visiting Professor,
> D.J. Sandoz, B.Eng., Ph.D. (Liverpool), C.Eng., M.I.E.E.

Reader,
> B.R. Varlow, B.Sc., Ph.D. (Manchester), C.Eng., M.I.E.E.

Senior Lecturers,
> P.L. Jones, M.Sc., Ph.D. (Manchester).
> R.S. Quayle, M.Sc. Ph.D.(Manchester), C.Eng., M.I.E.E.
> I.E.D. Pickup, M.Sc., Ph.D. (Manchester), C.Eng., M.I.E.E.
> E.W. Hill, B.Sc., Ph.D. (Nottingham), M.Inst.P.,C.Phys., M.I.E.E.E.
> C.D. Wright, B.Sc. (London), M.Sc. (Sheffield), Ph.D. (C.N.A.A.), M.S.P.I.E.

Lecturers,
> C.J. Hardy, B.Sc., Ph.D. (Manchester), C.Eng., M.I.E.E.
> David King, M.Sc., Ph.D. (Manchester), C.Eng., M.I.E.E.
> Roger Shuttleworth, M.Sc., Ph.D. (Manchester), A.M.I.E.E.
> D.J. Tait, B.Sc. (Kent), A.M.I.E.E.
> J.J. Miles, B.Sc. (Liverpool), M.Sc. (Salford), Ph.D. (Manchester), M.I.E.E.E.
> J.P. Oakley, B.Sc.,Ph.D. (Manchester), M.I.E.E., M.I.E.E.E., Grad I.M.A.
> C.G.M. Harrison, M.Sc., Ph.D. (Manchester), A.M.I.E.E.
> B. Al Zahawi, B.Sc., Ph.D. (Newcastle), A.M.I.E.E.
> Violet Leavers, B.Sc., Ph.D. (London).

External Lecturer,
> Peter Lilley, M.Sc. (Wales), Ph.D. (Manchester), C.Eng., M.I.E.E.

Honorary Visiting Lecturers,
> W.G. Chadband, M.Sc. (Birmingham), Ph.D. (Salford).
> Khetram Chandraker, M.A., Ph.D. (Sauger), C.Phys., M.Inst.P.
> K. Chanlaker, B.Sc. (Sauger), M.Sc., Ph.D.

Experimental Officers,
> I.W. Stutt.
> M.B. Stephens, B.Sc. (Manchester).

Honorary Fellows,
> T.E. Broadbent, M.Sc., Ph.D. (Manchester), C.Eng., F.I.E.E.
> G.F. Nix, M.Sc., Ph.D. (Manchester), C.Eng., M.I.E.E.
> D.E. Hesmondhalgh, M.Sc.(Manchester),Ph.D.(Belfast),C.Eng.,M.I.E.E.
> Dennis Tipping, B.Sc., Ph.D. (Manchester), B.A. (Open).

Appendix 3. List of first-degree graduates in Electrical Engineering and related subjects, 1915-1997

From the foundation of the Electro-technics Department in 1912 until 1968 students in the department's electrical engineering course (which was the only course until 1962) graduated in 'Engineering' and the word 'Electrical' did not appear on their degree certificates. Thus, technically, the electrical students obtained the same degree as their colleagues in civil and mechanical engineering, the only difference being their area of specialisation in the written work and laboratory. From the early 1920s it is easy to determine which students could truly be regarded as electrical engineers by noting the examination papers they took (information available from departmental records), but from 1912 to 1921 it is not so easy to differentiate. In the case of those early students the names which appear in the following list of graduates are those in whose work during the degree course the electrical subjects appear to predominate. In the first few years of the Electro-technics Department's existence the number of Electrical Engineering graduates was very small indeed, especially during the First World War when nearly all men above the age of 18 were called up into the Armed Forces. Most of the electrical teaching in these early years was for the benefit of civil and mechanical engineers and students from other Faculty of Science departments.

In the mid 1960s the title of the Department's main course was changed from 'Electrical Engineering' to 'Electronic and Electrical Engineering' and graduates of the Department's main course obtained this degree from 1968.

During the last 30 or so years the Department has offered other courses, namely:

3-year B.Sc. course in Physics and Electronic Engineering (shared between the Electrical Engineering and Physics Departments): There were graduates in this course from 1965 until 1992.

4-year B.Sc., M.Eng. course (industrially linked) in Electronic and Electrical Engineering (there have been graduates in this course from 1987). Graduates are awarded the double degree but only the B. Sc. part is classified.

Various other 3-year and 4-year courses (graduates from 1992).

These are named at the appropriate points in the lists which follow.

Note that the degree awarded at the end of 3-year courses was B.Sc. up to and including 1988. From and including 1989 it was changed to B. Eng. (excluding the *Physics & Electronic Engineering* course in which the degree remained B.Sc.) in response to the demands of national policy on engineering courses as formulated by the professional institutions and the Engineering Council.

Ordinary degrees are divided into divisions 1 and 2. Ordinary degree graduates in the following lists obtained a division 2 degree except where otherwise stated.

In accordance with University custom, a first name is included in the case of women graduates.

Appendix 3: Graduates in Electrical Engineering and related subjects, 1915-1997

3-year B.Sc. in Electrical Engineering:
(all graduates from 1915 to 1964 obtained this degree, both dates inclusive)

1915:
I: S. Brandt

1916:
Ord: A.N. Lansdell

1917:
Ord: C-T. Wu

1918:
Ord: H.L. Barretts

1919: None

1920:
I: W.J. Brown

1921:
I: F.S. Edwards II: E.F. Stephenson III: S. Edwards, A.E.C. Slater, W.A.L. Wolf.

1922:
II(1): J.W. Horner, H. de B. Knight. II(2): H.T. Aspinall III: T. Havekin, R.B. Johnson.
Ord: T.G. Francis, G. Marsden, H. Nattrass.

1923:
I: R.H. Evans II(1): S.H. Hemsley, H.C. Ogden. II(2): D.D. Evans, R.A.H. Sutcliffe.
Ord: P.I.H. Metcalfe

1924:
I: T. Boardman, E. Rushton.
II (not subdivided this year): J.T. Lea, V.M. Meswani, S.N. Panikkar, H. Shackleton.
Ord: B.H. Robertson, J.D. Warburton (div. 1).

1925:
I: T. Gill, W. Jackson, E. Swift.
II(1): F. Roberts Ord: D.L. Dixon, E. Morgan (div. 1), J. Smith.

1926:
III: E. Taylor Ord: G.E. Simpson, A.E. Starkey, I. Stewart.

1927:
I: C.F.J. Morgan III: C.V. Vinten-Fenton
Ord: W. Bamford, N.S. Nagpaul, D.F. Stewart, J. Wallace.

1928:
II(2): R.H. Dunn III: J.P. Wajnfeld
Ord: G.L. Armitage, P.E. Brockbank (div. 1), A.W.A. Dillamore, M. Green, S. Green, G.R. Kelkar.

1929:
III: S.V.G. Iyengar, J.H. Manekshaw. Ord: L. Raven, H. Robson.

1930:
I: H. Page II(1): W.H.R.A. Coates III: E.A. Jones, C. Stead.
Ord: R.N. Khosla, W.E. Marriott, R.W. Pearson, K.T. Thakur, J.S. Wason.

1931:
II(2): R. Garrett, J.L. Morrison. Ord: K.P.S. Narayana

Electrical Engineering at Manchester University

1932:
I: R.P. Desai, W.W. Sturge, F.C. Williams.
II(1): A. Raven, Beatrice Shilling.   II(2): Sheila E. McGuffie, R. Potts, S.W. Whiteman.
III: W.A.E. Gelder   Ord: M.D. Dubash, M.V. Paul.

1933:
I: R.F. Cleaver, A. Fairweather, F.H. Moon.
II(1): D.L. Armitage, H.B. Sedgfield.   II(2): J.S.C. Wheeler   III: W.L. Parkinson
Ord: E.W. Berth-Jones, J.A. Farrer, P.C. Guha, A.M. Raffael.

1934:
II(2): W.R.B. Frank, I.K. Paton.   III: E.J. Eaton   Ord: T.S. Ashe

1935:
I: K.B. Clayton, E.J. Wilkinson.
II(1): R.C. Jones, K.M. Sowerbutts, C.R. Taylor.   II(2): M.B. Wood   III: W.F. Taylor
Ord: C.W. Ricketts, H.B. Wood.

1936:
I: S. Jones, J.P. Wolfenden.   II(2): H. Gaskell, F. Hart.   Ord: J. Mackay, J.D.W. Taylor.

1937:
I: R.K. Beattie, C.F. Campbell, A.E. Chester, W. Heaton, F. Hornby.   II(2): A.F. Jackson
Ord: T.H. Chadwick, T. Hindle, E. Nicholson.

1938:
I: W. Beverley, P. Bowles, J. Highcock, E. Holland.   II(1): E. Moss, J.H. Porter.
II(2): W.A. Cowin   III: E.E. Sabbage   Ord: M.S. Ali

1939:
I: G.R. Brooks, R. Cooper, F. Horner, W.H. Morphet.
II(1): F. Duerden, R. Dunlop, B.N. Khosla, G. Saxon.
II(2): F.R. Finigan, W.H. Fisher, E.R. Pritchard.   III: S.A. Khan, N.M. Makin, J.R. White.
Ord: W.C. Crossley

1940:
I: K.M. Izzatullah   II(1): M.A.H. Khan, R.B. Quarmby, A.C.C. Robertson.
Ord: D.L. Coller

1941:
I: G.K. Armstrong, P. Dénes, T.A. Taylor.   II(1): P.V. Entwistle, E.G. Mallalieu.
Ord: D.G. Chalmers, W.E. Marrian, M.N. Sardana.

1942:
I: J.T. Allanson, P.A. Johnson, J.D. Qualtrough, W. Reddish, W. Rosenberg, J.R. Simmons.
II(1): C.E. Butler, H.M. Harrison, D.R. Nolan, G. Zahler.
II(2): E.K. Richardson   III: E. Farmer, E.H. Priestley, H. Ruth.
Ord: J. Barlow, J.E.P. Mills, K. Stephen.

1943:
I: R. Dunsmuir, J. Lamb, J.C. West.   II(1): J. Bodden, L. Rhoden.   II(2): T. Mulvey
III: R.H. Kelsall   Ord: E.W. Barton, E. Dobson, A.H. Grundy.

1943 (December: Conferred 1944):
I: A.E. Brierley, R. Croft, J.E. Makin, W.I. Plessner.
II(1): P.R. Allen, J. Beasley, R.J. Cooke, A.A. Dixon, F.H. Lomas, W. Roscow, R.G. Whitehead.
II(2): F.B. Todd   III: A.H. Hatton, P.H. King.
Ord: A.C. Bastable, D.B. Berry, W. Brittlebank, K.N. Dixon, E.R. Broadbent, P. McGlade,
   J.E.N. McInnes, R.J. Royds, I.M. Stringfield, J.K. Vose.

Appendix 3: Graduates in Electrical Engineering and related subjects, 1915-1997

1944 (December: Conferred 1945):
I: G. Berry, A.G.W. Edmunds, K.M. Entwistle, F.G. Heath, J.G. Powles, H. Rosbottom, M.R. Scott, D. Wray.
II(1): J.A.S. Hilditch, E. Rawlinson.   II(2): J. Hargreaves, J.W. Pollard, A.D. Sharp.
III: H. Nixon.
Ord: S.I. Ainsworth, H.K. Barker, M. Birkett, N. Bradley, H.P. Caldecott, N.S. Dean, W. Hockey, N. Lee, C.A. Marshall.

1945 (December: Conferred 1946):
I: J. Bird, A.J. Bourne, J. Wakley, F.M. Whitaker.
II(1): R.F. Chambers, R.S. Gow, W.C. Woodier.
II(2): A. Cormack, M.R. Cowan, G.L. Cracknell, P.A. Einstein, G. Frame, P.E. McKellan, D. Phillips, J.W. Thompson.
III: J de C. Baker, D.A. Broome, R. Powell.
Ord: A.R. Finniecome (div. 1), E.T. Warburton, A. Ward.

1946:
The degrees awarded were those earned in 1945 (see list above).

1947:
I: E. Boardman, P.F. Charnock, S. Elms, F.B. Somerville.
II(1): K. Knowles, D.B. Savage, J. Trickett, A. Williams.
II(2): J.S. Chadwick, W.R. Cheetham, H. Crawshaw, F. Harrison, G. Hughes, F.A. Manners, T. Scott.
III: R.B. Taylor   Ord: Margaret Makinson, S.K. Wallooppillai.

1948:
I: H. Aspden, H. Chippendale.   II(1): W.D. Carr, A.J. Graham, R.B. Haworth, J.G.W. West.
II(2): W. Derbyshire, J.C. Gladman, N.M. Pole, J.H. Riley.
III: T. Pratt, P.C. Rossides, H.E. Rubin.
Ord: J.E. Bailey, D.J.P. Byrd, Nora B. Kearle, T. Nuttall, R.H. Sykes.

1949:
I: D.K. Hewitt, J.G. Jones, E.R. Laithwaite, A.L. Stott, W.K. Taylor.
II(1): U.A. Aqqad   II(2): B. Harrison, J. Horne, J.A. Lodge, J.W. Machin, R.S. Paulden, J.S.B. Reynolds, A.H. Smith, A.J.L. Wright.   III: J. Pilling, C. Prescott, M.G. Seed.
Ord: G. Caldwell, F. Littler, N. Miller, J.E. Moorhouse, R.B. Rowley.

1950:
I: G.B.B. Chaplin, N. Eastman, R.L. Grimsdale, M.J. Somerville, D. Waltham.
II(1): T. Burgess, R.M. Foulkes, J.P.I. Tyas.
II(2): J.F. Delamere, G.T. Godier, D.H. Keene, J.H.W. Taylor, H. Turner.
III: J.D. Barritt, A. White.   Ord: R.H.C. Adams, E.P. Butt, J. King, E. Landless, K.G. Marwing, K.W. Morgan (div. 1), H.J. Palethorpe, C.B. Taylor.

1951:
I: F.G. Jenks
II(1): J.S. Bickerdike, G.S. Bracewell, P.J. Hurst, D.K. Partington, R. Shore, E.N. Toft.
II(2): R.N. Abbott, E.M. Adams, J.B. Baldwin, J.A. Bladon, J. Evans, A. Hicks, F.J. Ibbotson, P. Jackson, R. Postlethwaite.   III: R.M. Rear
Ord: A. Clayton, J.E. Dardier, R.M. Harrower, L. Jeavons, P.J. Oldridge, J. Procter.

1952:
II(1): T.E. Broadbent, R. Eatock, A.M. Marsden, T.E. Norris, J.K. Wood, H. Wroe.
II(2): F.A. Ambrose, D. Halton, E.S. Lee, A.J. Negus, J. Shelmerdine, N.J. Sylvester.
III: D.W. Whitehead   Ord: G.W. Bell, R.B.G. Benson, C. Dean, W.C.T. Hunter, D.A. Jones, J.G. Livesey, A.J.J. Moulam, D. Reid, J.A. Seabury, F.C. Summers, W.D. Worthington.

1953:
I: R.E. Hayes, H. Latham, R.W. Rowbottom.
II(1): D.F. Binns, B. Boardman. J.R. Foster, J.L. Leonard, W.T. Shelton.
II(2): G.B. Bourne, J. Edmundson, T.K. Hemingway, J. Hughes, G.O. Roscoe.
III: J.C. Hopkins, N. Sellers, N. Winterbottom.
Ord: P.F. Ainscow, W.A. Channing, I.F. Chrishop (div. 1), C.E. Duggan, E.B. Forsyth, J.D. Hansford, H.N. Iheukumere, H.K. Maddams, J.J. Weaver.

1954:
I: D. Gledstone, R.S. Stevenson, P. Wolstenholme.
II(1): P.J. Lawrenson, E. Rodwell. II(2): J.J. Hunt, J.C. MacKeand, D.J. Martin, D.C. Medford.
III: P. Balmer, J.B. Holroyd.
Ord: M.D. Charles, A. Clarkson, R.M. Dobson, J.C. Holmes, K. Hughes, K.A.J. Mapleston, B.M. Riley, J.F. Stringfellow, G.R. Swinn, A. Tomlinson, A. Williams.

1955:
I: M. Barraclough, M.J. Lanigan, R.S. Mamak, P.M. Patel, D. Tipping.
II(1): B.W. Dowd, D.C. Jeffreys.
II(2): K.J. Binns, W.E. Davies, J.F. Eastham, P.A. Jackson, N.L. Leece, N.T. Slater, B. Walker.
Ord: G.R. Gilbertson, I.S. Jackson, D. Orme, H.L. Tebicke, R. Woodward.

1956:
I: W. Roberts, F.I. Ross, A.H.A. Shlash.
II(1): D.M. Butler, E. Hall, W.E. Smith, M.J. Tyler, D. Zissos.
II(2): J.S. Armstrong, G.B. Bannister, P.A. Bradley, F.T. Connors, M. Flinders, A.J. Monks, J.W. Woodfield.   III: W.H. Condliffe, K.R.G. Garside.
Ord: S.M. Aboul-Timman, F. Catlow, M.I.H. Khan, D. Masheter, B. Mather, K. Mitchell, B.M. Sandford, G.E. Tracey, G.T. Waby (div. 1), J. Webster, I.W. Wright.

1957:
I: I.J. Cornish, C.S. Gaskell, F.A. Heys, C.H. Rowson.
II(1): M.R. Hill, A. Jebb, R.E. King, G.F. Nix, F. Shaw, G.H. Tomlinson, G.G. Weaver.
II(2): E.C.Y. Chen, A.D. Freeborn, F.W. Hadfield, P. Marlton, F.W. Senior, J.M. Shackleton, J.A. Turner, V.A. Williams, G.A. Wimbush.
III: S.S. El-Said, J.F.L. England, L.O.A. Seriki.
Ord: M.B.H. Alwan, S.H. Andrews, A.J. Durant, R.G. Hughes, H.C. Mangla, A. Owen, G.D. Parker, B. Quilliam, M.M.A. Rahman, J.T. Taylor, M.A. Wasfi, J. Williams.

1958:
I: K.F. Bowden, D.E. Hesmondhalgh.   II(1): W. Farrer, P.L. Jones.
II(2): R.D. Baker, P. Boyle, J. Dunsbee, W. Earle, D. Hodgson, S.P. Patience, E.J.M. Quirk, P. Reece, S. Simpkin, D.B. Watson, R.H. Wild, W.D. Willis.
III: A.D. Edwards, G.R. Gledhill, S.A.R.K. Hussain, M.A. Kelleher.
Ord: P.P.A. Calvert, H.R. Ghalib, G.H. Liew, B. Lott, D.D. Page, J.A. Parker, M. Pennington, C.I. Robertson, M.R.W. Silcock, B.K. Tandon, B.R. Wilkinson.

1959:
I: G.F. Turnbull
II(2): G.D. Claridge, D.G. Edwards, A.J. Faulkner, L.H. Fielder, D.J. Gardiner, I.J. Milne, M.H. Omar, J.W. Selwood, C. Stott, B.R. Taylor, D.C. Tedd, E.R. Wildridge.
III: E.R. Griffiths, G.C. Nnoli, R.J. Tee.
Ord: D. Briggs, C.H. Collett, B. Davis, N. Elston, C. Gillett, G.B. Higson, I. Jackson, G.P. Maycock, D. McNally, C.R. Newton, J.P. Nichols, M. Scferian, W.B. Sharpley, A. Simpson, T. Smith, A.C.J. Tamblyn, P.W. Torry, N. Tuntawiroon, H. Vickers.

Appendix 3: Graduates in Electrical Engineering and related subjects, 1915-1997

1960:
I: J.B. Gosling, D.F. Jesson, J.W. Robertson.
II(1): R.S. Broadfoot, L. Forth, A.E. Hill, M.G. Hunter, G.W. McLean, P. Mason, R.A. Newbury, J.A. Smethurst.
II(2): S.T.A. Addy, D.H. Copley, R.B. Dawe, H.E. Griffiths, A.D. Lockett, A.J. Martin, P.C. Page, J.R. Penketh, R. Smith, A. Whalley, O.P. Wright.
III: D.P. Edwards, J.E. Goodman, G.R.W. Horsfield, D.J. Tipple, S.D. Van.
Ord: D.G. Arnold, A.J. Boothman, E.A. Boxall, Judith M. Bradley, A.H. Brierley, G. Herring, W.A. James, R.S. Johnson, K.R. Jones, J.G. Kaye, R.N. Khana, B.J. Looker, M. Postlethwaite, M.I. Rackind, A.K. Ruffle, H.T. Sparrow, M.R.J. Wanner-Halder.

1961:
I: S.R.E. Dixon, M.R. Patel, K.P. Shambrook.
II(1): J.M. Birch, W. Broekhuizen, W.E. Freeman, K.A. Hill, W.R. Ingham, P. Lake, R.M. Moley, A.A. Shareef, C.J.V. Smith, J. Standeven, J.W. Townsend.
II(2): M. Bradford, T.S. Chubb, K. Davies, N.G. Depledge, R.A. Elmes, C.J.Ford, J.M. Garbutt, L.G.J. Goddard, T.H. Harrison, M. Haslam, E. Newton, M.S. Wright.
III: H. Chediak, A. Golding, H.D. Leefarr, K.A. Lowthian, L.A.M. Price, D.A. Smith.
Ord: J.W. Anderson, B. Bolton (div. 1), M.W. Curtis, A.J. Dawson, J.M. Fielding, A.S. Gahir, M.J. Halligan, B.W. Macarty, R. Neilson, D.W. Owen, N.C. Pickering, R.S. Pritchard, D.W. Shackleton, E. Ward, T. Whiteley.

1962:
I: F. Halsall, D.J. Kinniment.
II(1): W. Chisholm, B.W. Finnie, J.W. Green, S.H. Lavington, R.C. Prime, C. Robertshaw, R.J. Skerrett, B. Stott, D.E. Taylor, G-R.A. Tourzan, S.N. Vernon, M.J. Wood.
II(2): R.A. Addison, J.N. Anders, R.M. Barlow, D.H. Cashmore, C.T. Dargun, J.A.B. Dunk, N. Godinho, C. Hebden, P. Jakes, M.T. Jarvis, N.B. Jones, K.K. Kapur, R.H. Kelly, Stella Ravenson, G.A.J. Smith, D.H. Starbuck, B.P. Wainwright, H.W. Williams.
III: J. Armstrong, J. Bradbury, M.K. Malhotra, A.T.M. Mansuri, M. Osgerby, I. Vakilzadeh.
Ord: P. Arnold, A.K. Chatterjee, P. Clemerson, A.V. Clerk, C.R. Cowdy, G.L. Davies, U. Erdem, D.G. Fisher (div. 1), J.F. Harris, D. Hinton, G. Mofid, M.F. Pac-Soo, K. Panchmatia, B.A.M. Ross, K.A. Smith.

1963:
I: B.E. Jones, B.R. Varlow.
II(1): A. Bradwell, A.P. Chaddock, L.S. Chin, C.J. Hardy, B. O'Loughlin, D. Robinson.
II(2): J.R. Machin, D.M.W. Noble, T.A. Shaw, P.W.L. Simpson, J.C. Steele. III: N.Alivishahreza
Ord: D. Anderson, C.R. Bamforth, D.V. Bull, C. Goodman, M.A. Harris, G.W.F. Hulbert, D.T. Lalas, R.S. Mollard, E. Olcayto, J.S. Overton, C. Rigby, M.H. Shahin, M.J. Sharp, M.W. Whyman.

1964:
I: B. Gorman
II(1): T. Kelsey, K.A. McRae, S. Mihailovic, M. Phillips, R.J. Rhodes, D.M. Simpson, E.F. Taylor.
II(2): C.J. Balmforth, N.W.M. Cousins, D.P. Fathers, J.K. Haigh, P.J. Hyde, P.G. Nash, R.B. Smith, R. Verrill.
III: E.J. Abu-el-Timan, J. Aggrey-Mensah, P.H. Elliott, F.G. Irvin, M.T. Jackson, E.A. Middleditch.
Ord: H. Asad, P.S. Bannister, G.G. Bower, D.M. Browning, W.J. Butler, S.G.M. Camus, T.S. Chew, K. Chopra, Carol Creighton, N.T. Desai, P.D. English, J.B. Evans, M.S. Evans (div. 1), D.R. Gaskell, R.E. Gibbons (div. 1), M.S. Hasan, P.J. Horsley, M.J. Howard,

J.A. Maine, F.A. Moyes, M.J. Murtagh, A.D. Nicholls (div. 1), M.C. Patel, P.W. Stanley, S.J. Stephenson, E.K.A. Tchum, P.A. Wade (div. 1), G. Waterworth, V. Youel.

1965:
Electrical Engineering:
I: R.K. Cragg, D.Doyle, D.L. Pulfrey.
II(1): B. Altman, P.E. Bass, R.A. Bridges, J.H. Carlisle, J.E. Chatterley, I.F. Duff, A.P. McMahon, J.W.R. May, N.J. Stock, R.A. Willis.
II(2): A.J. Anderson, P.B. Hesketh, R. Kocache, M.J. Machin, P.J. Morley, J. Ormerod, J.P. Whitham, I. Woolley.
III: F.K. Al-Khawaja, R.L. Haywood, M.A. Shlash, W.H. Wombell, S.Z. Zahroun.
Ord: K. Abadani, K.T. Al-Kibasi (div. 1), O.Q. Bailey, J.P. Clegg, P.J. Francke, M.S. Kirvar, B. Medwell (div. 1), W.T. Ng (div. 1), G. Nowroozian, J.D.F. Selby, M.G.S. Sutton, J.S. Thorpe, P.W. Turner.
Physics and Electronic Engineering:
II(1): J.R. Moore   II(2): A.F. Dadds, N.H.C. Gilchrist.   III: E.A. Parr, M. Wintersgill, E. Wright.

1966:
Electrical Engineering:
I: D.A. Brash, R.W. Cripps, T.C. Harrington, D. King, I.E.D. Pickup.
II(1): T. Connolly, R.J. Cope, T.J. Lees, C.C. Myles, A.T.Y. Pearce, J.S. Pearson, G.B. Roper, M.A. Shallal, P.L. Sinclair.
II(2): S.R. Alwash, P.E. Battrick, S. Bodhisompon, A.L. Bragan, I.R. Coffey, B.R. Cooper, T.M. Jowitt, F.H. Marhoon, G.D. Richman, H.B. Saunders, J.M. Wood.
III: T.E. Andrews, R.D. Cottle, M. Counsell, R.E. Epworth, K.E. Marston.
Ord: J.A. Allenby, H.U. Ansari, R.J. Beshoory, J.W. Bolton, M. Bradley, H. Dodd, J. Eyre, R.N. Hollens, P.M. Levy, P.M. Overfield (div. 1), M.H. Said, D.A. Webster, M.R. Willetts.
Physics and Electronic Engineering:
I: N.P. Lutte   II(1): C.N. Brown, K.A. Moss, D.M.J. Seddon.
II(2): F.P. Coakley, K.N. Garad-Shenker.   III: D.R. Perkins   Ord: R. Allen

1967:
Electrical Engineering:
I: R. Fletcher, M.R. Higgins, G. Reed, P. Stamatiou, D.B. Terry.
II(1): J.M. Beaton, R.N. Biggs, G.M.S. Bland, S. Day, D.A. Edward, J.M. Kabwana, M.L. Sanderson, R.J. Taylor, G.B. Torry.
II(2): G.A. Abbas, B. Al-Kayssi, R.W. Appleby, I.H. Bainbridge, C.M. Blount, A.J.C. Rey, A.J. Rodrigues, R.A. Staton.
III: D.H. Balmforth, A.M. Hesketh, A. Sadre, T.M. Sembeguya, P. Sotiriou, S.L. Wong-Ng.
Pass: S.K. Dawood.   Ord: L.J. Brunton, P.W.T. Frisby, R.W. Gravestock, Y.F. Leung, D.H. Miller, R.J. Phipps, D. Pikoulis, I.C.D. Price-Smith, M.C.C. Rope, R.H. Shah, Z.K. Shaio, S.F. Sieh, J.A. Solly, P.A. Spencer, F.G. Tichborne.
Physics and Electronic Engineering:
I: M.B. Smith   II(1): G.P. Abraham, A.K. Boardman, K.J. Chadwick, J.E. Crampin, M.S. Francis, D.A. Graham, J. Pennington, G.T. Scullard.
II(2): P.M. Clare, M.T. Dent, R.L. Griffiths, D. Livingstone, R.M. Newbury, C.D. Pick, J.G. Ramage, G. Tyndall.
III: I.K. Barker, P.J. Campbell, J.F. Harrold, T. Matthews, A. Twyman, A.G. Williams.
Ord: E.C. Bartle, I.D. Chisholm, J.L. Hooper, M.K. Johnson, P.W.B. King, P.N.O. Plunkett, K.A. Starkey.

Appendix 3: Graduates in Electrical Engineering and related subjects, 1915-1997

1968:
Electronic and Electrical Engineering:
I: D.G. Baldwin, S. Hollock, S.E. Sutcliffe.
II(1): P.O. Akhionbare, B.F. Berry, M.E. Butcher, C-Y. Cheung-Yeung-San, C.W. Hothersall, L.B.C. Keitany, T.L. Lee, P.G. Long, D.G. Lovelock, R.N. Robinson, R. Russell, M.B. Skingley, A.J. Spivey.
II(2): D.A. Arnold, M.J. Bates, A.J. Clark, A.C. Evans, A.A.K.Jamil, J.P. Parrish, M.A. Usmani.
III: E.H. Ali, J.P. McCarthy. Pass: S.W. Hawley, Janet G. Richey.
Ord: S.W. Ainsworth, Maureen Brown, N.R. Clayton, A.J. Clynick, C.M. Eck, D.P. Franklin, C.E. Gidney, I.W. Lowrie, D.A. S. Mason, R. Mitchell, M.D. Moorhead, F. Newton (div. 1), D.A. Searby, Z.K. Shaio, J. Wareing.
Physics and Electronic Engineering:
I: S.A . Armitage, K.H. Bennett, R.A. Harris.
II(1): R.G.T. Bernard, M.J. Eccles, D.A. Edwards, A.K. Goldring, J.K. Hawkins, R.J. Hopkinson, C.S. Hughes, B. Priest, R.G. Proctor, L.A. Taylor, R.M. Upton, M.R. Weatherhead.
II(2): T.N. Carr, P.A. Denby, Heather Maggs, B. Polychronopulos, J.M. Read.
III: D.J. Range, V.J. Whitton.
Ord: D. Allen, A.R. Bryan, P.H. Durden, P.J. Gibson, S.N. Overington, I. Rowley, B.P. Wright.

1969:
Electronic and Electrical Engineering:
I: C.E. Dix, D.A. Grant, H.W. Thomas, M.S. Towers, C.M. Wong.
II(1): B.A. Bowles, M.E. Brigham, R.W. Edwards, W.G. Ferguson, F.C. Hulme, P. Liddell, B.K. Lord, C.C. Thomerson.
II(2): G.C. Aucott, F.D. Cave, N. Das, A.G. Dicker, K.F. Dinwoodie, E. Glasper, D.H. Mann, D.G. Podmore, D. Rutherford, M.R. Smith, J.P. White, D.J. Wittamore.
III: R.W. Curtis, R.K. Mehta, P.E. Yousif. Pass: A.P. Dennis, A.A.H.A. Janaby, P.K. Nelson.
Ord: A-S.M. Al-Mutlak, D.M. Butlin, M.A. Elwell, P.J. Garforth, C.R.J. Lillicrap, S.E. O'Neill, A.C. Patel, I Schofield.
Physics and Electronic Engineering:
I: J.R. Lewin, R.W. Mathieson, R.B. Phillips, C.H. Wong.
II(1): D.C. Brown, G.J. Crimes, S.N. Crisp, P.J. Hicks, P. Mason, R.C. Powell, P.W. Rivers-Latham, R.A. Smith, I. Watson, P.S. Wilson.
II(2): P.V.J. Adkins, J.R. Beckett, K.D. Brown, E.P. Clutten, D.W. Hadfield, O. Oner, M.A.B. Paessens, W.T. Phillips, J.L. Tattersall, S.R. Turner, J.S. Watts, A.R. Wooldridge.
III: K. Ashrafi, R.A. Bishop, J. Brotton, S.J. Delahunt, P.T. Kilgour, T.A. Shepherd, Z. Wojciechowski.
Pass: N.H. Payne, D.R. Stell, M.G. Vry. Ord: J. Chisholm, P.D.G. Milloy, K.R.B. Reed.

1970:
Electronic and Electrical Engineering:
I: K.C. Lo, R.S. Quayle. II(1): J. Barrick, J. Hiley, A.P.M. King, S.R. Lang, A. Sarson, P. Teather, D.K. Thomas, D.A. Urquhart, B.A. Walker.
II(2): D.J. Coates, S.B. Coster, M.A. Francis, S. Gorgorian, R. Hayward, R.J. Jessop, A. Kajiji, M. J. Lovell, J.P. Slinn, W-H. W. Yu.
III: I.W. Bettinson, D.J. Daniels, D.G. Fisher, N. Gregorian, G.E. Higginbottom, J. Prior, L. Shum. Pass: J.W. Friend
Ord: D.R. Cowbourne, D. Fischbach-Wright, M.F. Maltby, T.M. Parkinson.
Physics and Electronic Engineering:
I: M.H. Gilbert, R.B. Newman, C.R.C.B. Parker.
II(1): R. Brumer, P.J. Burrows, R.K. Donnelly, E. Evangelides, N. Gravill, E.C. Horton,

H.M. Jones, N.P. Knowles, M.J. Scholey.
II(2): G.V. Davies, J.T. Evans, M.L. Flowers, P.H. Griffiths, J.H.M. Hardy, N.F. Harrison, T.P. McElwee, J.P. Timms, R.G. Wassell.
III: R.E. Broughton, I.B. Deane, E.C.W. Lin, G.H.D. Powell, D.W. Robinson, I.S. Scales.
Pass: B.T. Goh, P.N. Leeson, G.R. Selvey. Ord: R. Bennison, B.K. Burridge, J.A. Connor, K.G. Martin, M.L. Nash, S.W. Pickering, M.E.A. Read.

1971:
Electronic and Electrical Engineering:
I: K.J. Allen, C.R. Eames, E.Y. Hadjiyannis, C.J. Hodge, H.M. McNaughton, K.N. Patel, L.C. Tong. II(1): A.D. Bracewell, D. Callaghan, W.C. Malham, D.W. Martin, W.H. Massey, K. Pacey, J.A. Smith, J.C. Wales.
II(2): R.A. Affleck, S.R. Anderson, G.H. Bailey, Mary Bower, D. Houghton, G.J. Low, N.I. Ojoo, P.S. Pilson, I.W. Pirie, C.S. Rawling, A.M. Solomons.
III: S.J. Appleton, R.C. Apps, A.J.B. Bott, W.L. Chew, S.E. Coan, E.B. Curtis, A. Eliasieh, B.J. Mullen, L.G. Smith. Pass: K.G. Lee, E.P. Roberts.
Ord: A-H. Jadidi, M-M. Jafar, D.A. Jones, B. Madubunyi, S-U-D. Qureshi, R.H. Round.
Physics and Electronic Engineering:
II(1): N.C. Butler, A.P. O'Leary, B. Wilson.
II(2): G. Cliff, K. Morgan, S.S. Nandra, D.H. Thomas, C. Wingrove.
III: D.P. Ward   Pass: A.D. Waters

1972:
Electronic and Electrical Engineering:
I: S.J. Gilliard, J.R. Hawke, R.W. McLintock, D.J. Webb.
II(1): M.K. Aiyewa, A. Behich, P.L. Binney, R. Chapaner, A.S. Chester, S.E. Contractor, N.A. Mirza, N.J. Richman, R. Shuttleworth, R.I. Thornborrow, S.R. Turner, D. Tyagarajah.
II(2): R.S. Compton, P. Creely, S.L. Matturi, C.J. Metcalfe, J.K. Parker, S. Porter, H. Shutt, D.J. Smith, R.M. Thomas, M.J. Waller.
III: J.J.W. Boundy, R.H. Jones, P.F. Kipling, P. Sephton, D.S. Tomlinson.
Pass: R.S.S. Jordan, M. Shakallis.   Ord: A.F. Graf, S.W. Shackleton.
Physics and Electronic Engineering:
I: J.R. Alexander, S.R. Paisley.
II(1): L. Di Marco, A. Gowland, J.S. Martin, P. Mosley.
II(2): D.J. Cunningham, C. Haynes, J. Jones, G.F. Miles, S. Rogers.
III: E.P. Burns, C.P.L. Conton, D.M. Harris, J.K. Li, A.S. Rodger, N.J. Smith, D.H. Warne.
Pass: P.B. Barstow, A. Bushell, T.J. Watson, D.S. Whiteway. Ord: A. Kyprianou, K.L. Stephens.

1973:
Electronic and Electrical Engineering:
I: T.M. Abram, S.F.G. Davies, J.H. Ho, J.Y. Huang, D.C. Markham.
II(1): E.N. Dayan, D.V. Evans, N. Goyal, P. Goyal, K. Hart, S. Hart, B.A. Mordecai, E.J. Simmonds, A.J. Spencer, W.J. Tildsley, C.K. Yeung Yam Wah.
II(2): T. Callaghan, A. Efthymiou, A.J. Flewitt, D.L. Goddard, H.M. Jones, D. Karageorgis, J.A. Kitchen, C.S. Nanayakkara, J.L. Odeny, J.J. Palmyre, R. Rogowski, A.W. Thomas, I.S. Trowbridge.
III: A.J. Brown, P.J. Burrows, C.J. Hamer, Mary M. Papadaki, R.D. Scott, M.C. Wong.
Pass: B.A. Mwasinga, J.J. Wood.   Ord: R.D. Gaiha, K.R. Hattersley, P.W. Smith.
Physics and Electronic Engineering:
II(1): I.D. Gatward, W.G. Hardy, K. Johnson, D.R. Owen.
II(2): G.R. Campbell, R.S. Hazeldine, T.R. Jackson, J.J.A. Wearing, S.P. Woolhouse.
III: A.D. Beecroft, J.A. Holden, G.R. Hunt.
Pass: M.A. Spencer, V. Sube, A.S. Wong.

Appendix 3: Graduates in Electrical Engineering and related subjects, 1915-1997

1974:
Electronic and Electrical Engineering:
I: M.W. Gregson, D.P.G. Hadfield, J.H. Quarmby, A.P. Russell, C.G. Tomlin, S. Wilson.
II(1): F. Aghdasi, D. Alemozaffar, A. Busse, M.D. Horwich, M. Owen, C.H. Pople, R.J. Rushton, N.S. Syrimis, T. Tabu, K.F. Thong.
II(2): P.M. Chatten, R.V. Davies, I. Lomas, R. Nair, S.W.C. Ng, A.J.R. Pleace, C.B. Puah, Susan E. Quick, J. Thorley.
III: P. Atkinson, A.P. Davis, J.J. Harden, D.H. Hill, J. Rose, N. Silou.
Pass: A. Jones, S.W. Wilkes. Ord: J.P. Hobson, K.D. Kaura, J.A. Kourtellis, K. Patel.
Physics and Electronic Engineering:
I: M.S. Abbott, S.M. Fensom, J.P. Foley, D.R. King.
II(1): J.M. Bilton, J.A. Green, A.I. Hawryliw, S.P. Holmes, A.J. Kerr, G.P. Lambert, G. Nocerino, A.J. Skepper.
II(2): A. Ghose, S. Hueber, C.T. Midgley, T.S. Smith, K.E. Thompson, J.R. Wells, A.J.R. Willis.
III: K.L. Bacriana, P.J. Harding, M.J. Smith, A.D. Taylor, J.J. Vrabel.
Pass: R.J. Bevan, J.F. Highcock. Ord: D.A. Forder

1975:
Electronic and Electrical Engineering:
I: F.A.M. Darwazeh, M.G. Mountis, J.N. Roach, W.O. Williams.
II(1): C.R. Coleman, P.A. Conton, L.E. Hall, R.D. Jelley, M.R. Walbridge.
II(2): I.A. Averill, J.W. Bridge, R.J. Burberry, S. Burns, N. Cohen, S.J. Lewis, K.V. Lim, J.B. Neeson, J.H. New, A.W. Paynter, M.J. Ratcliffe, J.P.A. Thurman, F.C. Verkroost, C.J.D. Woolmer, D.P. Yates.
III: S.S. Bancil, G.K. Orford, W.J. Russell, P.A. Watson. Pass: A. Apostolakos
Ord: S.E. Abolmaali, M.P. Chapman, P.T. Rees, D.B. Smith.
Physics and Electronic Engineering:
I: A.S. Bates, D.R. Cotton.
II(1): C.M. Buck, R.W. Deane, G.L. Hey Shipton, D.W. Holding, C. Papadakis, S.R. Taylor, D.P.N. Tong, P.J. Williams.
II(2): P.I. Bakewell, T.Q. Khan, D.C. McKee, K.A. Yates.
III: P. Burgess, P.G. Crossley, S.P. Houldsworth, R.M. Wadey, C.D. Wibberley, D. Wright.
Ord: W.D. Milburn

1976:
Electronic and Electrical Engineering:
I: D. Chew, T.J. Harding, D.S. Doyle.
II(1): S. Ahmadi, E.M.E. Brown, J.T. Gathariah, M. Haghighi, J.S. Harrop, S.L. Lim, C.F. Little, R.I. Muttram, C.P. Partassides, D. Pickering, J. Sanghera, S.H. Sia, B. Singh, L. Theobald, A. Venkatraman, R.T. Williams, D. Woods.
II(2): R.J.W. Ballantine, P.T. Davies, S.F. Deeny, C.J. Dunn, S. Goodwin, M.L. Graves, P.S. Heyes, G.A. Kourtellis, B. Nicholson, A.C. Orsten, J. Partington, P. Porter, K. Safarian, R.K. Shah, A.A. Shilbaieh, M. Tola, G.N.S. Zreikat.
III: G.W. Doherty, P. Grabham, W.G. Marsden, S. Swaddle, T.A. Treffner, W.G. Watkins, P.J. Wearon.
Pass: K.S.A. Athwal, E.N. Delis, A.A.A. Janahi, K. Skouros, P.A. Yiannouzis.
Ord: C.E. France, A. Gobbi, D.S. Jones (div. 1), V.V. Mangal.
Physics and Electronic Engineering:
I: M.R.D. Evans, K.L. Lee. II(1): M.H. Daish, I.J. Harvey, R.J. Salt, B. Saunders, J.R. Wood.
II(2): A.J. Daviel, S.P. Griffiths, S. Khan, G.D. Peck.
III: D.P. Culleton, A. Peabody, G.A. Perry. Pass: P.C. Lim Ord: T. Andrikopoulos

1977:
Electronic and Electrical Engineering:
I: M.I. Baines, R.J. Hodgson, B. Siddle, G.S. Virk.
II(1): T.A. Abboushi, N.P. Semmens, N.M. Snowdon, P.R. Tuson, M.R. Virani.
II(2): D.R. Atkinson, S. Baysan, M.V. Fitzgerald, M. Jacks, P.J. Kelly, A.P. Key, T. King, Y-F. Liew, J. Norbury, M. Shariatmadar, M.G. Walley, P. Whitcombe.
III: K. Bishop, A.N. Bowden, P.J. Siviter, N.N. Smith, E.M. Tindall, A.B.D. Worling, E.K. Yeo, P. Yousefpour, M.F. Yusuf.
Pass: I. Bradley, B.S. Glenn, D.W. Glover, J.D.S. Gunasekera, P. Hart, A.F. Hotten, S.D. Khalifeh, M.M. Masaud, Z.E. Mohamedbhai, A.M.A. Yacoob.   Ord: A.R. Nassar
Physics and Electronic Engineering:
I: P.N. Clarke, R. Thompson.   II(1): R.M. Connor, A.C.A. Hack, S.R. Laurenson, R.S. Lewis.
II(2): N.R. Adams, P.A. Brown, R.P. Tarpey, S. Terleckis.
III: S.R. Hutchinson, D. King, D.J. Law, J.R.M. Llewellyn, M.R. Sellars, D.M. Senior.
Pass: P.A. Cookson

1978:
Electronic and Electrical Engineering:
I: R. Haxby   II(1): G. Assaf, S.G. Bashiti, S.Y. Cheng, J.R.E. Cubbin, R.D. Hale, S.A.H. Khalfan, Z.L. Musallam, L.P. Saliba, Z. Sultani-Makhzumi, C.G. Tan.
II(2): S.R. Bhot, C-K. Chang, Y.C. Chen, J. Farnell, G.R. Fearnhead, D. Fletcher, S.G. Halliday, C.K. Hawkins, P.K.H. Ho, P.D. Macaulay, J. Matthews, P.J. McGrory, A.D.J. Muir, G. O'Neill, D.O. Omolo, R.P. Thomas.
III: M. Akram, G.D. Astarci, I.R. Blanchard, R.G. Chipalabela, H.R. Fattahi Kassili, S.N. Iyer, I.A. King, M. Mahbubian, L.G. Rustecki, P.N. Santamas, C.Y. Shee, J.J. Watts, D.E. Wells.
Pass: M.H.S. Akasheh, M.Akhtar, N.A.M. El Ghafari, P.F. Ennis, A.G. Godon, D.P.L. Mottershead, A.K. Pearson, Y.N.Y. Snobar, M. Tahir.   Ord: J.B.G. Fernandez
Physics and Electronic Engineering:
I: R. Carter   II(1): A.E. Collard, R.T. Howie, S.J. Levey, J.W. Morgan, P.G. Scotson.
II(2): K.I.H. Fasham, S.T. Glass, A.G. Ramli, P.M. Thompson.
III: J. Maddy, S-M. Yu.   Pass: B.P. Wilson

1979:
Electronic and Electrical Engineering:
I: K.L. Lai   II(1): M. Bullough, J.D. Crabtree, P.H. Martin, S.J. Mather, D.R. Slingsby, C.M. Tang, A.N. West.
II(2): K. Al-Saffar, H.K. Amin, Julie K. Brown, G.M. Camps, K.W. Chen, S. Cofinas, B. Eftekhari, M. Giotas, G.P. Gledhill, M.A. Innes, H. Lianeris, Heather Lobo, D. Lunn, I.R. Moledina, R. Monsefi, A.S. Rzadkiewicz, S.K. Saidi Shirazi, L.S. Yow.
III: H.A. Al-Kafrawy, C.J. Barnes, M.D. Button, B.M.C. Chow, I.M.H. El Ghawanmeh, M.H. Ghezelayagh, A.D. Grant, P.J. Hardy, C.J. Jackson, V. Math, S.K. Ng, S.P. Russ, D.M. Sumaili, C.W. Tam.
Pass: S. Ekmekcioglu, R.S. Knight, C.W. Loh, J.K. Nduli, S.B. Redshaw, S. Ryalls, W. Said, N. Scott.   Ord: I. Sheikh Yasin, I. Wilson, G. Zikos.
Physics and Electronic Engineering:
I: D.C. Chen   II(1): C.G. Harrison, P.J. Katzin, M.S. Piekarski, R.N. Turner.
II(2): L. Buckwell, K.M. Taylor.
III: D.R. Bolton, S.J. Chappell, M. Fill, M.G. Fisher, R.J. Parsons.   Pass: J. Evans, J. O'Dell.

1980:
Electronic and Electrical Engineering:
I: S.N. Iyer, Y.C. Leung, G.A.F. Ronan.
II(1): M.C.R. Chau, W.L.W. Cheng, G.W. Dawson, M. Hoggard, R.W. Hopper, C.O. Hung,

Appendix 3: Graduates in Electrical Engineering and related subjects, 1915-1997

P.J. Hutchinson, J.S. Lo, D.M. Page, J.F. Phiong, H.E.D. Talhamy, A.G. Tomlins, Caroline M. Waters, R.M.J. Whitaker, S.R. Young.
II(2): B. Abdul Aziz, Fatimah Ahmad, A.M. Antoniades, I. Arbuthnott, G. Ashikalis, I.P. Caldwell, K.H. Chong, H.S. Daniel, S. Glabedakis, M.F. Isaacs, G.W. Khor, P. Kontopoulos, R.J. Marsh, J.N. Mfune, K.L. Song, C. Subramaniam, T.H. Tan.
III: O. Baboomian, S.P. Boon, M. Bruck, J. Chebaclo, J.J.J. Chin, S.J. Chin, E.K. Dastur, S.E.M.A. Elamin, F.B. Geoghegan, H.K. Han, M.K. Rizaoglu, L. Yurtal.
Pass: A.H. Abang Taha, A.S. Ashour, G. Conaghan, D.S. Dissanayake, P.J. Kelly, C.A. Loh, L.M. Mbumwae, S.K. Ong, A. Patel, M. Rodgers.
Ord: G.S. Akasheh, F. Gerami, A. H. Mainwaring, P. McMahon, R.H. Taylor, S. Zomorrody.
Physics and Electronic Engineering:
II(1): V.A. Brashko, P.E. Cornforth, P.P. Skinner
II(2): P.I. Dowe, P.S. Forshaw, M.J. Hale, T. Oates.
III: J. Revie, E. Rossou.   Pass: J.A. Armitage, K. Luheshi.

1981:
Electronic and Electrical Engineering:
I: A.R.G. Bhaloo, G.L. Hirst, R.M. O'Sullivan, S. Sundram, R.A. Swann, F. Takawira, A.P. Varga, Hylda I. Wilson, T.W. Wong.
II(1): L.T. Aughton, R.S. Cameron, A.M. Chandley, A.J. Cooper, S. Fararooy, M.H. Goh, S.J. Linacre, R. Rao, J.C. Richardson, K.K. See, C.P. Sin, Catherine J. Threader.
II(2): D.W. Anderson, C. Barton, A. Chandegra, C.B. Conlon, J.E. Durrant, N. Elsby, L.C. Fung, P.R. Hernandez Herrada, M.S. Hitchcox, A. Khachik, H.K. Khor, H-S. Leow, J.K. Liang, F.M. Liau, G. Noonan, P.C. Phua, C.J. Powell, C.D. Ridley, S. Sanders, A.A. Sellak, G.R. Sherriff, Y.M. Siu, L.K. Tee, K.L. Teoh.
III: G. Bishop, M.R. Bull, J.D. Evans, H. Haji Rasouliha, K.P. Jones, M.A. Khayatan Mostafavi, W.W.S. Lam, S.V. Mabey, D. Mulenga, A. Ogucu, M. Tasniem.
Pass: D.B. Patel Ord: B.A. Akiwumi, B.A.I. Kabir, I. Mohammed, Anita V. Owen, M.A. Sacks.
Physics and Electronic Engineering:
I: H.C. Bhubata, D.R. Powell.   II(1): Hilary Davies, Joanne F. Goldberg, R.N. Kurji.
II(2): None III: J.L. Mather, P.S. Nagle, J.R. Ojeda Vina.   Pass: A. Puri

1982:
Electronic and Electrical Engineering:
I: P.G. Fenton, H.H. Ho, K.L.C. Lee, K.J. Lim, S.W. Mak, P.M. Muinde, W.A. Perera.
II(1): Clare S. Austin, J. Colgrave, D.M. Davenport, P.A. Drake, D.N. Farr, H.K. Lee, A.N. Mutisya, R.S.C. Potter, G.C. Townsend, Jenny W.C. Yap, L.W. Ying.
II(2): M.M.M. Alawi, J.H. Balmer, B. Bangui, H.A. Canavan, N.A. Cartledge, M.C.H. Chan, Joo H. Chong, Y.O.A. Chong, J.P. Diambra, J.S. Daniel, N. Dhala, D. Kelly, A.A.J. Lalji, P. Larkos, K.C.H. Leung, P.M.E. Mendis, P.M.C. Ng, K.W. Owen, M.R. Parker, D.M. Shiyukah, Siew L. Tan, A.G. Wilson, E.B. Zaier.
III: R.D. Clark, S.J. Clayton-Mitchell, R.G. Issaias, A.H. Mohamad Hussin, W.C. Moy, L. Panselinas, T.W. Poo, K.C. Sweeney. K. Tantivejkul, N-S. Tham, A.M. Wilkes.
Pass: K.J. Foo, M.J. Halliday, B.P. Jones, A.N. Winnett.
Ord: S.R. Gibson, H.B. Hogan, E.G. Maccariello, L.H. Sadiq, M. Samad, C.W. Savage, P.A.G. Shaw.
Physics and Electronic Engineering:
I: H. Brysh, R. Ettelaie, S.N. Lau, C.E. Smith, F.C.T. So.
II(1): K. Chandramouli, H.Y. Sze.   II(2): P.D. Carroll, S.R. Hunt, R.O. Kelly, D.M. Newell.
III: S.J. Hebbes, M.C. Lee.   Pass: A. Cope, H. Shirzad.

Electrical Engineering at Manchester University

1983:
Electronic and Electrical Engineering:
I: N. Cahill, P. Canavan, C.S.O. Choy, D.J. Cole, S.C. Goswami, J.R. Matthews, L.J. Richards.
II(1): O.O. Akinnola, M.B.A. Anderson, A.D. Bates, A.E.M. Fozard, J.N.C. Furnell, D.J. Hickey, Z.Y. Huang, B. Khamsehpour, Siu M.S. Lau, B.D. Lees, C.S. Leung, K.C. Lui, N.P. Main, A.J. McEntee, H.A.H. Mohammad, H.K. Muli, K.N. Patel, C.D. Psillides, M. Rabinowitz, N. Roughton, S.V. Savva, G.J. Welsh, H.P.W. Wightwick.
II(2): R.J. Anthony, N.R. Atkinson, P.M. Beck, J.N. Blennerhassett, S.R. Cooke, Christina M. Donaldson, R.D. Fitzgerald, A.S. Fries, R.D. Hull, Wendy A. Jones, T.S. Lee, K.Y. Leong, M. Longuet-Higgins, P. Mallinson, S.P. Manners, P.D. Martin, W.J. Mears, A.M. Mondoloka, S.P. Morgan, A. Morton, M.H. Noormohamed, D.J.E. Penny, A.J. Philiastides, M.J. Robinson, M. Sarshar, A.D. Shrier, M.E. Tung, N.H. Wong, J.S. Young.
III: S.J. Bentham, K.A.A. Jafar, Monica V. Koczan, A.C.K. Kwok, S.W.K. Ling, T.J. Mills, I.R. Minshall.
Pass: E.K.M. Chan, A.N. Foster, D.R. Moore, N.R.C. Russell, S.D. Tuson, S.T.K. Whittell.
Ord: A.W. Cooper, M.E.E. Summers, R.J. Webb.
Physics and Electronic Engineering:
II(1): M. Brown, N.J. Cartwright, J. Laing, A.M. McCullough, B.H. Seaton.
II(2): P.A. Haynes, A.B. Howlett, A.V. Ratcliff, T.W. Wan, L.H. Zhu.
III: P.H.A. Ashworth, M. Gordon, D.W. Heaton, M.H. Pearson, M.G. Quinlan, D. Rawsthorne.
Pass: M.J. Cummings, C.M. Owen, G.M. Pengilley.

1984:
Electronic and Electrical Engineering:
I: S.A.J. Bygrave, G.S.D. Farrow, K-C. Fung, M. Kipnis, R.P. Lancaster, C.K. Yau, C.C.K. Yeung.
II(1): N.A. Angelides, D.D. Buckley, D.S. Bye, J.A. Dutton, J.L. Ellis, S.M. Gough, C.M. Lo, A.H.K. Mak, M.G.W.C. Matley, T.I. McGibbon, I.A. Mobarek, P.G.J. Mullarkey, Geak H. Ng, C.M. Nixon, J.L.S. Palmer, A.S. Peake, M. Rice, A.M.D. Richardson, J.P. Smith, T.G. Tan, S.D. Williams, R.R. Wyrwas.
II(2): P. Armstrong, A.L.G. Brailsford, A.R. Carr, J. Chard, A.C.V. Collyer, Kathryn M. English, N.A. Holmes, D. Horridge, J.R. Jones, C.L.S. Lam, H.F. Lim, C.M. Liu, N.J.H. Mdaya, K.J. Perkins, S.R. Salter, M. Simpson, Polly M. Tang, D.R. Weatherill.
III: Man W.T. Au, Anuradha C. Banerjee, N.J. Borg Fenech, J.B. Child, D.F.C. Fuller, L. Hashim, R.J. Laver, V. Mavuba, S.N.C. Porritt, S.Y. Tang, S. Xuereb.
Pass: K.H. Kassam   Ord: A. Ahmad; F.C. Arnone (div. 1).
Physics and Electronic Engineering:
II(1): A.N. Clowes, A.L. Spray.
II(2): A.G. Casement, I.D. Clarke, M. Hamidian, R.M. Kerslake, P.J. Naybour, P. Parmar, C.P. Slater.
III: D.J. Callow, D.A. Kanaris, R.A. Odley, J.J. Slevin, M.E. Tipping, P.G. Whitehead.
Pass: R.J. Beaumont, M.A. Gregory, O.H. Harlow, M. Lloyd.

1985:
Electronic and Electrical Engineering:
I: S.L. Bradley, D.G. Evans, C.J. Kelley, S.C.P. Kwan, C.Y. Lau, D.A. Snowdon, H.M. Wong.
II(1): A. Brown, A.S. Challinor, Y. Choi, J.M. de Vile, K.W. Hicks, P.M. Ingram, A.M. Jones, G.J. Kemble, A.C. Kidger, D.B. Lamming, D.W. Leach, K.F.P. Ng, H.K. Patel, H.Y. Tam, L.S.S. Wong.
II(2): J.P. Biggs, R.G. Cliff, M.E. Cooper, P. Gath, W.M. Ho, Gillian L. Holmes, J.R. Kay, K.B. Ow, R.J. Robson, K.L. Stansfield, J.A. Stark, W.L. Thong, J.C. Wilkinson, P.Y. Yau.
III: M.C. Chan, M. Cowling, M.C. Hamilton, M.S. Hickman, C.D. Hutchinson, A.S. Leitch,

Appendix 3: Graduates in Electrical Engineering and related subjects, 1915-1997

Chali Tumelo, V.C. Turner.
Pass: M.P. Clough, N.M. Hedderick, C.T. Sit, I.J. Thompson.
Ord: G.A. Heeney, M.A. Maccariello; J.R. Tatchell (div. 1).
Physics and Electronic Engineering:
II(1): M. Ahmed, R.C. Robinson, I.D. Russell.
II(2): M.L. Bergman, J.O.A. Mosler, E.J. O'Donnell, S.M. Rudd, J. Wharmby.
III: R.W.M. Lok, D.G. Mathews, P.D. Morgan, A.R. Patel, S. Rodwell.
Pass: S.R. Dodd, Kathryn J. Howarth, K.K. Mak.

1986:
Electronic and Electrical Engineering:
I: S.E. Braznell, S.D. Cochrane, I.J. Hadfield, M.D.A. Hall, M.R.W. Manning, A.J. Reynolds, M.B. Roche, I.S. Tumber.
II(1): V. Amin, A.J. Ashill, T. Chan, W.K.A. Chung, M.P.R. Collins, P.J. Halliwell, H.R. Harji, A. Heaton, N.A.E. Heyes, R.T.C. Kwok, R.K. McCandless, M.B. Prior, P. Scaramangas, C. Stanfield, P. Thornber, M.K. Wong, E.H.T. Yung.
II(2): Y.L.R. Cheng, P. Fok, J.P. Francis, Sarah C. Harlow, H.F. Luker, S.H. Mak, G. Melrose, S. Ramsdale, T.E. Simmons, P.C.K. Tam, A.J.R. Taylor, S.C. Townley, S.O. Vital, B.P. Whittaker.
III: M.R. Burn, R.N. Cunliffe, H.A.H. Damji, D. Hatch, P.G.M. Lehmann.
Pass: O.A. Abaelu, P.H. Creighton, Jane Fitzsimons, A.S.M. Ma, D.M. Regan.
Physics and Electronic Engineering:
I: M. Moghisi, P.H.G. Smith.   II(1): S.A. Beck, R.N. Chappell, J.P. Keane, R.C. Wood.
II(2): J. Chan, Susan J. Davey, T.K. Lam, S.J. Potkins, K. Singh, C-K. So.
III: H. Cheng, D.A. Collyer, D.S. Cruickshank, S.A. Lark, M.C. Wood.   Pass: B.G. Reid.

1987:
Electronic and Electrical Engineering (3-year B.Sc.):
I: S. Brown, Helen D. Bunn, M.J. Crossley, K.K.M. So.
II(1): S.S. Ahmed, K.C. Chang, R.M. Coburn, S.G.T. Cooke, R.I. Crawford. Una C. Dunwoody, D.C. Feather, A.H. Fentiman, Y.C. Hui, H.Y. Lam, L.S.K. Mak, M.J. Mason, A.A. McNicol, S.L. Mifsud, R.L. Muir, W.K. Puplett, M.C. Roberts, D.L. Robinson, G.M. Roylance, M.B. Shaw, D.R. Smith, S.H. Smith, P. Tailor, Priscilla P. Tang, A.R.E. Temple Brown, I.D. Wigglesworth, S.C. Yu.
II(2): D.P. Anderson, G.S. Beer, G.I. Burrell, Chi L. Chao, R.N. Chapple, C-H. Cheng, J.R. Heston, M.J. Kelly, S.P. Kennedy, S.W.Y. Leung, M.C. Ngan, M.E. Preece, C.J. Ralph, P.R. Read. I.J.L. Russell, R.H. Wagner.
III: R.J.E. Amos, K.D. Cannon, B.S. Geary, H.A. Jafar, D.K.W. Li, D.J. Martin, M.W.H. Minchin, Eleanor K. Nabney, T. Patel, M.F. Sa, N.P. Taylor, R.D. Woods, W.C. Yeung.   Pass: A.B. Khan
Physics and Electronic Engineering:
I: M.B.H. Breese   II(1): A.C. Ford, N.D.G. Ford.
II(2): A.R. Brixton, S. Hussain, K. Lowe, S.J. Mather.
III: A.N. Fisher, J.R. Peacock.   Pass: J. O'Regan
4-year industrially-linked B.Sc., M.Eng. in Electronic and Electrical Engineering
(only the B.Sc. component is classified):
I: P.R. Miller, Katherine T. Neal, J.S. Thomas, D.F. Wolfe.
II(1): T.P. Beatson, R.J. Cryer, P.J. Eburne, Anne M. Eckersley, A.R. Flitcroft, R. Marciniak.
II(2): G. Crothers, K.J.F. Franklin, J. Hague, A.R. Wilkinson.

1988:
Electronic and Electrical Engineering (3-year B.Sc.):
I: S.D. Jolly, J.A. Siviter.
II(1): R. Baum, B.M. Beattie, B.L.C. Chiu, D.T.Y. Chou, H.S. Dabis, P.G. Davies, A. Drimiotis, J.P. Herron, R.P. Hira, C.G. Hodgson, L. Kalengo, M.A. Lambert, N.M. McKeeney, C.S. Ng, D. Nuttall, P.A. Porter, J. Shah, S.S. Shah, S.M. Shahrier, S. Sharma, C.K. Tang, G.B. Wilson.
II(2): S. Colston, J.H. Dyson, N.M. Faulkner, P.H. Garfield, P.J. Isherwood, K.M. Lai, C. Leung, S.R. Patel, C.J.Pepper, N.R. Powell.
III: P.S. Elliott, K.L. Lee, M.S. McLaughlin, K.C. Pang, S.C. Pun, S.C.I. Streat, A.J.R. Williams.
Pass: W.K.C. Chee, R.J. Friedlander, C.A. Herron, A. Mohamed, G.P. Salt.
Physics and Electronic Engineering:
I: M.S. Gill, C.W. Hill, S. Kapadia, J.C. Tracey, T. Wong.
II(1): D.J. Adams, L.O. Brown, M.P. Griffiths, P.A. Hester, B.J. Moss, A.P. Nisbet, Maria E. Winnett.
II(2): C.J. Cowell, C.N. Dente, B.P. McSweeney, A.J. Veitch, P.R. Williams.
III: J.R. Back, M.J. Fudge, M.J. Thornton.
4-year industrially-linked B.Sc., M.Eng. in Electronic and Electrical Engineering
(only the B.Sc. component is classified):
I: Catherine J. Thomas
II(1): Simone A. Abram, R.M. Bassham, E.C.J. Brady, B.J. McGrath, H. Singh.
II(2): C.F.Y. Lam, P.J. Linfitt, I.M. Overall, B.R. Pugh, S. Rana, M.I.F. Zakiuddin.

1989:
Electronic and Electrical Engineering (3-year B.Eng. from this year onwards instead of B.Sc. as previously):
I: P.J. Floyd, H.H.J.K. Li, T.A. Smith.
II(1): P.A. Chan, W.K. Chan, P.Y. Cheong, K.M. Cheung, C.M. Chileshe, B.P. Fowler, A.J. Fryett, R. Gurung, Anita L. Kelsall, A.S. Lok, Deborah Maloney, M.J. Mould, P. Stewart, K.C.B. Tam, Samantha J. Webber, L.G. Williams.
II(2): Dawn A. Acott, M.Z. Ahamed, J.A. Akabondo, C. Burrows, P.K. Cheung, P.J. Crozier, P.N.F. Hand, S. Harland, K.Y.K. Ho, A.C. Howard, G.S. Hudson, K.M. Kwan, G. MacKie, Catherine L. Meredith, D.C. Parmley, A.H. Ross, J.C. Szczerkowski, A. Tuaima, J.M. White.
III: E. Hardy, C.K. Jones, C.R. Maddock, P. Meadows, G. Moorthy, M.S. Rana, S.S. Velji, W.S. Wong, M. Wooldridge.
Physics and Electronic Engineering:
I: D.A.M. Smith   II(1): D.M.W. Ho, D.L. Jackson.
II(2): F.S. Abdoo, T.J. Griffiths, S.R. McKiernan, D.P. Nixon, R.J. Perks.   III: S. Coates
4-year industrially-linked B.Sc., M.Eng. in Electronic and Electrical Engineering
(only the B.Sc. component is classified):
I: P.Z. Bernat, P.A. Cain, A.A. Costantino, P.J. Cox, A.N. Fletcher, Amy S. Geschke, T.J.A. Jones, I. Lowrey, R.H. O'Hara, N.J. Stubley, P.S. Thompson.
II(1): G.S. Humber, R.J. McCarthy, M.R. Proctor.   II(2): S.J. Burrows, M. Hales.
B.Sc. pass: J.C.D. Smith

1990:
Electronic and Electrical Engineering (3-year B.Eng.):
I: Sze-Yin Y. Chan, P.P.F. Hackney, P.Y. Tam, K.N. Yeo.
II(1): R.J. Bibb, C.C. Chong, Lisa A. Drew, T.H. Hou, D.J. Hudson, A.H. Mak, J.W. Rothnie, C.D. Smith, P. Smith, S.K.P. Tong, H.M.F. Wong.
II(2): R.C. Bvulani, Karina H.Y. Chan, H. Chauhan, T. Cho, A. Clews, S.W. Gaskell, S.J. Gibbs, S. Grundy, M.P. Hatch, N.J. Houghton, S. Iqbal, R.N. Johnson, E. Kaplanellis, E. Mailey, G.L. Melnick, S.A. Sporton, H.D. Umney, R.N. Watts, T.E. Wharton, Y.K. Yu.

Appendix 3: Graduates in Electrical Engineering and related subjects, 1915-1997

III: O.A.G. Adewumi, A.K. Bangar, S. Blagburn, R.J.E. Connolly, D. Doll-Steinberg, J. Hillman, C.R.N. Hull, P.A. Hurst, A.B. Lee, S.W. Leung, C.N. Muyanda, Betty M. Mwangi, A.K. Patterson, C.R. Slade, C. Thompson.   Pass: V.K. Awumee

Physics and Electronic Engineering:
I: N. Hepworth, M.J. Shaw, R.E. Vialls.   II(1): J. Babad, S.M. Shah   II(2): M.J. Cooper
III: S.F. Bell, P. Gordon, N. Henderson, W.A. O'Neill.   Pass: G.R. Coney, D. Sleath.

4-year industrially-linked B.Sc., M.Eng. in Electronic and Electrical Engineering
(only the B.Sc. component is classified):
I: J.N. Binnie   II(1): Linda A. Beaton, M.C. Bort, D.J. Faria, R.N. Hobson, C. Hurst, G. Keeble, L.A. Renforth, A.G. Robins, W.J. Wass, G.A. Westlake, M.D. Wood.
II(2): I.H. Aitken, L.H. Burley.

1991:
Electronic and Electrical Engineering (3-year B.Eng.):
I: S.K. Cheng, A. Lanitis, C.S. Liew, H.P. Ngien, J.E. Posner, W.A. Powell, C.G. Soh.
II(1): R.M. Brown, D. Carey, I.P. Cooper, J.R. Durant, K.W. Lam, D.J. Lovett, P.C.P. Mak, S.J. Oliver.
II(2): M.J. Carson, M.J. D'Netto, A. Hayim, G. Hussain, P. Janabalan, V.H. Joseph, S.K.V. Kwan, K.W. Ross, M.R. Rubenstein, K.S. Sung, M. Thomas, D.M.P.Z. Tseu, M.T. Wong, D. Zhu.
III: A.J. Abd Rahim, J.A. Callaghan, M.W. Chauhan, S. Enrique, R.A. Jones, P.C. Morrison, M. Munkanta, N.A. Qureshi, A. Riat, R. Tamin, C.M.O. Wade, W.T. Wong.
Pass: W.G. Elliott, Z.B. Husin, P.B. McKeown, B.M. Sepiso.

Physics and Electronic Engineering:
I: C.J. Brown   II(1): P.J. Bearpark, Helen D. McCormick, K. Tangri.
II(2): N.C. Heaney, A. Warbrick.

4-year industrially-linked B.Sc., M.Eng. in Electronic and Electrical Engineering
(only the B.Sc. component is classified):
I: M. Akingbade, A.L. Gillett, B.P. James, J.M. Pearce, Monique S. Sachdeva.
II(1): A.F. Atkins, S. Ballantyre, R.J. Bruggisser, J.P. Connell, A.D.R. Granick, M.E.P. Keohane, P.J. Langslow, A.R. Main, Catherine J. Nayler, Rowena L. Sweatman.   II(2): R.G. Murley

1992:
Electronic and Electrical Engineering (3-year B.Eng.):
I: P. Michael
II(1): J.B. Gibson, V.K.V. Hee, G.J. Martindale, C.T.L. Phua, P.S. Royston, E.E. Ryott, G.R. Walker.
II(2): L. Brown, W.M. Cheung, K.B. Cockett, N.A. Dato Hj Abd Aziz, G.C. Douglin, Katherine M. Duncan, S.A.D. Fung, T.Y. Fung, C.D. Furgusson, D. Gardner, A. Iqbal, J. Jamaludin, R. Kaspin, C.J. Latherman, Parmjeet K. Nazran, G.P.D. Owen, K.N. Pang, G. Stoker, D. Wan, Y.K. Wong, Swee E. Yeoh.
III: S. Abu Bakar, S. Al-Hiddabi, D.E. Boddington, H.B. Hashim, M.P.J. Leung, D.V. Mackie, S.N. Sheikh Ismail, A. Sipan, Andrea M. White.
Pass: T.M.H. Abu-Hassan

Physics and Electronic Engineering:
Pass: S.C. Fisher, M. Prasad.

3-year B.Eng. in Electronic Engineering Science:
I: A.J. Perring   II(1): A.H. Brough

4-year M.Eng. in Electronic and Electrical Engineering (integrated European programme):
I: H.P. Jones   II(1): A.R. Hall, G.F. Holland, P.N.P. Sidebotham.

4-year industrially-linked B.Sc., M.Eng. in Electronic and Electrical Engineering
(only the B.Sc. component is classified):

Electrical Engineering at Manchester University

I: I.G. Cowsill, S.R. Cumpson, R.W. Friel, A.P. McIlwaine, S.A. Potts, J.A. Robinson, A.J. Smith.
II(1): V. Devlin, G.S. Francis, A.E. Glasgow, C.G. Harron, F.E. John, P.A. McManamon, Claire M. O'Grady, S.L. Vickers, D.J. Wilson.   II(2): A.P. Davies

1993:
Electronic and Electrical Engineering (3-year B.Eng.):
I: K.A. Beg, D.I. Greenspan, P.W. Nutter, P.A. Stevens.
II(1): J.M. Bates, Y.Y. Bham, T.V. Chong, F.N. Kaiser, C.C.A. Liu, I.D. Locker, S.T.E. Marley, J.N. Mufugi, B.P. Pang, R.J. Vlemmiks, R.S. Wilkhu.
II(2): P.D. Bennett, Geraldine C. Betts, F.H. Bt Ahmad, J. Cocking, A.J. Davies, M.D. Hossain, F. Kasujja, T. Lobo, N.A. Morby, S. Navaneethakumar, L. Seaman, B.J. Walsh, K.L. Wu.
III: R. Abd Rahman, K.S. Basra, A.P. Cummings, C. Dunham, N. Giannakogiorgos, A.C. Hirst, A.S. Md Yusoff, A. Salleh, E.M. Sambi, M. Slowey, D. Williams, N.F. Yarwood.
3-year B.Eng. in Electronic Engineering Science:
I: S. Gaskell   II(2): J.D. Clark   III: S. Chaudry, D.J. Clayton.
3-year B. Eng. in Electronic Systems Engineering:
III: A.D. Ingham
4-year B.Eng. in Electronic and Electrical Engineering (industrial programme):
II(1): R.J. Craig, A.M. Southworth.
4-year M.Eng. in Electronic and Electrical Engineering (integrated European programme):
I: Joanne M. Kirkman   II(1): C.W. Coppard, J.R. Hyde.
4-year M.Eng. in Electronic Engineering Science (integrated European programme):
II(1): M.D. Darragh
4-year industrially-linked B.Sc., M.Eng. in Electronic and Electrical Engineering
(only the B.Sc. component is classified):
I: Charlotte M. Attenborough, N. Deville, M.J. Gill, Mary-Claire Kennedy, R.J. Morton, A.W. Preston, S. Rahilly, A.C. Scheele, P.D. Seccombe, A. Stringfellow, J.D. Taylor.
II(1): C.D.M. Andrew, P.C. Atkinson, J. Bodger, Elaine E. Greig, S. McLean, M.J. Turner, C.J. White.   II(2): M.A. Khan, J.J. Turner.

1994:
Electronic and Electrical Engineering (3-year B.Eng.):
I: P.R. Aggarwal, A.A. Al-Maqrashi, A.C. Burrows, Z.M. Datoo, S.L.F. Wong, J. Wright.
II(1): O.M.D. Al-Saleem, T. Boyle, N.M. Cheung, R.P. Fenton, M.I. Hartley, D.A. Howarth, S.W. Malpass, G. Moore, D.T. Smith, S.P. Walmsley, M.M. Yazid.
II(2): S.P. Allen, D.M. Chanda, J. Corr, P.D. Cropper, S.K. Derby, M. Hopper, M.J. Juliano, C. Kaputu, T. Lyrintzis, F. Mwale, D. Nathanael, M.E. Peterson, G.S. Probyn, J.J. Ridgway.
III: M.S. Ahmad, K. Ajayi-Obe, H. Alias, J. Chipulu, J.C.Q. Cox, M. Hussain, S.Z. Idris, R.A. Jack, N.A. Lloyd, T.C. Lungu, Nicola J. Phillips, J.M. Silungwe.
Aegrotat: S.S. Al-Mahrooqi   Pass: A. Chilenka, A.G. Mohamad, M.S. Samad.
3-year B.Sc. in Electronic Engineering Science:
II(1): A.S. Milne   II(2): R.J. Bagnall   III: N.S. Sutherland, G.B. Young.
3-year B. Eng. in Electronic Systems Engineering:
II(1): M. Melbourne   II(2): J. Byron   III: N.J. Bryant, S.J. Clough, M.D. Jemmeson, J.V. Kulasingham.
4-year M. Eng. in Electronic and Electrical Engineering:
I: A.J. Polley, P.M. Zimmerman.   II(1): Hannah Grimshaw, M.N. Hallett, D.J. Marton.
II(2): Sarah Archampong, S.D. Knightley.
4-year M. Eng. in Electronic and Electrical Engineering (integrated Japanese programme):
II(2): D.M. Lamont
4-year M.Eng. in Electronic Engineering Science (integrated European programme):
II(1): Nicola J. Cartlidge

Appendix 3: Graduates in Electrical Engineering and related subjects, 1915-1997

4-year industrially-linked B.Sc., M.Eng. in Electronic and Electrical Engineering
(only the B.Sc. component is classified):
I: N.E. Brown, J. Clark, T.D. Daldry, C.D. Hewitt, P.D. Parry, W.M. Yates.
II(1): J.A. Gillespie, C.J. Harding, M. Harrison, C. Knowles-Spittle, P.E. Thompson, R.A. Veitch.
II(2): D.S. Auty, M. Batey, W.M. Macauley.

1995:
Electronic and Electrical Engineering (3-year B.Eng.):
I: L. Eugene, C.H. Siew.
II(1): D.C. Chang, T.L. Chen, Helen Dixon, Shona Fewster, S. Hassan, Y.K. Ho, W.K. Leung, B.T. Lim, C.Y. Lim, C.H. Ling, J.C. Mann, S.M. Wang.
II(2): A.K. Abd Rahman, H. Abdul, Vania L. Bastajian, B.C. Beh, Yesim Celtik, K.K. Chick, J. Chikota, E. Haque, K.K. Lam, M.D. Plant, P.J. Ravenscroft, Lay P. Tan, T. Wahib, M. Zadoroznyj.
III: M.H. Abdul Samad, P.M. Higgins, A. Mehan, M.I. Mulenga, C.E. Pinkstone, B.S. Salasini, D.S. Smith, S.N. Syed Mohd.
Aegrotat: M.J. Griffin
Pass: O. Abd Latiff, M.Z.B. Abdullah, P. Chonya, H.S. Dhadwal, N. Salah Adam, S. Suntharesan, M.O.O. Yusoof.
3-year B.Eng. in Electronic Engineering Science:
II(2): A.E. Mustafa   III: A. Croshaw
3-year B. Eng. in Electronic Systems Engineering:
II(1): S.A. Suleman, M.H. Yung.   II(2): W.H. Chan, P.M. Glasgow, M.B. Phokoletso.
III: A.N. Elgergawi, H.S. Labana.   Pass: R. Ashworth, Tania S. Baker.
4-year M.Eng. in Electronic and Electrical Engineering (integrated European programme):
II(1): Catherine Askam, K.L. Chan, S.P. Connolly, W.M. Leong.   II(2): Alison R. Dimsdale
4-year M. Eng. in Electronic and Electrical Engineering (integrated Japanese programme):
I: N.R. Purcell   II(2): P.S. Wimalasiri
4-year industrially-linked B.Sc., M.Eng. in Electronic and Electrical Engineering
(only the B.Sc. component is classified):
I: G.A. Bartlett, S.J. Hardingham, D.K. McCoy, J.M. Wise.
II(1): M.J. Bennett, D.J. Guest, J.C. Pringle, S. Williams.   II(2): D.J. Roulston, A.E. Shields.

1996:
Electronic and Electrical Engineering (3-year B.Eng.):
I: C.A. Chew, T.B. Koh, P.J. Merewood, E.O.P. Omiyi, K.Y. Ong, J.K. Tsao.
II(1): H.K. Ang, B.H. Cheng, C. Chilumbu, K.W. Chng, C.K. Chu, D.M. Courtney, B. Essien, C. Holly, N. Howroyd, B. Kabaso, C.C. Lee, J.G. Malkin, M. Mzyece, K.W. Ngeow, T.C. Por, P.A. Sampson, D.J. Saul, N.K. Southern, Y.H. Tan, S.C. Yip.
II(2): M.A. Abd Rahman, J.R.H. Arkell, J.D. Banda, G.J. Bradly, Paula Caddick, K. Chiramanee, D.R. Clarke, B.R. Cooling, R.G. Hartshorne, J.G. Hook, A.J. Hussain, S. Hussain, R. Karani, N.T. Larkins, H. Mgdob, P.R. Millgate, M.J. Pearson, A. Peddie, N.R. Ramoo, P.D. Royall, S.P. Sandles, J.R. Shacklock, K.H. Tse, M.W. Westrap, M.R. Williams.
III: M.R. Abdul Rahman, C.H. Chingumbe, T. Ellis, T.D. Evans, C.R. Howells, G.S. Jones, Gemma M. McDonnell, D. Mintah, A.M. Mohamed Munawir, Z. Mohammad, G. Nikolakakis, S.R. Roberts, A. Syed, M.J. Would, K. Yargici.
3-year B.Eng. in Electronic Systems Engineering:
I: D. Williams   II(1): P. Mulligan, K.W. Williams.
II(2): M.R. Bryce, P. Dawkins, A.L. Jordan, Louise C. Moore, J-L. M. Pellet.
III: A.K. Abdul Rahman, A. French, R. Matthews.

3-year B.Eng. in Electronic Engineering:
II(1): S. Betab, A.R. Buddhev, H.G. Chia, M. Evans, Annie P. Lata, W.K. See.
II(2): N.J. Deacon, G.R. Ellison, J. Pappas.   III: S. Amani Gheshlaghi, R. Tekchandani.
4-year B. Eng. in Electronic and Electrical Engineering (industrial programme):
II(1): M.J. Ayub
4-year M.Eng. in Electronic and Electrical Engineering (integrated European programme):
I: T.M. Chapman, N. O'Connor, D.J. Wiper.
II(1): G.F. Gillison, G.R. Heyes, J.M. Plimmer.   II(2): G.R. Huggett
4-year M. Eng. in Electronic and Electrical Engineering (integrated Japanese programme):
II(2): J.C. Msuya
4-year M. Eng. in Electronic Systems Engineering (integrated Japanese programme):
II(1): S.W. McManus
4-year industrially-linked B.Sc., M.Eng. in Electronic and Electrical Engineering
(only the B.Sc. component is classified):
I: M.D. Burden, I.P. Roper.   II(1): S.E. Martin, D.P. Murphy.

1997:
Electronic and Electrical Engineering (3-year B.Eng.)
I: K.C. Chan, M. Larizadeh, K.K. Law, D.A. Liu, L. Lukama, R. Maseko, B.B. Philip, S.J. Wong, E.B. Zyambo.
II(1): C.K. Choi, A.B. Ismail, M. Kandeke, E. Manyonganise, M.T. Mikal Tasya, T. Nicholls, K.U. Nzewi.
II(2): W.J. Biggs, S.E. Booth, Victoria J. Broughton, E. Chisanga, M.A. Hilmy, F.J. Lyons, Razlina Muhammad Sabri, C.Y. Ng, K.I. Ramli, C.J. Wheeldon, S.H. Woo.
III: C. Bilimoria, D.S. Cox, N. Sampatas, G.S. Sekhon, M.M.A. Shehada.
Pass: N. Hussain, R.H. Patel.
Electronic Engineering (3-year B.Eng.)
I: K. Zulhisham   II(1): C. Brown, R.J. Brown, M.D. Crutch, Rezawiti Ismail.
II(2): A.H. Ahmed, G. Arnold, G.J. Clarke, Z.A. Maz Iskandar, D. Mirza-Hekmati, Md H. Mohd Hasrulnizam.
III: M.C. Branford, Karen L. Dabrowski, T. Frost, S. Grey, A. Ingram, I.R. Mellor, D.J. Pennington, N.M. Turtell.   Pass: M.J. Ashdown, B.D. Powell.
Electronic Systems Engineering (3-year B.Eng.)
I: C.H. Guo   II(1): K.D. Maifoshis   II(2): M. Ahmed, C. Dimitrakakis, I.K. Hunter.
III: A. Kashim, T. Kokkinias, M.P. Parker, S. Pymm, S. Riaz.   Pass: R.G. James, A.J. Jenn.
4-year B.Eng. in Electronic and Electrical Engineering (industrial programme):
I: C.W. Chong   II(2): A. Buscombe
4-year M.Eng. in Electronic and Electrical Engineering (integrated European programme):
I: R.W. McKellar, V. Petrovic.   II(1): C.A. Moore
4-year M.Eng. in Electronic and Electrical Engineering (Japanese programme):
I: R. Webb   II(1): Elene Y.L. Lee
4-year M.Eng. in Electronic and Electrical Engineering (Singapore programme):
II(2): D.J. Pugh
4-year B. Eng. in Electronic Engineering (industrial programme): I: M.D. Grist
4-year M.Eng. in Electronic Engineering (integrated European programme): I: I. Ali
4-year M.Eng. in Electronic Systems Engineering (integrated European programme):
II(1): Kirsty E. Stevenson   II(2): S.J. Olima
4-year M.Eng. in Electronic Systems Engineering (Japanese programme):   II(1): F. McConville
4-year industrially-linked B.Sc., M.Eng. in Electronic and Electrical Engineering
(only the B.Sc. component is classified): I: T.M. Passingham, A. Sud.

Appendix 4. Numbers of first-degree graduates in Electrical Engineering and related subjects and their degree classifications, year by year, 1915-1997

Year	I	II(1)	II(2)	III	Ord div.1	Ord div. 2	Pass	Total
1915	1	-	-	-	-	-	-	1
1916	-	-	-	-	-	1	-	1
1917	-	-	-	-	-	1	-	1
1918	-	-	-	-	-	1	-	1
1919	-	No graduates		-	-	-	-	0
1920	1	-	-	-	-	-	-	1
1921	1	1*	-	3	-	-	-	5
1922	-	2	1	2	-	3	-	8
1923	1	2	2	-	-	1	-	6
1924	2	4*	-	-	1	1	-	8
1925	3	1	-	-	1	2	-	7
1926	-	-	-	1	-	3	-	4
1927	1	-	-	1	-	4	-	6
1928	-	-	1	1	1	5	-	8
1929	-	-	-	2	-	2	-	4
1930	1	1	-	2	-	5	-	9
1931	-	-	2	-	-	1	-	3
1932	3	2	3	1	-	2	-	11
1933	3	2	1	1	-	4	-	11
1934	-	-	2	1	-	1	-	4
1935	2	3	1	1	-	2	-	9
1936	2	-	2	-	-	2	-	6
1937	5	-	1	-	-	3	-	9
1938	4	2	1	1	-	1	-	9
1939	4	4	3	3	-	1	-	15
1940	1	3	-	-	-	1	-	5
1941	3	2	-	-	-	3	-	8
1942	6	4	1	3	-	3	-	17
1943	3	2	1	1	-	3	-	10
1944	4	7	1	2	-	10	-	24
1945	8	2	3	1	-	9	-	23
1946	4	3	8	3	1	2	-	21
1947	4	4	7	1	-	2	-	18
1948	2	4	4	3	-	5	-	18
1949	5	1	8	3	-	5	-	22
1950	5	3	5	2	1	7	-	23

* In 1921 and 1924 Class II degrees were not subdivided into II(1) and II(2).

Year		I	II(1)	II(2)	III	Ord div.1	Ord div.2	Pass	Total
1951		1	6	9	1	-	6	-	23
1952		-	6	6	1	-	11	-	24
1953		3	5	5	3	1	8	-	25
1954		3	2	4	2	-	11	-	22
1955		5	2	7	-	-	5	-	19
1956		3	5	7	2	1	10	-	28
1957		4	7	9	3	-	12	-	35
1958		2	2	12	4	-	11	-	31
1959		1	-	12	3	-	19	-	35
1960		3	8	11	5	-	17	-	44
1961		3	11	12	6	1	14	-	47
1962		2	12	18	6	1	14	-	53
1963		2	6	5	1	-	14	-	28
1964		1	7	8	6	3	26	-	51
1965	EE:	3	10	8	5	3	10	-	39
	PE:	-	1	2	3	-	-	-	6 [45]
1966	EE:	5	9	11	5	1	12	-	43
	PE:	1	3	2	1	1	-	-	8 [51]
1967	EE:	5	9	8	6	-	15	1	44
	PE:	1	8	8	6	-	7	-	30 [74]
1968	EE:	3	13	7	2	1	14	2	42
	PE:	3	12	5	2	-	7		29 [71]
1969	EE:	5	8	12	3	-	8	3	39
	PE:	4	10	12	7	-	3	3	39 [78]
1970	EE:	2	9	10	7	-	4	1	33
	PE:	3	9	9	6	-	7	3	37 [70]
1971	EE:	7	8	11	9	-	6	2	43
	PE:	-	3	5	1	-	-	1	10 [53]
1972	EE:	4	12	10	5	-	2	2	35
	PE:	2	4	5	7	-	2	4	24 [59]
1973	EE:	5	11	13	6	-	3	2	40
	PE:	-	4	5	3	-	-	3	15 [55]
1974	EE:	6	10	9	6	-	4	2	37
	PE:	4	8	7	5	-	1	2	27 [64]
1975	EE:	4	5	15	4	-	4	1	33
	PE:	2	8	4	6	-	1	-	21 [54]
1976	EE:	3	17	17	7	1	3	5	53
	PE:	2	5	4	3	-	1	1	16 [69]
1977	EE:	4	5	12	9	-	1	10	41
	PE:	2	4	4	6	-	-	1	17 [58]
1978	EE:	1	10	16	13	-	1	9	50
	PE:	1	5	4	2	-	-	1	13 [63]
1979	EE:	1	7	18	14	-	3	8	51
	PE:	1	4	2	5	-	-	2	14 [65]
1980	EE:	3	15	17	12	-	6	10	63
	PE:	-	3	4	2	-	-	2	11 [74]

On this page, EE = Electrical or Electronic & Electrical Eng; PE = Physics & Electronic Eng.

Appendix 4: Numbers of graduates and their degree classifications, 1915-97

Year		I	II(1)	II(2)	III	Ord div.1	Ord div.2	Pass	Total
1981	EE:	9	12	24	11	-	5	1	62
	PE:	2	3	-	3	-	-	1	9 [71]
1982	EE:	7	11	23	11	-	7	4	63
	PE:	5	2	4	2	-	-	2	15 [78]
1983	EE:	7	23	29	7	-	3	6	75
	PE:	-	5	5	6	-	-	3	19 [94]
1984	EE:	7	22	18	11	1	1	1	61
	PE:	-	2	7	6	-	-	4	19 [80]
1985	EE:	7	15	14	8	1	2	4	51
	PE:	-	3	5	5	-	-	3	16 [67]
1986	EE:	8	17	14	5	-	-	5	49
	PE:	2	4	6	5	-	-	1	18 [67]
1987	EE(3yr):	4	27	16	13	-	-	1	61
	EE(4yr):	4	6	4	-	-	-	-	14
	PE:	1	2	4	2	-	-	1	10 [85]
1988	EE(3yr):	2	22	10	7	-	-	5	46
	EE(4yr):	1	6	6	-	-	-	-	13
	PE:	5	7	5	3	-	-	-	20 [79]
1989	EE(3yr):	3	16	19	9	-	-	-	47
	EE(4yr):	11	3	2	-	-	-	1	17
	PE:	1	2	5	1	-	-	-	9 [73]
1990	EE(3yr):	4	11	20	15	-	-	1	51
	EE(4yr):	1	11	2	-	-	-	-	14
	PE:	3	2	1	4	-	-	2	12 [77]
1991	EE(3yr):	7	8	14	12	-	-	4	45
	EE(4yr):	5	10	1	-	-	-	-	16
	PE:	1	3	2	-	-	-	-	6 [67]
1992	EE(3yr)†:	2	8	21	9	-	-	1	41
	EE(4yr)†:	8	12	1	-	-	-	-	21
	PE:	-	-	-	-	-	-	2	2 [64]
1993	EE(3yr)†:	5	11	14	15	-	-	-	45
	EE(4yr)†:	12	12	2	-	-	-	-	26 [71]
1994	EE(3yr)†:	6	13	16	18	Aegrotat:1	-	3	57
	EE(4yr)†:	8	10	6	-	-	-	-	24 [81]
1995	EE(3yr)†:	2	14	18	11	Aegrotat: 1	-	9	55
	EE(4yr)†:	5	8	4	-	-	-	-	17 [72]
1996	EE(3yr)†:	7	29	33	20	-	-	-	89
	EE(4yr)†:	5	6	2	-	-	-	-	13 [102]
1997	EE(3yr)†:	11	12	20	18	-	-	6	67
	EE(4yr)†:	8	4	3	-	-	-	-	15 [82]
Totals:		385	781	859	497	21	417	152	
					+ 2 aegrotat:			Overall total:	3114

EE(3yr)†: The total of the graduates in the various 3-year courses for the year.
EE(4yr)†: The total of the graduates in the various 4-year courses for the year.
The figures in square brackets are the total number of graduates for the year in all courses.
For the purpose of these figures the 285 graduates in Physics and Electronic Engineering are regarded as graduates of the Electrical Engineering Department.
Of the above 3114 graduates 96 were women, 84 of whom graduated between 1979 and 1997.

Appendix 5. Papers on electrical subjects published by Owens College staff, 1851-1900

This is a list of papers published by staff and research students of the college on subjects that to a greater or lesser extent relate to electrical engineering; i.e. all the published papers that were concerned with electricity, magnetism, etc. It is as complete as it has been possible to ascertain and it is doubtful whether much has been missed. All these papers, with the exception of three from Professor Lamb (Mathematics Department) came from the Engineering Department (a few) or the Physics Department (many). None of the earliest scientific research at Owens (1851- 1869) by Frankland, Williamson, Sandeman, Roscoe, Clifton and Jack had anything to do with magnetism or electricity.

After 1900 the papers become too numerous to quote.

Engineering Department

The following papers were all by Osborne Reynolds:

The tails of comets, the solar corona, and the aurora, considered as electrical phenomena. Manchester Lit. and Phil. Soc. Memoirs, Series 3, Vol. 5, Session 1870-1871.

On cometary phenomena. Manchester Lit. and Phil. Soc. Memoirs, [3], **5**, Session 1871-72.

On an electrical corona resembling the solar corona. Manchester Lit. and Phil. Soc. Memoirs, [3], **5**, Session 1871-72.

On the electro-dynamic effect which the induction of statical electricity causes in a moving body: This induction on the part of the sun a probable cause of terrestrial magnetism. Manchester Lit. and Phil. Soc. Memoirs, [3], **5**, 1872.

On the electrical properties of clouds and the phenomena of thunderstorms. Manchester Lit. and Phil. Soc. Proc., **12**, Session 1872-1873.

On the bursting of trees and objects struck with lightning. Manchester Lit. and Phil. Soc. Proc., **13**, Session 1873-74.

On the surface-forces caused by the communication of heat. Philosophical Magazine, **48**, November 1874.

On the forces caused by the communication of heat between a surface and a gas; and on a new photometer. Proc. Roy. Soc., **24**, 1876; Phil. Trans., 1877.

On the principle of the electromagnet constructed by Mr. John Faulkner. Manchester Lit. and Phil. Soc. Proc., **15**, 1875.

Physics Department

Balfour Stewart. *Results of the monthly observations of dip and horizontal force made at the Kew Observatory, from April, 1863, to March, 1869, inclusive.* Proc. Roy. Soc., **18**, 1870.

J.H. Poynting. *Arrangement of a tangent galvanometer for lecture-room purposes to illustrate the laws of the action of currents on magnets and of the resistance of wires.* Manchester Lit. and Phil. Soc. Proc., **18**, 1879.

Balfour Stewart. *On the variations of the diurnal range of the magnetic declination as*

Appendix 5: Papers on electrical subjects by Owens College staff, 1851-1900.

recorded at the Prague Observatory. Proc. Roy. Soc., **27**, 1878.
Balfour Stewart and W. Lant Carpenter. *On the inequalities of diurnal range of the declination magnet.* Proc. Roy. Soc., **28**, 1879.
Balfour Stewart and W. Dodgson. *Note on the inequalities of the diurnal range of the declination magnet as recorded at the Kew Observatory.* Proc. Roy. Soc., **28**, 1879.
Balfour Stewart and M. Hiraoka. *A comparison of the diurnal range of magnetic declination as recorded at the observatories of Kew and Trevandrum.* Proc. Roy. Soc., **28**, 1879.
Balfour Stewart and W. Dodgson. *An analysis of the recorded diurnal ranges of magnetic declination, with the view of ascertaining if these are composed of inequalities which exhibit a true periodicity.* Manchester Lit. and Phil. Soc. Memoirs, [3], **8**, 1881.
Balfour Stewart. *On the connection between the state of the sun's surface and the horizontal intensity of the earth's magnetism.* Proc. Roy. Soc., **34**, 1882.
Balfour Stewart. *Preliminary report to the solar physics committee on the comparison for two years between the diurnal ranges of magnetic declination as recorded at the Kew Observatory, and the diurnal ranges of atmospheric temperature as recorded at the observatories of Stonyhurst, Kew and Falmouth.* Proc. Roy. Soc., **34**, 1882.
A. Schuster. Experiments on the discharge of electricity through gases. *Bakerian Lecture.* Proc. Roy. Soc., **37**, 1884.
A. Schuster. *Über die Entladung der Electricität durch Gase.* Wied. Ann, **24**, 1885.
A. Schuster. *On comparing and reducing magnetic observations.* Brit. Assoc. Report (Aberdeen), 1885.
A. Schuster. *On Helmholtz's views on electrolysis and on the electrolysis of gases.* Brit. Ass. Report (Aberdeen), 1885.
Balfour Stewart and W. Lant Carpenter. *Note on a preliminary comparison between the dates of cyclonic storms in Great Britain and those of magnetic disturbances at the Kew Observatory.* Proc. Roy. Soc., **38**, 1885.
Balfour Stewart and S.J. Perry. *A comparison of certain simultaneous fluctuations of the declination at Kew and at Stonyhurst during the years 1883 and 1884.* Proc. Roy. Soc., **39**, 1885.
Balfour Stewart and others. *Three reports of the committee appointed for the purpose of considering the best means of comparing or reducing magnetic observations.* Brit. Assoc. Reports, 1885, 1886, 1887.
Balfour Stewart. *On the cause of the solar diurnal variations of terrestrial magnetism.* Phil. Mag., [5], **21**, 1886.
Balfour Stewart and W. Lant Carpenter. *On a comparison between apparent inequalities of short period in sun spot areas and in diurnal declination ranges at Toronto and Prague.* Proc. Roy. Soc., **40**, 1886.
H. Holden. Measurements of the magnetic induction and permeability in soft iron. Manchester Lit. and Phil. Soc. Proc., **26**, 1886.
A. Schuster. *On the diurnal period of terrestrial magnetism.* Phil. Mag., [5], **21**, 1886.
A. Schuster. *On the best means of comparing and reducing magnetic observations.* Brit. Assoc. Report (Birmingham), 1886.
W.W. Haldane Gee. *On a comparison magnetometer.* Brit. Assoc. Report, 1887.
C.H. Lees and R.W. Stewart. *On electrolytic polarisation.* Manchester Lit. and Phil. Soc. Proc., **26**, 1887.

A. Schuster. *Experiments on the discharge of electricity through gases.* Proc. Roy. Soc., **42**, 1887.

A. Schuster. *Conduction of electricity through gases.* Brit. Assoc. Report (Manchester), 1887.

W.W. Haldane Gee and W. Stroud. *On a null method in electro-calorimetry.* Brit. Assoc. Reports, 1877 and 1888.

W. Haldane Gee, H. Holden and C.H. Lees. *Experiments on electrolysis and electrolytic polarisation.* British Association Report (Manchester), 1887.

W.W. Haldane Gee and H. Holden. *Experiments on electrolysis: Part I: Change of density of the electrolyte at the electrodes. Part II: Irreciprocal conduction.* Phil. Mag., [5], **25** and **26**, 1888.

H. Holden. *A method of calculating the electrostatic capacity of a conductor.* Manchester Lit. and Phil. Soc. Memoirs, [4], **1**, 1888.

A. Schuster. Article on "Volta". Encyclopedia Brittanica, 9th Edn., 1888.

A. Schuster. *The diurnal variation of terrestrial magnetism.* Phil. Trans., 1889.

A. Schuster. *The passage of electricity through gases.* Brit. Assoc. Report (Newcastle-on-Tyne), 1889.

A. Schuster. *On Joule's explanation of the varying dips of magnetic needles of different length.* Manchester Lit. and Phil. Soc. Memoirs, [4], **3**, 1889.

A. Schuster. *The disruptive discharge of electricity through gases.* Phil. Mag., [5], **29**, 1890.

A. Schuster. *The discharge of electricity through gases.* Bakerian Lecture, Proc. Roy. Soc., **47**, 1890.

A. Schuster. *Discharge of electricity through gases.* Letter to *Nature*, **42**, 1890.

A.T. Stanton. *The discharge of electricity from glowing metals.* Proc. Roy. Soc., **47**, 1890.

A. Schuster. *The influence of bending of magnetic needles on the apparent magnetic dip.* Phil. Mag., [5], **31**, 1891.

A. Schuster. *On summer lightning.* Manchester Lit. and Phil. Soc. Memoirs, [4], **5**, 1891.

A. Schuster. *Electrical Notes, I.* Phil. Mag., [5], **32**, 1891.

A. Schuster and A.W. Crossley. *On the electrolysis of silver nitrate in vacuo.* Proc. Roy. Soc., **50**, 1892.

A. Schuster. *On primary and secondary batteries in which the electrolyte is a gas.* Brit. Assoc. Report (Edinburgh), 1892.

A. Schuster. *The present condition of mathematical analysis as applied to terrestrial magnetism.* Report of Chicago Meteorological Congress, 1893.

J.R. Ashworth. *A method of measuring dynamo efficiency.* Letter to *Electrician*. **30**, 1893.

A. Schuster. *On the construction of delicate galvanometers.* Brit. Assoc. Report, 1894.

J. Frith. *An analysis of the electromotive force and current curves of a Wilde alternator under various conditions.* Manchester Lit. and Phil. Soc. Memoirs, [4], **8**, 1894.

W. Gannon. *On copper electrolysis in vacuo.* Proc. Roy. Soc., **55**, 1894.

A. Schuster. *On an oak tree struck by lightning.* Manchester Lit. and Phil. Soc. Memoirs, [4], **8**, 1894.

A. Schuster. *A suggested explanation of the secular variation of terrestrial magnetism.*

Appendix 5: Papers on electrical subjects by Owens College staff, 1851-1900.

Brit. Assoc. Report (Oxford), 1894.
J.R. Ashworth. *Methods of determining the efficiencies of dynamos and motors.* College Prize Essay. Electrical Plant, 1894-5.
A. Schuster. *Electrical Notes, II.* Phil. Mag., [5], **39**, 1895.
A. Schuster. *On some remarkable passages in the writings of Benjamin Franklin.* Manchester Lit. and Phil. Soc. Memoirs, [4], **9**, 1895.
A. Schuster. *Atmospheric electricity.* Roy. Institut. Proc., 1895.
A. Schuster and W. Gannon. *A determination of the specific heat of water in terms of the international electric units.* Phil. Trans., **186**, 1895.
A. Griffiths. *Some experiments with alternating currents.* Phil. Mag., [5], **39**, 1895.
J. Frith. *On the true resistance and on the back electro-motive force of the electric arc.* Manchester Lit. and Phil. Soc. Memoirs, [4], **9**, 1895.
A. Schuster. *Note on Mr. Frith's paper On the true resistance and on the back electro-motive force of the electric arc.* Manchester Lit. and Phil. Soc. Memoirs, [4], **9**, 1895.
A. Schuster. *On some experiments with Lord Kelvin's portable electrometer.* Brit. Assoc. Report (Ipswich), 1895.
J.R. Ashworth. *Electroscopes in lecture.* Letter to Nature, **51**, 1895.
A. Schuster. *On electrical currents induced by rotating magnets and their application to some phenomena of terrestrial magnetism.* Terrestrial Magnetism, **1**, 1896.
J. Frith. *The effect of wave form on the alternate current arc.* Phil. Mag., [5], **41**, 1896.
J. Frith and C. Rodgers. *On the resistance of the electric arc.* Phil. Mag., [5], **42**, 1896.
J.R. Ashworth. *Methods of making magnets independent of changes of temperature.* Proc. Roy. Soc., **62**, 1897.
J.R. Ashworth. *Discharge of electricity by phosphorus.* Letter to Nature, **55**, 1897.
A. Schuster. *Electrical notes, III.* Phil. Mag., [5], **43**, 1897.
A. Schuster. *On the constitution of the electric spark.* Brit. Assoc. Report (Toronto), 1897.
A. Schuster. *On the application of terrestrial magnetism to the solution of some problems of cosmical physics.* Brit. Assoc. Report (Bristol), 1898.
J.R. Ashworth. *The construction of magnets of constant intensity.* Terrestrial magnetic conference; Brit. Assoc. Report, 1898.
A. Schuster. *On the interpretation of earth current observations.* Terrestrial Magnetism, **3**, 1898.
A. Schuster. *On the interpretation of earth current observations.* Brit. Assoc. Report, 1898.
A. Schuster. *On a simple method of obtaining the expression of the magnetic potential of the earth in a series of spherical harmonics.* Brit. Assoc. Report, 1898.
A. Schuster. *Die magnetische ablenkung der kathodenstrahlen.* Wied. Ann., **65**, 1898.
A. Schuster. *On the possible effects of solar magnetisation on periodic variations of terrestrial magnetism.* Phil. Mag., [5], **46**, 1898.
J.R. Ashworth. *The construction of magnets of constant intensity.* Brit. Assoc. Report (Bristol), 1898.
J.R. Ashworth. *Dynamo efficiency.* Letter to *Electrician*, **40**, 1898.
C.H. Lees. *On the resistance between opposite sides of a quadrilateral one diagonal of which bisects the other at right angles.* Manchester Lit. and Phil. Soc. Memoirs, [4], **44**, 1899.

A. Schuster and G. Hemsalech. *On the constitution of the electric spark.* Phil. Trans., 1899.

C.H. Lees. *On the conductivities of certain heterogeneous media for a steady flux having a potential.* Phil. Mag., [5], **49**, 1900.

A. Schuster. *On the periodogram of magnetic declination at Greenwich.* The "Stokes" Volume of Camb. Phil. Soc. Trans., 1900.

R. Beattie. *The spark length of an induction coil.* Phil. Mag., [5], **50**, 1900.

R. Beattie. *Note on a possible source of error in the use of a ballistic galvanometer.* Phil Mag., [5], **50**, 1900.

C.H. Lees. *On the production of electrolytic copper.* Letter to Nature, **61**, 1900.

Mathematics Department

H. Lamb. *On ellipsoidal current sheets.* Phil. Trans., 1887.

H. Lamb. *On the principal electric time-constant of a circular disc.* Proc. Roy. Soc., **42**, 1887.

H. Lamb. *On the reflection and transmission of electric waves by a metallic grating.* Proc. Lond. Math. Soc., **29**, 1898.

Appendix 6. Robert Beattie's early published papers in Manchester 1899-1916

The Wehnelt interrupter. Letter to Electrician, **42**, p. 732, 1899.
The spark length of an induction coil. Phil. Mag., [5], **50**, pp. 139-148, 1900.
Notes on a possible source of error in the use of a ballistic galvanometer. Phil. Mag. [5], **50**, pp. 575-579, 1900.
The hysteresis of nickel and cobalt in a rotating magnetic field. Phil. Mag., [6], **1**, pp. 642-647, 1901.
Note on the length of the break-spark in an inductive circuit. Phil. Mag., [6], **2**, pp. 653-658, 1901.
Test coils for alternating-current wattmeters. Electrician, **48**, pp. 818-819, 1902.
The normal saturation of alternator fields. Electrician, **49**, pp. 479-480, 1902.
An electric quantometer. Electrician, **50**, pp. 383-385, 1902.
Methods of measuring capacity with alternating currents of complex waveform. Electrician, **61**, pp. 531-534, 1908.
Measurement of iron losses by short, straight strips. Electrician, **62**, pp. 136-139, 1908.
The elementary theory of the quadrant electrometer. Electrician, **66**, 12th August, 1910, pp. 729-733, + correspondence, 821-822, 910-911, 1034-1035.
Harmonic analysis diagrams. Electrician, **67**, 11th June, 1911, pp. 326-328 and 370-372, + correspondence, pp. 344, 471.
An extension of Fischer-Hinnen's method of harmonic analysis. Electrician, **67**, 8th September, 1911, pp. 847-850.
Simplified electrometer theory. Electrician, **69**, pp. 233-234, 1912.
An electrostatic quantometer. Electrician, **69**, p. 457, 1912.
Effective capacity of a quadrant electrometer when used as a quantometer. Electrician, **70**, 683-685, 1913.
Compensated electromagnet as standard of field strength. Electrician, **73**, pp. 929-930, 1914.
A permeameter for straight bars. Electrician, **76**, 3rd March, 1916, pp. 781-782.

Joint papers:

R. Beattie and P.M. Elton: *Differential ballistic methods of measuring hysteresis losses.* Electrician, **63**, pp. 299-301, 1909, and pp. 341-343, 1909.
R. Beattie and H. Gerrard. *Note on the use of Kapp's apparatus for measuring permeability of iron under high magnetising forces.* Electrician, **64**, 18th February, 1910. 7pp.
R. Beattie and H. Gerrard. *Magnetic Hysteresis at the temperature of liquid air.* Electrician, **65**, 23rd December, 1910, pp. 411-415.
R. Beattie and H. Gerrard. *A method of measuring permeability by means of alternating currents.* Electrician, **67**, 22nd December, 1911. 7pp.

Appendix 7. "Technical Electricity" - an examination paper of the Honours Physics course, June 1904

TECHNICAL ELECTRICITY

(Three hours)

1. State what is meant by the leakage factor of an induction motor, explaining carefully its exact relation to the stator and rotor leakage factors.

A 500 volt 3-phase induction motor has a leakage factor of ·0375, and takes a magnetising current of 3 amperes. Neglecting losses, calculate the maximum power which it can develop, and its maximum power factor.

2. A single-phase alternator drives an exactly similar machine as a synchronous motor, the latter developing 100 H.P. From the following data calculate (a) between what limits the voltage of the motor must lie, (b) when the action of the motor will be most stable.
Voltage of generator = 1000.
Resistance of generator armature = 1 ohm.
Inductance of generator armature = ·05 henry.
Resistance of line = 2 ohms.
Inductance of line negligible.
Efficiency of motor at 100 H.P. output = 85%.
Frequency = 50.

3. If with the field normally excited the full working pressure be suddenly applied to the terminals of a direct current motor armature at rest, show that it will run up to speed according to the law

$$n = N(1 - e^{-\frac{t}{T}})$$

where n is the speed at time t, N is the normal speed, and T is a time-factor proportional directly to the armature resistance, and inversely to the the induction factor.

4. Explain how the power factor of an inductive circuit may be raised to unity, and prove that the possibility of doing this depends entirely on the absence of harmonics in the supply voltage.

5. Prove that, except at light loads, the small secondary terminal pressure drop of a good single-phase alternating current transformer, working on a non-inductive load, may be represented by an expression of the form

$$A + BI + CI^2$$

where I is the secondary current, and A, B, C are constants which can be obtained from measurements of the unloaded transformer. (contd.

Appendix 7: "Technical Electricity" - a Physics examination paper, June, 1904

6. A 54-pole star-connected 3-phase generator, to run at 110 r.p.m. and deliver 520 amperes at 500 volts, has a winding distributed over 2 slots per pole per phase, with 1 conductor per slot. Calculate as accurately as you can, without further data, the useful flux required at no load, and estimate how many ampere-turns you would allow to compensate for armature reaction at full load with $\cos \phi = \cdot 75$.

7. Write a short account of recent developments in alternating current commutator motors.

8. Give the theory of some form of direct-reading phasemeter suitable for use on a 3-phase balanced system.

9. If the cost of buying and laying a cable can be represented by £$(a + bV + cS)$ per mile where V is the voltage, S is the sectional area of the copper, and a, b, c are constants, show how, assuming a load factor of unity, you would determine the best voltage to adopt for transmitting electrically a given amount of power to a distance, (1) on the basis of a fixed current density, (2) allowing a fixed percentage drop per mile.

10. In the wattmeter method of measuring alternating current power, determine the magnitude of the error arising from (a) the presence of self-induction in the fine-wire circuit, (b) the method of connecting up the pressure coil on one side or other of the current coil. State how you would test a wattmeter for error due to (a), and show that the error due to (b) is a maximum when the apparent power measured is equal to

$$\frac{E^2}{\sqrt{Rr}},$$

where E is the voltage and R, r are the resistances of the current coil and of the fine-wire circuit respectively.

Appendix 8. M.Sc. and Ph.D. projects carried out by first-degree graduates of the Electrical Engineering Department from its foundation until 1946.

Name	B.Sc.	M.Sc.	Ph.D.	Project
A) To 1930 (Chapter 8)				
F.S. Edwards	1921	1939	-	Calibration of sphere spark gaps for voltage measurements up to 1MV at 50Hz, and the design of a portable impulse generator for 240kV. (Work carried out externally at Metropolitan-Vickers Electrical Co.)
H. de B. Knight	1922	1947	-	Production of a thyratron for use in a particular class of high-power radar modulator.
R.H. Evans	1923	1928	-	Stresses produced in reinforcing bars by the loading of concrete T beams. (Work done in the Engineering Dept.)
H. Shackleton	1924	1937	-	Measurement of the network fault resistance of electricity supply networks. (Work done externally with the Manchester Corporation Electricity Supply Company.)
F. Roberts	1925	1927	-	Errors in the valve voltmeter.
W. Jackson	1925	1926	-	Experimental determination of the amplification efficiency of modern high-frequency transformers. (He obtained his D.Sc. in 1935.)
C.F.J. Morgan	1927	1928	-	Determination of the harmonic components of an alternating emf wave form.
C.V. Vinten-Fenton	1927	1934	-	Investigation into the torsional properties of all the British Standard Sections for rolled beams, angles and channel irons. (Work done in the Engineering Department.)
P.E. Brockbank	1928	1929	-	Phenomena accompanying "back water suppressor" in hydro-electric plant. (Work carried out in collaboration with the Engineering Department.)
H. Page	1930	1940	-	The design of a ring aerial system for broadcasting.
B) To 1938 (Chapter 9)				
F.C. Williams	1932	1933	-	Efficient detection of the audio frequency signal from a radio frequency carrier by a thermionic valve.
Beatrice Shilling	1932	1933	-	Experimental investigation of the temperature at several points of the piston of a diesel engine under operating conditions. (Work carried out in the Engineering Department.)
R.F. Cleaver	1933	1936	-	Investigation into the accuracy of high-frequency current and voltage measurements over the range 1.5 to 10 MHz.
A. Fairweather	1933	1934	-	Stability and performance of a high-frequency amplifier.
F.H. Moon	1933	1935	-	Frequency conversion in superheterodyne reception.
C.R. Taylor	1935	1945	-	Work relating to two investigations concerned with the public supply of electricity. (Work done externally with Enfield Cables.)
W.F. Taylor	1935	1940	-	Investigation of the operating temperature of pistons in a compression engine fitted with sleeve valves and "Vortex" cylinder heads. (Work done in the Engineering Department.)
J.P. Wolfenden	1936	1937	-	Voltage regulation of a mercury-arc rectifier.
R.K. Beattie	1937	1938	-	Investigation of a new system of modulation for use in radio.
A.E. Chester	1937	1938	1942	M.Sc. An investigation of wide-band amplifiers. Ph.D: Conditions affecting the extraction of electrons from cold metals at high electric fields; breakdown in vacuum.

Appendix 8: M.Sc. and Ph.D. projects in Electrical Engineering to 1946

P. Bowles	1938	1939	-	Development of wire strain gauges. (Work carried out in the Engineering Department.)
J. Highcock	1938	1939	-	Investigation into the performance of diode valves which has been suggested by analytical studies of the "shot" effect in these valves.
W.A. Cowin	1938	1941	-	Investigation into the viscous forces opposing the movement of ions and polar molecules in solid dielectric materials.

C) To 1946 (Chapter 10)

R. Cooper	1939	1941	1946	M.Sc: The electrical properties of polythene [then a new material] for use in electric cables. Ph.D: Electrical breakdown of gases when subjected to recurrent impulse voltages of short duration.
F. Horner	1939	1941	-	Dielectric parameters using cavity resonators.
G. Saxon	1939	1942	-	Dielectric properties of salt solutions; various microwave measurements.
R.B. Quarmby	1940	1941	-	Attenuation and phase-shift constants in polythene cables.
P. Dénes	1941	1943	-	Amplification of small low-frequency voltages by the modulation of a carrier oscillation.
T.A. Taylor	1941	1944	-	High-frequency electrical measurements with cavity resonators.
W. Reddish	1942	1945	-	Measurement of dielectric properties at high frequencies; high permittivity ceramics.
W. Rosenberg	1942	1944	-	An improved cosmic ray recorder. (Work carried out in collaboration with the Physics Department.)
R. Dunsmuir	1943	1945	1946	M.Sc: Various aspects of work on waveguides. Ph.D: Radiation from a rectangular waveguide projecting normally through a plane metal sheet; radiation from open ended waveguides and E.M. horns.
J. Lamb	1943	1945	1946	M.Sc: Various aspects of work on waveguides. Ph.D: Work on waveguides with respect to dielectric materials; attenuation of supersonic waves in liquids.
K.W. Plessner	1944	1945	(1947)	M.Sc: Measurement of permittivity, power factor, etc. in dielectric materials at frequencies of 600 and 1200 MHz. Ph.D: Attempt to check the theoretic prediction of H. Fröhlich and others that the intrinsic electric strength of dielectric materials should increase as the thickness is reduced.
K.M. Entwistle	1945	1946	(1948)	Investigation of the frictional forces in metals by the measurement of damping in vibrating bars. (Work carried out in the Metallurgy Department.)
J.G. Powles	1945	1946	(1948)	Cavity resonators; investigations on high permittivity ceramics at microwave frequency.

Ph.D.s in brackets are outside the time period covered in the chapter.
The following were not first-degree graduates of the department but were staff members:

J. Higham	1911	1921	-	The flickering of incandescent lamps on alternating current circuits.

(Mr. Higham graduated in the Physics Department under Rutherford, specialising in electro-technics.)

R. Feinberg	-	1942	-	Investigation of a thermionic valve invertor circuit for static conversion of direct current to alternating current.

(Dr Feinberg obtained his doctorate at Kahlsruhe before the war.)

N.B. After 1946 the Department's M.Sc. and Ph.D. degrees become too numerous to quote individually.

Appendix 9. Syllabuses of the Electrical Engineering Lecture Courses, early 1930s.

Only the syllabuses of the electrical courses are given here; not those of the lectures in mathematics, physics and mechanical and civil engineering which comprised the remainder of the course.

Electrotechnics: Second year - Professor Beattie.
A. Alternating currents.
1. Ideal alternator. Heteropolar.
2. Types of generator: homopolar, heteropolar; ring, disc, drum. Winding configurations. Spreading, Spread Factor.
3. Waveshape: Harmonic Analysis.
4. A.C. Measurements. Frequency; tuned reed. Waveshape: galvanometers; oscillographs. Peak, mean, RMS values.
5. Vectorial representation.
6. Algebra and trigonometry of periodic quantities. Products, modulation, summation. Phase difference. Three voltmeter method.
7. Alternating current circuits.
8. Inductance: toroid, long solenoid, chokes, overhead lines. Energy of magnetic field. Force between surfaces.
9. Circuits with various combinations of inductance, capacitance and resistance. Time-constants. Oscillatory circuit. Skin effect.
10. Power and power measurement. Wattmeter. Power factor.
11. Complex Circuits: series and parallel; equivalents; resonance and tuning; circle diagram; filters; complex quantities; mutual inductance; transformer, untuned and tuned.
12. Application of complex method to power problems. Exponential operator.
13. Alternator (contd.) Load characteristics; parallel running; behaviour as motor; vector diagram; stability and instability.
14. Transformer. Ratio loading; non-inductive and inductive vector treatment; winding resistances; leakage.

B. Generators and motors.
1. Dynamical induction: magnitude and direction of emf: Fleming's right hand rule. Homopolar and heteropolar: conductor mounting - radial and parallel. Faraday's Disc.
2. Heteropolar machines: voltage fluctuations; armature windings: spiral, drum, wave. Commutator. Core construction: smooth and slotted. End connections - evolute. Eddy losses. Equalising rings.
3. Field magnets: magnetic circuit. Salient and non-salient poles; leakage; ampere turns; shunt and compound windings.
4. Armature reaction: curves; tip-saturated, slotted pole, counter mmf (Ryan).
5. Commutation: use of extra winding (Swinburne) and interpoles. Linear, over, under and resistance commutation.
6. Series dynamo. Characteristics: regulation; topped and diverted field; uses of the dynamo; boosting and battery-charging; stability; external characteristics and magnetic circuit.
7. Shunt dynamo. Characteristics. Critical shunt and external resistances; armature reaction; compound dynamo characteristics.
8. Efficiencies of shunt and series dynamos.
9. Motors. Principle of reversibility; speed and torque.
10. Shunt (separately excited) motor; mechanical characteristics; Series motor: traction. Compound motor: steady speed.
11. Combinations of motors and dynamos.
12. Starting. Speed control methods: rotary convertor; rheostat, tapped field; shifting armature; shifting brushes; varying armature turns; use of two motors - series parallel; Ward-Leonard.

Appendix 9: Syllabuses of the Electrical Engineering lecture courses, early 1930s

13. Efficiencies.
14. Special uses: meters; balancing; motor generators and 3-wire systems; compounding; boosters; automatic; with batteries.
15. Dynamos (contd.). Interconnection. Output and armature dimensions. Temperature rise: heat tests; overload; practical rules. Efficiency tests: direct; indirect; back-to-back (regenerative; Hopkinson); transmission dynamometer; calibrated motor; mechanical brakes - rope, Brony. Soames; electrical brakes - dynamo, rocking field, rotating field; Field's Method.
16. Loss measurements: total losses - Threlfall; stray losses - Swinburne, Kapp, Kapp-Housman, Hopkinson; variations thereon.

Higher Theory and Design of Electrical Machines and Apparatus.
Third year - Harold Gerrard.

1. Properties of alternators.
   Load/voltage characteristic. Armature reaction: fixed and rotating (double frequency) components. Use of wave method. Cross-magnetisation.
   Vector diagram. Ordinary and synchronous impedances and their measurement.
   Voltage regulation measurement: wattless current method (Blondel and Poitier).
   Voltage control: Methods: rheostatic, induction regulator, compounding, excitation shunt (Parsons), Westinghouse, Heyland, Tirrill/BTH, Oemstead/Westinghouse.
2. Power Measurement
   Dynamometer wattmeter; electrometer (Beattie); hot-wire (Irwin).
3. Alternator (contd.)
   Waveform distortion due to flux-distribution, armature teeth, armature reaction on load. Analysis. Remedies.
   Non-sinusoidal waves: circuit behaviour; selective resonance; effect of iron.
   Power: Power factor.
   Circuit calculations: symbolic method: wattmeter problem.
4. Mutual induction: transformer.
5. Synchronous motor: vector diagram; load, long line; circle diagram; excitation, power and stability. Uses: rotary converter, power factor correction, phase-advancing. Hunting. Coupled motor/alternator. Parallel running of alternators. Synchronisation; stability.
6. Transformer
   Instruments; current; power. Vector diagrams. Effect of load, iron losses and leakage. Equivalent circuit.
   Tests: Kapp, open- and short-circuit.
   Parallel operation; connection to mains.
7. Polyphase systems/comparison.
   Advantages for generators, motors and transmission. Dependence of alternator output on number of phases. Spread factor; windings; star-delta connections; distribution factor.
8. Induction motor.
   Slip; loss; torque; power; characteristics; vector diagram; revolving fields; conditions at no-load and standstill; Heyland diagram; crawling.
   Starting: Squirrel cage; phase-wound; Lewis.
   Speed control: frequency-changing; pole-changing; rheostatic; cascade (Kramer); cascade (two motors). Hunt motor.
   Power factor improvement: transformer analogy; rotor lead; use of phase advance; Leblanc; Leblanc improved (2-phase) - application to 3-phase case; Scherbius (Brown-Boveri) system.
   Power-factor improvement on 3-phase systems: Leblanc, phase advancers, Walker (MV-Westinghouse), vibrator, Swinburne, Kapp.
   Improvement of ordinary induction motor: rotor feed, D.C., 3-phase.
9. A.C. Commutator motor.
   Shunt type: Speed control: Scherbius, AEG.
   Motors as generators: Induction, commutator, synchronous (Heyland). Series type; compensation, circle diagram; losses; power factor.

10. Repulsion motor.
   Rotating field: commutation; vector diagram; starting; Déri brush connection; circle diagrams. compensated motor (Latour). Shunt motor (Atkinson).
11. Rotary convertor
12. Electrical Design.
12.1 Closed coil drum armature windings.
   Rules: Lap, Simplex, Multiplex. Wave, Simplex, Multiplex.
   Symmetry: Equalising rings.
   Commutation: Linear, over, under.
   Separate excitation: Rosenberg dynamo; with batteries for train lighting.
   Self-excitation: arc-welding generator.
12.2 Magnetic circuit
   Calculation of dimensions.
   Determination of paths of lines of force: methods; least reluctance, Hay/Hele Shaw; Douglas: Carter.
   Air-gap reluctance; Still; Carter; Ossanna; Baillie.
   Effects of fringing; teeth; ventilating ducts.

Generation, Transmission and Distribution of Electrical Energy.
Third Year - Joseph Higham.

Economics of power transmission and distribution.
1. Load variation. Charging of consumers.
2. Station costs: fixed and variable. Unit cost.
3. Allocation of fixed cost amongst different classes of consumer. Methods: diversity factor, peak responsibility, load scheme, multiple plant methods, Quinan, Moore.
4. Tariffs: flat, restricted, fixed, multi-part (Hopkinson).
5. Metering: maximum demand indicators; Wright, Fricker, Lincoln. Summation metering; paralleling, impulse, remote reading; Midworth.
6. Number of units and location of generating plant.
   Basic factors; load, load losses, power, unbalance. Power factor correction.
7. Power and energy measurement. Wattmeters; eddy currents. Watt-hour meters.
   3-phase systems: Star, delta, balanced, unbalanced. Maximum demand; Sperti-Blacksmith, Ball, Aron, Landis & Gyr Trivector.
8. Distribution economics. Cable for least cost. Siting of substations, Carruthers' model. Feeders: boosted, unboosted.
9. Power transmission.
9.1 D.C., High voltage: Constant current, transverter, mercury vapour rectifiers, thyratrons.
9.2 A.C. (Short line). Regulation, efficiency, voltage drops; 3-phase case, Ferranti effect; locus diagrams. Star/mesh transformations.
9.3 A.C. (long line). Rigorous solution.
   Regulation, efficiency, charging current; measurement of leakage and conductor resistances; 3-phase case. Locus diagrams. Constant voltage transmission.
   Capacity of synchronous phase modifier.
   Transformer; effect on line constants; equivalent circuit.
10. Wave propagation on lines.
   Physical concept. Interpretation of line equation; forward travelling and reflected waves; velocity and attenuation.
   Heaviside condition for no distortion; minimum attenuation. Loading: continuous, lumped. Relative directions of electric and magnetic forces and motion. Fleming and corkscrew rules.
   Energy. Poynting vector; superposition; impure waves; reflection and transmission: general theory.
11. Symmetrical short-circuits.
   Alternators. Polyphase systems: method of symmetrical components. Inductance and capacitance of lines and cables.
   Proximity effect. Internal, external and mutual linkages.

Appendix 9: Syllabuses of the Electrical Engineering lecture courses, early 1930s

3-phase line; transposition.
12. Switchgear.
Mechanical forces. Ampère's rule. Laplace's equation. Finite wire; infinite parallel wires.
Inductance method (energy). Spiral spring; infinite parallel wires; solenoid; circular wire; helix; finite parallel wires.
Switch crossbar; perpendicular and inclined conductors.
13. Protective gear. Lines and alternators.
Similar devices are used for protection against external faults (lines) and internal faults (alternators). "Standby" protection provided by "back-up" alternators.
14. Fire. Protection by modern closed-circuit air cooling systems. Oxygen supply is limited and they involve air cleaning, dry and wet.
15. Line and alternator faults.
Line fault due to earthed neutral can be dealt with by earthing transformer; zig-zag reactor; Petersen coil.
Alternator faults can be due to earth (frame); between phases; between lines in same phase; due to excitation failure; due to faulty prime-mover.
Protective devides include: overload trip relays; direct, battery, A.C. instantaneous, delayed (Buchholz); inverse time (Statter).
Reverse power relays: wattmeter (Everett Edgcumbe), differential (BTH); core-balancing (Ferranti-field); balanced current/voltage (Merz-Price); self-balancing (Beard); circulating current (McColl). Combinations: differential and reverse-power (McColl); balanced voltage and balanced core (Ferranti-Hawkins). "Pilot" systems: screened (Beard-Hunter); "mock pilot"; self-compensating (BTH); "split" pilot (Reyrolle); "Translay" (Met-Vick); Pilotless; 4-core (Callender-Hunter).

Higher Frequency Currents. Third Year - Frank Roberts.

A. Thermionic valves.
1. Thermionic emission.
"Work function". Types of filament: tungsten; thoriated tungsten; oxide-coated. Operating temperature. Vacuum; residual gas; gettering. Collecting electrodes.
2. Diode.
Space charge; planar and cylindrical electrodes; "soft" tubes.
3. Triode.
Action of grid. Equivalent diode. Penetration and magnification (amplification) factors. Characteristic curves. Mutual conductance. Voltage amplification; effect of anode resistance. "Short path" valves.
Secondary emission: Dynatron. Inter-electrode capacitance: disadvantageous effect in radio frequency amplification. "Neutrodyne" circuit and screen-grid valve. Parallel operation.
4. Tetrodes.
Types: space charge grid; screen grid: pliodynatron; variable μ; beam
5. Pentode.
6. Hexode.
B. *Circuits. transmission and reception.*
1. Oscillatory circuit.
Decrement. Energy transfer. Condenser leakage.
Resonance. Series and parallel. Tapped core.
2. Coupled circuits.
Mesh coefficients. Transformer-coupled amplifier. Equivalent circuit. Frequencies for maximum amplification. Optimum power output.
High frequency (HF) amplifier: neutralisation.
Low frequency (LF) amplifier: requirements for uniform response: Causes of distortion. Filters.
3. Instability of H.F. amplifier.
Condition for self-oscillation.
4. Oscillation generator.
Conditions for inception and for sinusoidal output. Various circuits.

5. Modulation and detection.
6. Amplitude modulation.
   Carrier. sidebands. Single sideband transmission. Signal inversion. Privacy. Methods: choke, grid and absorption.
7. Detection.
   Types: square law (small signal); linear (power); peak. Methods: small signal anode bend; small signal leaky or cumulative grid; power anode bend; peak diode.
8. Superheterodyne reception.
9. Automatic volume control.
   Types: simple, delayed, amplified.
C. *Electromechanical instruments.*
   Vibration galvanometer. Analogy with coupled circuit. Butterworth circuit. Telephone receiver. Moving coil speaker.

Higher Pure Electricity and Magnetism. Third year - Frank Roberts.

1. Electromagnetism.
   Electrostatics; magnetostatics (magnetism); electrodynamics. Reciprocal relationships: magnets in motion/electricity at rest. Electricity in motion/magnets at rest.
   Interpretation in terms of stress in a medium (ether): electric force/stress; magnetic force/change in stress.
2. Maxwell's laws (experimental).
   Torsion balance; Coulomb's law (inverse square). Unit of charge. Cavendish. Electrostatic system of units.
3. Electrostatic field.
   Intensity. Line of force. Potential. Equipotential surfaces.
4. Mathematical expressions of the inverse square law.
   Potential a scalar function - conservation of energy. Continuous distributions.
5. Gauss's Theorem and corollaries.
6. Cavendish proof of the inverse square law.
7. Equations of Poisson and Laplace. Physical meaning.
8. Deductions from the inverse square law.
9. Tubes of force. Coulomb theorem. Electric doublet.
10. Conductors and condensers.
    Spherical and cylindrical conductors; spatial and surface intensities. Spherical, cylindrical and plate condensers; electrical and mechanical forces; energy; dielectrics.
11. Systems of conductors.
    Unique relationship between charge distribution and individual charges and potentials. Principle of superposition. Green's reciprocation theorem and corollaries. Coefficients of potential, capacity and induction. Mechanical forces.
12. Electrometers.
    Types: Attracted disc; quadrant; Crompton; Hofmann; "Duant"; String.
13. Dielectrics and specific inductive capacity.
14. General theory of stresses in a medium at rest.
    Energy; Maxwell's displacement theory.
15. Determinants.
    Simple theory, six theorems. Generalised Kronecker delta; Einstein summation convention; linear equations; Cramer's method.
16. Electrolysis.
    Chemical definitions; Faraday's laws; mechanism.
    Polarisation; dissociation; ionic migration; Hittorf ratio; Kohlrausch equation.
    Thermodynamics and reversible cells. Standard cells: Daniell, Weston, Clark. Concentration cells.
17. Seebeck, Peltier, Thompson Effects. Laws of intermediate metals and temperatures. Power: Thermodynamics and Thompson effect. Thermoelectric diagram. Neutral temperature.

Appendix 10. Laboratory experiments done by Electrical Engineering students, late 1940s and early 1950s

D.C. machines laboratory:

First year:
Voltage characteristics of separately-excited shunt-wound and compound-wound d.c. generators.
Tests with a 'three brush' car dynamo.
The direct current series motor.
The ballistic galvanometer and the measurement of field strength.
The measurement of insulation resistance.
The running down test.
Experimental analysis of armature losses in a direct current dynamo or motor.
Measurement of the leakage factor of a dynamo or motor.
Photometry and illumination.
The metadyne.
Ammeter readings on various wave forms.
Separately-excited shunt-wound and compound-wound d.c. motors.
The Kelvin double bridge.
The potentiometer.
Ewing's hysteresis tester.

Second year:
Determination of the B-H curve of iron by a ballistic method.
Fusing currents.
Determination of the characteristics of two element or diode valves.
Testing a d.c. ampere-hour meter.
Determination of the characteristics of triode valves and measurements of the internal impedance and amplification factor.
Measurement of the hysteresis loss in iron by a ballistic method.

A.C. machines laboratory:

Second year:
Measurement of inductance and capacitance.
Measurement of mutual inductance.
Vector diagrams, e.m.f. and current.
Resonance in a series circuit.
Resonance in a parallel combination of inductance and capacitance.
Characteristics of a single-phase alternator.
Transformer tests.
Measurement of hysteresis and eddy current loss in sheet iron.
Three-phase induction motor tests.
The three-phase synchronous motor.
Waveform analysis: arithmetic method.
Resonance in coupled circuits.
Three-phase induction motor tests.
The three-phase synchronous motor.

Electronics laboratory (second year):
Richardson's law of thermionic emission and the three halves power law.
Valve characteristics.
Characteristics of a cathode-ray tube.
Cathode-ray tube as a tracing instrument.

The cathode follower.
The tuned-anode oscillator.
The pentode
The hexode frequency changer.
Multivibrators.
Time bases.
Resonance measurements.
Negative feedback.
Measurement of the frequency response of a single-stage triode amplifier.
Characteristics of a gas-filled triode.
Rectification with gas-filled triodes.
The glow discharge tube.
The photocell.

Third year electrical laboratory (prior to and overlapping the start of the project):
The Shrage motor.
The non brush-shifting variable-speed, three-phase, shunt-type, commutator motor.
The compensated single-phase series motor.
The repulsion motor.
General high-voltage measurements.
The six-phase grid-controlled mercury-arc rectifier.
The high-voltage Schering bridge.

The **project** began about Christmas of the final year, overlapping with the last 'set' experiments.

Experiments in the Engineering Department:
Strength of materials laboratory (first year):
Tensile tests on mild steel specimens using a Buckton machine and a Cambridge extensometer.
Ewing's extensometer for a rod.
Comparison of the moduli of elasticity of mild steel and aluminium.
Ewing's extensometer for a wire.
Transverse test on cast iron specimens.
Determination of the modulus of rigidity of a mild steel bar.
Relationship between torque and angle of twist for a brass bar.
Tests using a Jackman oil-pressure Brinell hardness machine.
Avery Rockwell - type hardness machine.
Vickers pyramid hardness testing machine.
Tests with the Izod impact machine.
To check the calibration of the two gauges provided.
Complete torsion test on a steel bar.
Determination of the elastic constant for the material of a close-coiled helical spring (static method).
Experiments on a centrally loaded beam.
Tests on a cantilever beam.
Experiments on a beam loaded by overhanging weights.
Tests on a 100-ton Buckton testing machine.
Tests on an Avery 30/12 ton universal dial indicating testing machine.

Drawing office classes: 3 hours per class on Saturday mornings for one term.

Heat Engines laboratory (second year):
Calorific value of coal using Darling's calorimeter.
Calorific value of fuel oil using a Mahler-Cook Bomb calorimeter.
Calorific value of coal gas (Orsat apparatus) and flue gas analysis.
Testing arrangements (engine trials) of four engines: National Gas Engine, Mirrlees Diesel Engine, Corliss Tandem Compound Steam Engine, and Musgrave Zoelly Turbine and Boiler.

Appendix 11. Student and graduate Engineering Societies at the University

The **Engineering Society** was founded in 1883. It is, and always has been, a Society run for students by students, and very large numbers have enjoyed and benefited from its social and other activities. At various periods over the years a magazine containing news and high-quality articles of technical and non-technical interest has been produced, and hundreds of visits to engineering works and other venues have been arranged.

In 1977 a group of Electrical Engineering students requested that an **Electrical Engineering Society** be formed, the reason being that the link between Electrical Engineering and the other branches within the university had by then become rather tenuous and the students felt it would be advantageous to have a society which addressed their interests more directly. A constitution was drawn up and approved by the University Registrar, and the Society was duly formed. It operated successfully for several years. However, on the creation of the Manchester School of Engineering in 1994 the various branches of Engineering were once again drawn more closely together and it was decided that a separate society for Electrical Engineering students was no longer necessary. Consequently the Society was merged with the Engineering Society which then once again, as in years gone by, served students in all branches of Engineering.

The **Osborne Reynolds Society** was founded in 1955 to enable Engineering graduates of the University to keep in touch with the various engineering disciplines at the university and with each other. There are social and other activities from time to time and a Newsletter is published periodically. Any Engineering graduates (including Electrical Engineering) interested in the possibility of joining the society should contact:

The Secretary
Osborne Reynolds Society
The Manchester School of Engineering
The University of Manchester
Oxford Road
Manchester M13 9PL

Index

This index includes the names of people mentioned in the Preface and in the text of the 15 chapters.

Page numbers in italics refer to photographs or (in the case of John Owens) an illustration.

Titles (Dr, Professor, Sir, etc.) are not ascribed to Owens College or Manchester University staff in the index as the titles usually changed as the individual's career progressed.

Commercial firms are included only if they had a strong connection with Electrical Engineering at Owens College or the University, or made donations.

Abbosh, F.G., 223, 237, *239*, 263
Alderman, J.K., 186, 187, *193*
Alexander, Samuel, 58
Allanson, J.T., 159, 161, 169
Allen, Jack, 31, 34, *141*
Allen, John, 186, 187, *193*
Al-Zahawi, B., 278, 282
Andrade, E.N. da C., 67
Andrews, Colin, 273
Antonoff, G.N., *73*
Arkwright, Sir Richard, 3
Ashton, C.C., 135
Ashton, Thomas, 26, 27, 28
Ashworth, J.R., *64*
Aspinall, David, 211, 212
Aspinall, H.T., 122
Asquith, Herbert, 107
Atherton, D.P., 215
Atkinson, S.G., *64*
Auckland, D.W., 223, 237, *239*, 246, 247, 264, *271*, 272, 275, 278, 282, 287
Ayrton, Professor W.E., 42

Bain, Alexander, 22
Baird, J.L., 140,
Baker, H. Wright, 118, 124, *141*
Baker, Mrs H. Wright, see Drew, Kathleen M.
Baldwin, C.T., 154
Baldwin, S.T., 184
Balfour, Arthur, 107
Bamber, James, 159
Bamford, Harry, 34
Barraclough, Michael, 193
Barton, Mr, 180

Basterfield, Alderman George, 139
Bateman, H., *73*
Bateman, J.F., 4
Baumbach, Otto, 83
Beakley, J., *64*
Beale, Maurice, 267
Beard, J.R., *64*
Beatles, The, 220
Beattie, Elizabeth, née Watt (Robert Beattie's mother), 44
Beattie, Janet (Robert Beattie's infant daughter), 112
Beattie, Janet E. (Mrs Janet Paterson, Robert Beattie's daughter), 112, 141, 143, 144, 150, 151-152
Beattie, Robert, senr. (Robert Beattie's father), 44
Beattie, Robert, ix, x, 26, 43, 44, 46, 57, 58, 60, *64*, 68, 72, *73*, 74, 75, 76, 80, 83, 87, 88, 90, 91, 94, 99, 100, 110, 111, 112, 115, 116, 117, 119, 120, 124, 126, 127, 130, 131, 136, 140, 141, *141*, 143, *143*, 144, 146, 149, 150, 151-152, 223, 277, 285
Beattie, Mrs Robert (née Janet Kyd), 90, 112, 150, 151
Beattie, Robert Kyd (Robert Beattie's son), 112, 140, 141, 142, 143, *143*, 144, 147, 150, 151-152
Beatty, Admiral Sir David, 106, 107
Beatty, Lady, 106
Beaumont, J.W., 50, 52, 189, 197
Beebe, J.R., 263
Bell, A.H., *64*
Beyer, C.F., 23, 33, 76

Index

Beyer, Peacock & Co., 23, 33
Binney, Edward, 4, 5
Blackett, P.M.S., 135, 146, 147, 162, 165, 166, 171, 180
Boardman, Ellis, 179
Bohr, N.H.D., 67, 68
Borodowsky, W., *73*
Boswall, Mary, see Jackson, Mrs Willis
Boswall, R.O., 146
Bowden, K.F., 211, 212
Bowles, Percy, 148
Boyd, C.A., 267, *271*, 281
Bradshaw, Lawrence, *64*
Bragg, W.H., 104, 105, 107, 108, 109, 117
Bragg, W.L., 117, 119, 124, 125
Braudo, C.J., 154
Brett, Lieut. G.H., 105
Bridge, Philip, 263
British Insulated Callender's Cables Ltd., 216, 248
British Westinghouse Electric and Manufacturing Co. Ltd., 44, 52, 54, 99
Broadbent, Marguerite, 253
Broadbent, T.E., 187-188, 193, 197, 198, *199, 200*, 211, 216, 231, 237, *239*, 245, 247, 253, *271*
Brockbank, P.E., 127
Bromilow, Joseph, 141
Brooker, R.A., 204, 206, 208, 211, 212
Brown, S.G., 105
Buccleuch, 7th Duke of, 102
Buchanan, D.G., 223, 238
Buckingham, Frederick, 187
Bunsen, R.W., 14, 16
Burke, J.B.B., *64*
Butterworth, Stephen, 72

Cambridge Scientific Instrument Co., 82
Camm, Donald, 243
Carnegie, A.Q., 105
Chadwick, James, 67, 72, *73*, 101, 120
Chalmers, B.J., 273
Chaplin, G.B.B., 193
Chapman, Sydney, 118, 119, 125
Charnley, John, 138
Cheetham, J.F., 75
Chester, A.E., 116, 141, 142, 147, 158, 162
Child, J.M., 124
Chloride Electric Storage Co., 55
Chopin, Frédéric, 9

Chrishop, I.F., *200*
Churchill, W.S., 107
Clark, Sir James, 18
Clarke, Margaret E., *271*
Clay, Thomas, 28
Cleaver, R.F., 131, 134, 135
Clegg, Neville, 75
Clegg, W.W., 223, 238, *239*, 247, 265, *271*, 281
Clifford, G.H.W., 124
Clifton, Charles, 33
Clifton, R.B., 17, 18, 19, 20, 21, 36
Coates, W.H.R.A., 123
Cobden, Richard, 10, 33
Cochran, James, *64*
Cockroft, John, 158, 209
Cole, D.J., 263
Collier, John, 60
Collins, G.C., 154, 184
Colton, Robert, 71
Cook, C.W., 82
Cook, Gilbert, 71
Cooke, W.F. (Wheatstone's collaborator), 4, 22
Cooper, G.A., 128, 130, 133, 136, 140, 141, 142, 146, 156, 159, 160, 164, 166, 186, 196, 197, 226, *239*, 242-243, 247
Cooper, Ronald, 148, 149, 156-163, 165, 166, 167, 169, 170, 193, 211, 216, 223-224, 237, *239*, 244, 246, 247, 248, 264, 286
Cooper, W. Mansfield, 229-230, 231, 241
Core, T.H., 37, 38, 40, 41, 42, 46, 60, *64*
Cornish, R.J., 131
Cotton, Henry (known as Harry), 72, 131
Cowin, W.A., 147
Cowling, T.G., 159
Craggs, J.D., 185
Crank, John, 135
Crankshaw, H.M., *64*
Crippen, Dr H.H., 94
Crompton Electrical Co., 41
Crompton, Samuel, 3
Crookes, William, 35
Crossland, Sir Bernard, 273
Crossley, A.W., *64*
Crossley, W.J., 54
Cunningham, M.J., 223, 235, *239*, 247, 263, *271*, 278, 279, 281
Curtis, C.D., 273

343

Dalton, John, 2, 3, 4, 6, 9, 58
Dancer, J.B., 8
Darbishire, O.V., *64*
Darwin, C.G., 67, 70, *73*, 75
Daskalakis, Costas, 269
David, A.E., 138
Davidson, Robert, 22
Davies, John, 6
Davy, Sir Humphry, 2
Dean, H.R., 119
Delépine, H.G.S., 113
Dénes, Peter, 153, 154, 168
Devonshire, 7th Duke of, 28, 47
Dewhurst, C.B., 46, 59, 60, 71, 119, 123, 125
Diamond, Jack, 194
Dickie, A.C., 119
Dixon, H.B., 119
Donnachie, Alexander, 272, 276
Douce, J.L., 215
Dowson, Duncan, 275
Drew, Denis, *200*
Drew, Kathleen M. (Mrs Wright Baker), 124
Dryden, S., 167
Duffield, W.G., *64*
Dunkerley, Stanley, 60, 69, 70
Dunn, R.H., 123
Dunsmuir, Robert, 161, 162, 166, 167, 169
Dunstan, E.M., 211, 212
Durnford, John, 154
Dye, D.W., 131
Dylan, Bob, 220

Eastham, J.F., 211, 214
Eccles, N., *73*
Eddington, J.S., *64*
Edwards, C.A., 117, 119
Edwards, D.A., 223
Edwards, D.B.G., 179, 181, *182*, 185, *193*, 208, 211, 212
Edwards, F.J., 186, *193*
Edwards, F.S., 122, 127, 216
Egerton, Lord Francis, 3
Elizabeth, H.M. the Queen (now the Queen Mother), 197
Elliott, C.T., 211, 237
Elton, P.M., *64*
Engels, Friedrich, 9
Entwistle, K.M., 162, 187, 273, 274

Erwood, A.F., 269, *271*, 282
Evans, E.J., *73*, 75
Evans, H.R., 273
Evans, R.H., 123, 127
Eve, A.S., 101, 109, 110

Fairbairn, William, 4, 20
Fairweather, Alan, 130, 133, 134, 135, 142, 147
Faraday, Michael, 2, 3, 18, 22, 69, 134, 289
Farmer & Brindley, 110
Farrell, P.G., 246, 247, 249, 253, 259, *260*, 267, 270, *271*, 272, 273, 275, 276, 278, 281, 287
Faulkner, George, 4
Faulkner, John, 10
Faulkner, John (early research worker on electromagnets), 35
Feinberg, Raphael, 148, 152, 153, 161, 168, 184, 185, 188
Ferranti Ltd., 41, 44, 180, 181, 183, 184, 185, 202, 203, 204, 205, 206, 207, 208, 209, 210, 251
Ferranti, S.Z. de, 80, 131
Fiddes, Edward, 87
Finniston, Sir Montague, 256, 257
Fleming, A.P.M., 99, 146
Fleming, Professor J.A., 44
Florance, D.H., *73*, 101
Fortesque, C.L., 166
Foster, J.F., 13
Frankland, Edward, 12, 13, 14, 15, 16, 19, 277, 285
Fraser, Captain, 108
Frith, Julius, *64*, 124, 125

Gambling, W.A., 273
Gannon, William, 42, 43, *64*
Garfitt, Dorothy E.M., 185
Garner, C.D., 273
Gaskell, Elizabeth C., 9
Gauss, K.F., 22
Gee, W.W. Haldane, *64*
Geiger, Hans, 62, 67, 68, 70, 72, *73*, 75, 82, 101, 117
George IV, H.M. King, 69
George V, H.M. King, 107
Gerrard, Harold, x, 72, *73*, 75, 76, 83, 88, 91, 94, 99-111, 114, 115, 116, 117, 120,

344

Index

124, 125, 126, 127, 128, 130, 139, 147, 149, 153, 168, 184, 185, 187, 193, *193*, 194, 195
Gerrard, Percy, 114
Gibson, A.H., 60, 71, 119, 123, 124, 131, 132, *141*, 149, 152, 186
Gill, Thomas, 123, 124
Gladstone, Murray, 27
Gladstone, W.E., 27
Glazebrook, R.T. (later Sir Richard), 101
Gledson, Arthur, 186, *233*
Good, I.J., 176
Gough, T., *64*
Gourley, S.F., 223, 238
Grassot, M.E., 68
Greengrass, Mary E., *64*
Greenwood, H.C., *73*
Greenwood, J.G., 12, 13, 15, 18, 27, 277
Gregory, R.D., 273
Grieg, J., 136
Griffiths, Albert, 42, 46, *64*
Griffiths, James, *64*
Grime, R.E., *64*
Grimsdale, R.L., 193
Grindley, J.H., 59
Grundy, Eric, 180
Gurd, J.R., 272, 273

Hall, H.E., 273
Hall, W.B., 194
Hallé, Charles, 9
Halsbury, Lord (3rd Earl), 183
Hardy, C.J., 223, 238, *239*, 251, *254*, 265, *271*, 278, 281
Hardy, D.R., 185, 188, 193, *200*, 215, 216, 243
Hargreaves, James, 3
Harker, J.A., *64*
Harlan, Frank, 185
Harris, M.B. (Vice-chancellor), 276
Harris, R.A., 223
Harrison, C.G.M., 265, 269, *271*, 278, 282
Harrison, Eleanor, see Mrs Edward Stocks Massey
Hart, Michael, 273
Hartog, P.J., *64*
Hartree, D.R., 126, 132, 135, 162
Hartree, Mrs D.R., 132
Harty, H. Hamilton, 105
Harvison, R.E., 179

Harwood, W.A., *73*
Havekin, Thomas, 123, 153, 154, 184, 185-186, 187, 194
Hawke, J.R., 223, 264
Haxby, Roger, *239*
Hayton, J.W.P., *64*
Heaviside, Oliver, 131
Helmholtz, H. von, 39
Hemsalech, G.A., *64*
Henry, D.C., 125, 164
Henry, T.A., 125, 164
Henry, Thomas, 2
Herbert, A.M., *64*
Herbert, Rev. William, 4
Hesmondhalgh, D.E., 193, 211, 214, 235, *239*, 247, *254*, 261, *271*, 282
Hewitt, Ronald, 186
Hewlett, Edith L., *64*
Heymann, F.F., 185
Heywood, James, 9
Higham, Charles J, 120
Higham, Joseph, x, 72, 120, 125, 126, 127, 130, 136, 137, *141*, 142, *143*, 147, 149, 153, 155, 168, 171, 184, 185, 186, 187, 188, 189, *193*, 193-194, 195, 197, 202
Highcock, John, 142, 147
Hill, E.W., 265, *271*, 278, 281
Hirst, Harry, *64*
Hitler, Adolf, 148
Hodgkinson, Eaton, 4
Hoffman, G.R., 193, 211, 223, 226, 238, *239*, 245, 246-247, 248, 251, 265
Holme, Edward, 4
Holmes (electric lighthouse), 22
Hopkinson, Alfred, 50, 74, 75, 88, 89, 99
Hopkinson, Alice (John's daughter), 51
Hopkinson, Charles, 51
Hopkinson, Edward, 51, 52, *64*, 99
Hopkinson, Mrs Evelyn (John's wife), 51
Hopkinson, Alderman John, 50, 51, 52
Hopkinson, Mrs (Alderman Hopkinson's wife), 51
Hopkinson, John, 41, 50-52, *53*, 54, 74, 75, 76, 94, 99
Hopkinson, Lena (John's daughter), 51
Hopwood, F.L., 105, 108
Horne, M.R., 194
Horner, Frederick, 159, 160, 169
Horner, J.W., 122
Hoyle, David, *64*

345

Hughes' telegraph, 22
Hutton, R.S., 57, 58, 62, *64*, 68, 69, 70, 75, 76
Hutton, Mrs R.S., see Schuster, Sybil

Ingham, John, 135
Ionides, Captain A.G., 109
Irvine, J.M., 272
Iyer, S.N., *239*

Jack, William, 36, 37
Jackson, A.F. (Alex), 141
Jackson, Brian, 216
Jackson, J.D., 31, 35, 272, 276
Jackson, J.M., 132
Jackson, Willis, 123, 124, 127, 143, 144, 146-149, 152-163, 166-171, *171*, 175, 223, 244, 285, 286, 287
Jackson, Mrs Willis (née Mary Boswall), 146
Jackson, W.H., *64*
James, R.W., 131
Jánossi, Lewis, 162
Jeffreys, D.C., 211, 238
Jellicoe, Admiral J.R., 107
Jevons, W., 126
Jones, B.E., 223, 235, 237, *239*, 263
Jones, Eric, 124
Jones, E.A., 123
Jones, F. Wood, 163
Jones, Henry Longueville, 9
Jones, P.L., 193, 211, 238, *239*, 251, *271*, 278, 281
Joule, J.P., 4, 5, 6, 7, 8, 9, 23, 58

Kastner, L.J., *141*
Kay, William, *73*, 88, 112, 113, 121, 169
Kelvin, Lord, see Thomson, William
Kennedy, Sir Alexander, 71
Kennedy & Donkin, 71
Kennington, T.B., 52
Kenyon, G.H., 128, 132, 240
Kilburn, Tom, *Front cover*, 72, 135, 173-185, *179*, *182*, 188, *193*, 204, 205, 207, 208, *208*, 210, 211, 212, 223, 244, 286, 287
King, David, 223, 238, *239*, 265, *271*, 278, 281
Kingdon, Francis, 38
Kinniment, D.J., 211, 212, 223

Kinoshita, S., *73*
Kirchhoff, G.R., 38
Knight, H. de Boyne, 122, 127
Knowles, Vincent, 243
Kyd, Janet, see Beattie, Mrs Robert

Laithwaite, E.R., 185, 188, 193, *193*, 194, 195, 204, 211, 212, *213*, 214, 235, 286
Lamb, Horace, 36, 40, 59, 117, 119, 125
Lamb, John, 161, 162, 166, 167-170
Lander, C.H., 71
Lang, W.H., 136
Langworthy, E.R., 38, 76
Lanigan, M.J., 193, *208*
Latham, Herbert, *200*
Launder, B.E., 273
Lautsberry, W.C., *64*
Lavington, S.L., 176, 177, 205, 206, 209, 231
Lawrenson, P.J., 287
Leavers, Violet F., 267, 278, 281
Léclanché's primary cell, 23
Lees, C.H., 42, 46, 58, 60, *64*
Leigh, John, 5
Leonard, J.L., 215
Lilley, Peter, 223, 238, *239*, 265, 271, 278, 281
Liszt, Franz, 9
Litting, C.N.W., 171, 184, 185, 186, 188, *193*, 196, 211, 238, *239*, 241, 247
Littler, T.S., 154, 187
Livesley, R.K., *193*
Lockspeiser, Sir Ben, 180, 181
Lodge, Oliver, 100
Lord, James, *64*
Lovell, A.C.B., 165, 187
Lustgarten, J. (later Langton, J.L.), *64*
Lutte, N.P., 223, 237

McCaig, Malcolm, 162
Maccall, W.T., *64*, 131
McCormick, Joseph, 140, 141, 160, 186
McGrory, P.J., *239*
McGuffie, Sheila E., 128, 134
McLean, G.W., 193, 211, 214, 235, *239*, 247, 261, *271*, 272
Makinson, William, 139, 140
Makower, Walter, *64*, *73*
Marchant, E.W., 99
Marchington, J.M., 137, 138

Index

Marsden, Ernest, 67, 72, *73*, 112
Marsh, Arthur, 175
Mason, C.M., 71, 123, 125, *141*, *193*, 194
Mason, William, *64*
Massey, Alderman, 76
Massey, Charles, 76
Massey, Edward Stocks, 76, 77, 78, 79, 80, 87, 88, 125, 144, 146, 174, 175, 241
Massey, Mrs Edward Stocks (née Eleanor Harrison), 77
Massey, H.S.W., 136
Mather, William, 51, 52
Mather & Platt Ltd., 41, 44, 51, 55, 87, 99
Matheson, J.A.L., 128, 132, *141*, 186, 187, 194, 240
Mathieson, Robert, 186, *193*
Matteui's mica capacitor, 22
Maxwell, J. Clerk, 22, 39
Meek, J.M., 154, 287
Mendelssohn, Felix, 9
Merrifield, C.S., 275
Merriman, J.H.H., 245
Metcalfe, J.S., 275
Metropolitan-Vickers Electrical Co. Ltd., 44, 134, 135, 146, 147, 154, 156, 157, 168, 185, 202, 207, 216, 242
Michaelson-Yeates, P.G., 247
Middleton, B.K., 246, 247, 259, 261, 265, *271*, 278, 281, 287
Middleton, Commander, 108
Miers, H.A., 88, 117
Miles, J.J., 265, *271*, 278, 281
Millar, J.B., 32, 34, 46, 59, 60
Millikan, R.A., 289
Montague, Peter, 272, 273, 275, 276
Moody, N.F., 173
Moon, F.H., 134, 135
Moorby, W.H., 46
Moore, S.A., 275
Mordell, L.J., 125-126, 163
Morgan, C.F.J., 126, 127
Morris, Derrick, 206, 211, 212
Morse, Samuel, 22
Moseley, H.G.J., 67, *73*, 75, 101, 104, 113
Moullin, E.B., 134, 142, 146, 161
Mucklow, G.F., 124, 131, *141*
Mudge, J.L., 223, 238

Naylor, G.A., 133
Needham, R.M., 273

Newbery, Edgar, 109, 111
Newman, M.H.A., 176, 181
Nicholls, Agnes (Mrs Hamilton Harty), 105
Nicholson, Miss, 135
Nicholson, Sir Robin, 273
Nield, William, 18
Nix, G.F., 193, 211, 214, 235, *239*, 247, 261, *271*
Noble, S.W., 173
Normington, A.C., 153, 154, 184
Nuttall, J.M., 67, 72, *73*, 125, 140, 162

Oakes, W.G., 161, 167
Oakley, J.P., 269, *271*, 278, 282
O'Farrell, Timothy, 267, 281
Oliver, Martin, 157
O'Neill, F.H.H., *64*
O'Sullivan, J.K., 186, *193*
Owen, Robert, 2
Owen, Thomas, 157
Owens, John, 9, 10, *10*, 26, 28, 30, 76

Paganini, Nicolò, 9
Page, Harold, 123, 127
Paget, Sir Richard, 102, 106
Parsons, Charles, 78
Partridge, Lieut. G.F., 109
Paterson, Janet E., see Beattie, Janet E.
Payne, R.B., 211
Pear, T.H., 119
Percival, Thomas, 2
Petavel, J.E., *64*, 68, 69, 70, 71, 117, 118, 119, 124
Pickard, R.M., 223, 238, *239*, 247, 251, 265, *271*
Pickford, Fred, 60, 71
Pickup, I.E.D., 223, 235, *239*, 261, *271*, 278, 282
Piggott, L.S., 174, 185, *193*, 211, *239*
Planté's secondary cell, 22
Playfair, Lyon, 4, 7
Plessner, K.W., 161, 162, 167
Plowman, J.C., 179
Poll, D.I.A., 275
Popplewell, Cicely M., 183, 203, 211, 212
Porter, Arthur, 132, 135
Powles, J.G., 161, 162, 167, 169, 170
Poynting, J.H., 38
Princess Royal, H.R.H., 259, *260*

Pring, J.N., *64*, *73*, 75
Prinz, D.G., 184
Prowse, W.A., 167
Pye, F.W., 103
Pyman, F.L., 119

Quarmby, R.B., 160
Quayle, R.S., 223, 238, *239*, 251, 265, *271*, 278, 281

Radford, Katharine, *64*
Rankine, W.J.M., 32-33
Rankins, A.O., 109
Raven, Arthur, 123
Rawcliffe, John, 235
Rawlinson, Eric, 185
Rayleigh, Lord (3rd Baron), 39, 54
Read, F.H., 275
Reddaway, Messrs. F., 52
Reddish, Wilson, 161, 166, 167, 169
Reece, A.J.B., 247
Rees, David, 176
Reid, S.R., 273
Reynolds, Osborne, senr., 30, 31
Reynolds, Osborne, 23, 26, 30-36, *34*, 40, 46, 59, 60, 69, 70, 83, 128, 175, 276, 290
Reynolds, Mrs Osborne (Osborne Reynolds' first wife), 31
Richardson, Charles, 121, 128, 140
Riding, Geoffrey, 212
Riley, Clifford, *64*
Ritson, F.J.U., 173
Robbins, Lord L.C., 199, 220, 221
Roberts, Frank, 123, 127, 128, 130, 131, 134, 137-139, 140, 142
Roberts, Mrs Frank, 131, 139
Robertson, Andrew, 71
Robinson, A.A., 179, *182*, 183, 203
Robinson, H.R., 67, 72, *73*
Robinson, John (early 19th century inventor), 23
Robson, B.T., 273
Rochester, G.D., 162
Rolls, Hon. C.S., 70
Röntgen, W.K. von, 39
Roscoe, H.E., 16, 17, 18, 19-21, 26, 27, 31, 37, *64*, 277, 285
Rose, Lieut. H.C., 105
Rosenberg, Waldemar, 153, 154

Rossi, Roberto, *73*
Round, H.J., 106
Royce, Sir Henry, 70, 164
Royds, Thomas, *64*
Rushton, Eric, 123, 124
Russ, Sidney, 62, *64*, *73*
Rutherford, Ernest, 63, 67, 68, 69, 70, 72, *73*, 74, 75, 99, 100, 101, 102, 103, 104, 106, 111, 112, 113, 117, 120, 125, 285
Ryan, Commander C.P., 102-107
Rybner, Joergen, 136

Sale, F.R., 272, 273
Sandeman, Archibald, 12, 13, 17, 18-19, 36
Sandeman, Edward, 123
Sandoz, D.J., 246, 247, 263
Saunders, S.M., *64*
Saunderson, Constance M., *64*
Sawyer, W.W., 168
Saxon, Godfrey, 161,
Saxton, J.A., 160
Schilling (telegraph), 22
Schuster, Arthur, ix, 38, 39, 40, *40*, 41, 42, 43, 46, 49, 50, 51, 52, 54, 57, 58, 60, 62, 63, *64*, 67, 69, 70, *73*, 80, 102, 125, 285, 290
Schuster, Mrs Arthur, 62, *64*
Schuster, Leonard, *64*
Schuster, Sybil (Arthur Schuster's daughter, later Mrs R.S. Hutton), 57, 70
Schwartz, A., xi
Scott, A.J., 12, 13, 14, 15, 277
Sedgwick, Adam, 7
Shackleton, Sir Ernest H., 131
Shackleton, Herbert, 123, 124, 125, 127
Sheldon, R.A., *64*
Shilling, Beatrice, 128, 133, *133*, 134, 170-171
Shuttleworth, Roger, 223, 238, 264, *271*, 278, 282
Siemens, Werner von, 23
Simon, Henry, 53
Simpson, G.C., 62, *64*
Singer, K.E., 273
Sinnot, Mr, 108
Slade, R.E., *73*
Smee, J.F., 184, 185
Smith, F. Graham, 272
Smith, H. Bombas, 119

Index

Smith, I.M., 275
Soddy, Frederick, 67
Solomons, A.M., 263
Somerville, M.J., 193, 215, 216
Spooner, J.R., 186, *193*
Stamford, N.C., 154, 159, 160
Standing, Basil, 186, 187, *193*
Stanley, Peter, 272
Stansfield, Harry, 62, *64*, *73*, 75
Stanton, Arthur, 62
Starkey, A.E., 123
Stead, Clifford, 123
Steinthal, E.T., *64*
Stelfox (tinsmith), 120
Stewart, Balfour, 37, 38, 40
Stewart, C.P., 23
Stewart, Ian, 123
Stoney, G.G., 119
Stopes, Marie C.C., 62
Stopford, J.S.B., 152, 165, 288
Sturgeon, William, 4, 5, 6, 7, 22
Sumner, F.H., 211, 212
Sutherland, G.A., 132, 149, 168
Swan, Joseph, 22
Swift, Edgar, 123
Sykes, Fred, 186

Tait, D.J., 267, *271*, 278, 281
Taylor, Alice J., *64*
Taylor, E.F., 223, 238, *239*, 265
Taylor, G.I., 70
Taylor, John (of Mather & Platt), 99
Taylor, T., *73*
Taylor, T.A., 160, 161
Thomas, Dylan, ix
Thomas, G.E., 179, 181, *182*, 185, 193, *193*
Thomas, H.W., 223, 237, *239*, *252*, 263
Thomasson, J.P., 50
Thompson, F.C., 117, 124, 163
Thompson, Silvanus P., 111
Thomson, J.J., 32, 35, 38, 59, 63, 67, 117
Thomson, William (Lord Kelvin), 8, 159
Threlfall, Sir Richard, 102
Tipping, Dennis, 193, 211, 214, 231, 235, *239*, 247, 261, *271*, 282
Tipping, Mr, 158
Tizard, Sir Henry, 180
Todd, J.R., 132
Tolanski, Samuel, 162

Tomlinson, G.R., 272, 275
Tootill, G.C., 175, 176, 179, 180, 183
Tout, T.F., 119
Townsend, J.M., 211, 237
Trinci, A.P.J., 273
Tudor Accumulator Co., 55
Tuomikoski, T., *73*
Turing, A.M., 176, 181, 194-195, 204
Turnbull, G.F., 211, 237
Turner, J.A., 211, 238
Turner, P.J., 263
Tyrwhitt, Commodore Sir R.Y., 107

Uttley, A.M., 173

Varley, C.F., 23
Varlow, B.R., 223, 237, *239*, 253, 264, *271*, 278, 282
Vaughan, Arnold, 186
Victoria, H.M. Queen, 30, 44
Vinten-Fenton, C.V., 123, 127

Wade, P.A., 223, 237
Walker, Alderman, 131
Walker, Miles, xi, 99
Walmsley, H.P., 72
Walsh, Joan E., 272
Ward, Dr Adolphus, 60
Ward, Bernard, 186
Ward, Gladys, see Williams, Mrs F.C.
Waterhouse, Alfred, 28, 29, 33, 35, 50, 59, 70, 164, 286, 290
Watson, Stanley, 137, 139
Watt, Elizabeth, see Beattie, Elizabeth
Watteville, C.F. de, *64*
Weber, W.E., 22, 39
Weiss, Frederick, 88
Wells, D.E., *239*
West, J., *64*
West, J.C., 155, 171, 184, 185, 186, 188, *193*, 197-198, 214, 215, 234, 286, 287
Whaley, F.W., *64*, *73*
Whalley, C.A., 161
Wheatstone, Sir Charles, 4, 22, 23
White, Arthur, 88, 121
White, Margaret, *73*
Whitehead, J.R., 173
Whitworth, Sir Joseph, 5, 23, 33, 76
Wilberforce, Professor L.R., 103
Wilde, Henry, 22, 23, 55, 69

Wilkes, M.V., 135
Williams, F.C., *Front cover*, 72, 129, 130, 132, 134, 135, 139, 140, 141, *141*, 142, *143*, 147, 148, 149, 152, 171, 173-185, *179*, *182*, 186, 188, 192, 193, *193*, 197, 202, 204, 205, 210, 211, 212, 213, *213*, 214, 220, 223, *224*, *225*, 231, 235, 236, *236*, *239*, 240-244, 248, 261, 285, 286, 287, 288
Williams, Mrs F.C. (née Gladys Ward), 140
Williamson, W.C., 3, 12, 13, 277
Wilson, David, 112
Wilson, G.W., 46, 59
Wilson, R., *73*
Wilson, William, *64*, *73*
Winpenny, Clifford, 181
Wolfenden, J.P., 142, 147

Wood, A.B., 74, 100-108, 110, 111, 115
Wood, G.W., 4
Wood, John, 72
Wood, J.K., 216
Woodyatt, W.N., 247
Woolley, Hermann, 75
Woolner, T., 10
Wright, C.D., 265, *271*, 278, 281
Wrigley, J.S., 124
Wroe, Harold, 188, 216

Xydeas, C.S., 246, 247, 261, 269, 270, *271*, 272, 278, 282, 287

Young, F.B., 108, 126
Young, R.J., 273

Zussman, Jack, 273

Ted Broadbent entered the Electrical Engineering Department of Manchester University as an undergraduate student in 1949, and won the Ashbury Scholarship in 1950. After graduating in 1952 he stayed on to do research, financed by Matthew Kirtley Senior and Edmund Travis Engineering Scholarships, and obtained his M.Sc. in 1953 and Ph.D. in 1955. Following a two year spell at Ferranti Ltd as a research engineer he returned to the department as a Lecturer in 1957 and has been associated with it ever since, as Senior Lecturer (1966), Senior Lecturer and Tutor (1969) and Honorary Fellow (since his retirement in 1990). As a staff member he lectured in power systems, high-voltage engineering and a variety of other subjects and his research specialities were high-voltage technology and electrical breakdown in gases. His interest in the history of the department is of long standing and he has been collecting information on the subject for many years.